Computational Aspects of
Polynomial Identities

Research Notes in Mathematics

Volume 9

Computational Aspects of Polynomial Identities

Alexei Kanel-Belov
Hebrew University of Jerusalem, Israel
Moscow Institute of Open Education, Russia

Louis Halle Rowen
Bar-Ilan University, Ramat-Gan, Israel

CRC Press
Taylor & Francis Group
Boca Raton London New York

CRC Press is an imprint of the
Taylor & Francis Group, an **informa** business

AN A K PETERS BOOK

First published 2005 by A K Peters, Ltd.

Published 2018 by CRC Press
Taylor & Francis Group
6000 Broken Sound Parkway NW, Suite 300
Boca Raton, FL 33487-2742

© 2005 by Taylor & Francis Group, LLC
CRC Press is an imprint of Taylor & Francis Group, an Informa business

First issued in paperback 2019

No claim to original U.S. Government works

ISBN 13: 978-0-367-44650-5 (pbk)
ISBN 13: 978-1-56881-163-5 (hbk)

Visit the Taylor & Francis Web site at
http://www.taylorandfrancis.com

and the CRC Press Web site at
http://www.crcpress.com

Library of Congress Cataloging-in-Publication Data

Rowen, Louis Halle.
  Computational aspects of polynomial identities / Alexei Kanel-Belov, Louis Halle Rowen.
    p. cm. -- (Research notes in mathematics ; 9)
  Includes bibliographical references and index.
  ISBN 1-56881-163-2
  1. PI-algebras. 2. Polynomial rings. 3. Combinatorial analysis. I. Kanel-Belov, Alexei,
1963- II. Title. III. Research notes in mathematics (Boston, Mass.) ; 9.

QA251.R687 2004
512'.4--dc22

                                                                        2004053474

This work was carried out under the auspices of the
Israel Academy of Science Center for Excellence
at Bar-Ilan University.

Contents

Preface

An *identity* of an associative algebra A is a noncommuting polynomial that vanishes identically on all substitutions in A. For example, A is commutative iff $ab - ba = 0$, $\forall a, b \in A$, iff $xy - yx$ is an identity of A. An identity is called a *polynomial identity* (PI) if at least one of its coefficients is ± 1. Thus in some sense PIs generalize commutativity.

Historically, PI-theory arose first in a paper of Dehn [De22], whose goal was to translate intersection theorems for the Desarguian plane to polynomial conditions on the underlying division algebra D, and thereby classify geometries that lie between the Desarguian and Pappian axioms (the latter of which requires D to be commutative). Although Dehn's project was only concluded much later by Amitsur [Am66], who modified Dehn's original idea, the idea of PIs had been planted.

Wagner [Wag37] showed that any matrix algebra over a field satisfies a PI. Motivated by an observation of M. Hall [Ha43] that the polynomial $(xy - yx)^2$ evaluated on 2×2 matrices takes on only scalar values, Kaplansky asked whether matrix algebras have nonidentities that take on only scalar values; in 1972, Formanek [For72] and Razmyslov [Raz72] discovered such polynomials on arbitrary $n \times n$ matrices. This led to the introduction of techniques from commutative algebra to PI-theory, culminating in a beautiful structure theory with applications to central simple algebras, and (more generally) Azumaya algebras.

Since a representation of an algebra is a homomorphism to a matrix algebra, the degree of the representation is tied to the identities of $n \times n$ matrices, and PI-theory has close connections to finite dimensional representation theory. In particular, one of our main objects of study are *representable* algebras, i.e., algebras that can be embedded into an algebra of matrices over a suitable field.

PIs also can be viewed as the atomic universal elementary sentences satisfied by algebras. Consider the class of all algebras satisfying a given set of identities. This class is closed under taking subalgebras, homomorphic images, and direct products; conversely, any such class of algebras is called a *variety* of algebras. Varieties of algebras were studied in the 1930s by

Birkhoff [Bir35] and Mal'tsev [Mal36], thereby linking PI-theory to logic, especially through the use of constructions such as ultraproducts.

In this spirit, one can study an algebra through the set of all its identities, which turns out to be an ideal of the free algebra, called a T-*ideal*. Specht [Sp50] conjectured that any such T-ideal is a consequence of a finite number of identities. Specht's conjecture turned out to be very difficult, and Kemer's solution [Kem87] (in characteristic 0) is a tour de force that involved most of the theorems then known in PI-theory, in conjunction with several new techniques such as the use of superidentities.

Another very important connection, discovered by Regev, is a way of describing identities of a given degree n in terms of the group algebra of the symmetric group S_n. This led to the asymptotic theory of codimensions, one of the most active areas of research today in PI-theory.

While the interplay with the commutative structure theory was one of the main focuses of interest in the West, the Russian school was developing quite differently, in a formal combinatorial direction, often using the polynomial identity as a tool in word reduction. The Iron Curtain and language barrier impeded communication in the formative years of the subject, as illustrated most effectively in the history of Kurosh's problem, whether or not finitely generated (i.e., affine) algebraic algebras need be finite dimensional. This problem was of great interest in the 1940s to the pioneers of the structure theory of associative rings—Jacobson, Kaplansky, and Levitzki—who saw it as a challenge to find a suitable class of algebras which would be amenable to their techniques. Levitzki proved the result for algebraic algebras of bounded index, Jacobson observed that these are examples of PI-algebras, and Kaplansky completed the circle of ideas by solving Kurosh's problem for PI-algebras. Shirshov, in Russia, saw Kurosh's problem from a completely different combinatorical perspective, and his solution was so independent of the associative structure theory that it also applied to alternative and Jordan algebras. (This is evidenced by the title of his article, "On some nonassociative nil-rings and algebraic algebras," which remained unread in the West for years.)

A similar instance is the question of the nilpotence of the Jacobson radical J of an affine PI-algebra A, demonstrated in Chapter 2. Amitsur had proved the local nilpotence of J, and had shown that J is nilpotent in some cases. There is an easy argument to show that J is nilpotent when A is representable, but the general case is much harder to resolve. By a brilliant but rather condensed analysis of the properties of the Capelli polynomial, Razmyslov proved that J is nilpotent whenever A satisfies a Capelli identity, and Kemer [Kem80] verified that any affine algebra in characteristic 0 indeed satisfies a Capelli identity. Soon thereafter, Braun found a characteristic free proof that was mostly structure theoretical, employing a series of reductions to Azumaya algebras, for which the assertion is obvious. Even today, this dichotomy between the two approaches can be seen readily in the literature.

Both approaches are unified in the perspective of algebraic geometry. Whereas algebraic varieties are the subsets of a given space that are solutions of a system of algebraic equations, i.e., the zeroes of a given ideal of the algebra $F[\lambda_1, \ldots, \lambda_n]$ of commutative polynomials, PI varieties are collections of algebras *all* of whose subsets yield 0 when substituted into a given set of noncommutative polynomials. Thus, the role of ideals of $F[\lambda_1, \ldots, \lambda_n]$ in commutative algebraic geometry is analogous to the role of T-ideals of the free algebra. Hilbert's Basis theorem says every ideal of $F[\lambda_1, \ldots, \lambda_n]$ is finitely generated as an ideal, so Specht's conjecture is the PI-analog viewed in this light.

The introduction of noncommutative polynomials vanishing on A intrinsically involves a sort of noncommutative algebraic geometry, which has been studied from several vantage points, most notably the coordinate algebra, which is an affine PI-algebra. This approach is described in the seminal paper of Artin and Schelter [ArSc81].

Starting with Herstein [Her68] and [Her71], many expositions have already been published about PI-theory, including a book [Row80] and a chapter in [Row88, Chapter 6] by one of the coauthors (relying heavily on the structure theory), as well as books and monographs by leading researchers, including Procesi [Pro73], Jacobson [Jac75], Kemer [Kem90], Razmyslov [Raz89], Formanek [For91], Bakhturin [Ba91], Belov, Borisenko, and Latyshev [BBL97], Drensky [Dr00], and, recently, Drensky and Formanek [DrFor04] (which appeared while this book was in press). Nevertheless, we feel that some of the important advances in the last 20 years, largely combinatoric, still remain accessible only to experts (at best), and this has limited the exposure of the more advanced aspects of PI-theory to the general mathematical community.

Our main objective is to describe these breakthroughs in full, starting with Shirshov's theorem and carried forward so successfully by his school. The motivating result is Kemer's solution of Specht's conjecture in characteristic 0; the first six chapters of this book are devoted to the theory needed for its proof, including the featured role of the Grassman algebra and the translation to superalgebras (which also has considerable impact on the structure theory of PI-algebras). From this point of view, the reader will find some overlap with [Kem91] and, more recently, Iltyakov's monograph [Ilt03]. Although the framework of the proof is the same as for Kemer's proof, based on what we call the *Kemer index* of a PI-algebra, there are significant divergences; in the proof given here, we stay more within the PI context, moving away from the structure of subdirectly irreducible finite dimensional algebras. This approach enables us to develop Kemer polynomials for arbitrary varieties, as a tool for proving diverse theorems in later chapters, and also lays the groundwork for analogous theorems that have been proved recently for Lie algebras and alternative algebras. ([Ilt03] treats the Lie case.) Our primary goal is to present a full proof of Kemer's solution to Specht's conjecture (in characteristic 0) as quickly and intelligibly as we can, so for the most part we choose the easier proofs. (The notable exception is Shirshov's original proof of his theorem, which is proved three times, because Shirshov's theorem is so fundamental to the text. However, the first proof that we provide is new and considerably shorter than the others.) Nonetheless, this is all relative, and Kemer's theorem remains quite difficult. In Chapter 7, we present counterexamples to Specht's conjecture in characteristic p, as well as their underlying theory.

Other topics are delayed until after Chapter 7. For example, we postpone the finer considerations of Shirshov height until the end of Chapter 9. Additional topics include Noetherian PI-algebras, Poincaré–Hilbert series, Gelfand-Kirillov dimension, the combinatoric theory of affine PI-algebras, the ideals of identities ("T-ideals"), description of multilinear identities in terms of representation theory, and trace identities. In the process, we also develop some newer techniques, such as the "pumping procedure." Asymptotic results are considered more briefly, since the reader should be able to find them in the forthcoming book of Giambruno and Zaicev.

Since most of the combinatorics needed in these proofs do not require structure theory, there is no need for us to develop many of the famous results of a structural nature. But we felt these should be included somewhere in order to provide balance, so we have listed them in Section 1.6, without proof and with a different indexing scheme (Theorem A, Theorem B, and so forth). The proofs are to be found in most standard expositions of PI-theory.

Although we aim mostly for direct proofs, we also introduce technical machinery to pave the way for further advances. One general word of caution is that the combinatoric PI-theory often follows a certain Heisenberg principle—complexity of the proof times the manageability of the quantity computed is a constant. One can prove rather quickly that affine PI-algebras have finite Shirshov height and satisfy a Capelli identity (thereby leading to the nilpotence of the radical), but the bounds are so high as to make them impractical for computations. On the other hand, reasonable bounds are now available for these quantities, but the proofs become highly technical.

Our treatment largely follows the development of PI-theory via the following chain of generalizations:

1. Commutative algebra (taken as given)

2. Matrix algebras (references quoted)

3. Prime PI-algebras (references usually quoted)

4. Subalgebras of finite dimensional algebras

5. Algebras satisfying a Capelli identity

6. Algebras satisfying a sparse identity

7. PI algebras in terms of the Razmyslov-Zubrilin identity

8. PI-algebras in terms of Kemer polynomials (the most general case)

The theory of Kemer polynomials, which is embedded in Kemer's proof of Specht's conjecture, shows that the techniques of finite dimensional algebras are available for all affine PI-algebras, and perhaps the overriding motivation of this book is to make these techniques more widely known.

Another recurring theme is the Grassman algebra, which appears first in Rosset's proof of the Amitsur-Levitzki theorem, later as the easiest example of a finitely based T-ideal (generated by the single identity $[[x_1, x_2], x_3]$), still later in the link between algebras and superalgebras, and finally as a test algebra for the counterexamples in characteristic p.

During the writing of this book, Alexei Kanel-Belov was visiting Bar-Ilan University and was also affiliated with the International University Bremen.

We are indebted to Amitai Regev, one of the founders of the combinatoric PI-theory, especially in connection with representations of the symmetric group S_n. Besides generously making his notes available to us, Regev provided a great deal of help in preparation for this book, and also suggested various improvements in the exposition. The organization of this book is according to his suggestion of proceeding as quickly as possible to solving Specht's problem, and then filling in related material. We also are grateful to Uzi Vishne for providing valuable comments on the various draft versions of the manuscript, and to Allan Berele for his suggestions. Ilya Bogdanov supplied us with Kuzmin's paper (in Russian) on the lower bound for the Nagata-Higman theorem, cf. Exercise 12.9. Also we would like to thank Lance Small for suggesting to Belov the problem of finite presentation of Noetherian PI-algebras.

We would like to thank Leonid Bokut for suggesting this project, and Klaus Peters for his friendly persistence in encouraging us to complete the manuscript. Also thanks to Miriam Beller for her invaluable assistance in helping put together the manuscript, and to Charlotte Henderson at A K Peters for her patient help at the editorial stage.

Enumeration of Results

The text is subdivided into chapters, sections, and at times subsections. Thus, Section 9.4 denotes Section 4 of Chapter 9; Section 9.4.1 denotes subsection 1 of Section 9.4. The results are enumerated independently of these subdivisions. Except in Section 1.6, which has its own numbering system, all results are enumerated according to chapter only; for example, Theorem 6.13 is the thirteenth item in Chapter 6, preceded by Definition 6.12. The exercises are listed separately at the end. When referring in the text to an exercise belonging to the same chapter we suppress the chapter number; for example, in Chapter 9, Exercise 9.12 is called "Exercise 12," although in any other chapter it would have the full designation "Exercise 9.12."

Symbols Commonly Used in Text

Note: Due to the finiteness of the English and Greek alphabets, some symbols have multiple uses. For example, in Chapters 2 and 9, μ denotes the Shirshov height, whereas in Chapter 4 and 6, μ is used for the number of certain folds in a Kemer polynomial. We have tried our best to recycle symbols only in unambiguous situations.

Page	Symbol	Meaning
1	$M_n(A)$	the algebra of matrices over A
1	e_{ij}	matrix units for the algebra of matrices
1	δ_{jk}	the Kronecker delta
1	$\mathbb{Z}/n$	the integers modulo n
1	$[a,b]$	the ring commutator $ab - ba$
1	S_n	the symmetric group
1	$\operatorname{sgn}(\pi)$	the sign of the permutation π
1	$\mathcal{M}\{X\}$	the word monoid on the set of letters X
2	$C\{X\}$	the free C-algebra on the set of letters X
2	$f(x_1,\ldots,x_m)$	the polynomial f in indeterminates $x_1,\ldots,x_m$
2	$f(A)$	the set of evaluations of a polynomial f in an algebra A
2	$\deg f$	the degree of a polynomial f
3	f.g.	finitely generated (used for modules)
3	$\operatorname{Cent}(A)$	the center of an algebra A
5	$\Delta_i f$	the multilinearization step of f in x_i
6	$\tilde{s}_n$	the symmetric polynomial in n letters
6	$\operatorname{id}(A)$	the set of identities of A
6	$A_1 \sim_{\mathrm{PI}} A_2$	A_1 and A_2 satisfy the same identities
8	s_t	the standard polynomial (on t letters)
8	c_t	the Capelli polynomial (on t letters)
9	πf	the left action of a permutation π on a polynomial f

Page	Symbol	Meaning
10	$f_{A(i_1,\dots,i_t;X)}$	the alternator of f with respect to the indeterminates $x_{i_1},\dots,x_{i_t}$
10	$\tilde{f}$	the symmetrizer of a multilinear polynomial f
11	$F[\Lambda]$	$= F[\lambda_1,\dots,\lambda_n]$, the commutative polynomial algebra
13	G	the Grassman algebra, usually in an infinite set of letters
13	$e_1, e_2, \dots$	the standard base of the Grassman algebra G
13	G_0	the odd elements of the Grassman algebra G
13	G_1	the even elements of the Grassman algebra G
17	det	determinant
18	tr	trace
22	ℓ_i	the left replacement operator
22	$\mathfrak{r}_j$	the right replacement operator
23	$D(x_1,\dots,x_n)$	the identity of algebraicity
24	$J(A)$	the Jacobson radical of an algebra A
43	$|w|$	the length of a word w
43	$\bar{w}$	the image of a word w in $A = C\{a_1,\dots,a_\ell\}$, under the canonical specialization sending $x_i \mapsto a_i$
44	$\mu = \mu(A)$	the Shirshov height of an affine PI-algebra A
46	$\beta(\ell, k, d)$	the Shirshov function for an affine algebra $A = C\{a_1,\dots,a_\ell\}$ of PI-degree d; any word of greater length must either be d-decomposable or have some nonempty subword repeating k times
46	u^∞	the infinite periodic hyperword with period u
48	$\beta(\ell, k, d, h)$	the Shirshov function for a given hyperword h evaluated on A
52	$\hat{A}$	the trace ring of a representable algebra A
58	$\delta_j(A)$	the Razmyslov-Zubrilin trace operator
62	$\ker \varphi_a$	the obstacle to integrality of degree n, of an element a
63	$M_{\psi;f}$	the A-submodule of $C\{X\}$ generated by f and the Razmyslov-Zubrilin trace operators
72	$\delta(xv)$	the cyclic shift
96	t_A	the dimension of the semisimple part of a f.d. algebra A
96	s_A	the nilpotence index of the Jacobson radical of a f.d. algebra A

Page Symbol   **Meaning**

Page	Symbol	Meaning
96	$\bar{F}$	the algebraic closure of the base field F
97	$A = R_1 \oplus \cdots \oplus R_q \oplus J$	the decomposition of a f.d. algebra A over an algebraically closed field, as the direct sum of its matrix parts and the Jacobson radical
102	$\tilde{f}_{X_1,\dots,X_\mu}$	the μ-fold alternator of a polynomial f
102	$\beta(W)$	the first Kemer invariant of a PI-algebra W
103	$\gamma(W)$	the second Kemer invariant of a PI-algebra W
103	index(W)	$= (\beta(W), \gamma(W))$, the Kemer index of a PI-algebra W
103	index(Γ)	the Kemer index of a T-ideal Γ
103	$\omega(\Gamma)$	a third invariant of the PI-algebra W
106	$\tilde{f}_{\mathcal{A}(I_1)\dots\mathcal{A}(I_s)\mathcal{A}(I_{s+1})\dots\mathcal{A}(I_{s+\mu})}$	the μ-fold multiple alternator
125	$\Gamma(U_i)$	the set of all values of all polynomials in Γ, evaluated on the algebra U_i
131	$T(I')$	for $I' \subseteq I$ in S_{2n}, defined in $\mathbb{Z}[S_{2n}]$
141	$F[S_n]$	the group algebra
141	$M_\sigma(x_1, \dots, x_n)$	$= x_{\sigma(1)} \cdots x_{\sigma(n)}$, the monomial of a permutation
142	$\sigma M_\pi(x_1 \dots, x_n)$	$= M_{\sigma\pi}$, the left action of a permutation on a monomial
142	$M_\sigma(x_1 \dots, x_n)\pi$	$= M_{\sigma\pi}$, the right action of a permutation
144	p^*	the Capelli-type polynomial obtained from a polynomial p
146	$\lambda \vdash n$	$\lambda = (\lambda_1, \dots \lambda_k)$ is a partition of n
147	f^λ	the number of standard tableaux of shape λ
147	$R_{T_\lambda}^+$	$= \sum_{p \in R_{T_\lambda}} p$
147	$C_{T_\lambda}^-$	$= \sum_{q \in C_{T_\lambda}} \operatorname{sgn}(q) q$
147	e_{T_λ}	$= \sum_{q \in C_{T_\lambda}} \sum_{p \in R_{T_\lambda}} \operatorname{sgn}(q) qp$, the semi-idempotent of a tableau
149	λ^+	partition branched up
149	μ^-	partition branched down
149	$\chi^\lambda \uparrow$	the induced character
151	$g_d(n)$	the number of d-good permutations in S_n
151	$h_{i,j} = h_{i,j}(\lambda)$	$= \lambda_i - j + \lambda_j' - i + 1$, the hook number
152	$T^n(V)$	the homogeneous polynomials of degree n
152	$E_{n,k}$	the endomorphism ring of $T^n(V)$
153	$A(n,k)$	$= \varphi_{n,k}(F[S_n])$

Page Symbol   **Meaning**

153   $H(k, 0; n)$ those partitions of n contained in the infinite k-strip

154   Γ_n $= \mathrm{id}(A) \cap V_n$

155   $c_n(A)$ the codimension $\dim_F (V_n/\Gamma_n)$

161   $H(k, \ell; n)$ $= \{\lambda \vdash n \mid \lambda_{k+1} \le \ell\}$, the (k, ℓ)-hook

164   $L(y_1^{d_1} \cdots y_k^{d_k})$ multilinearization operator

170   $F\{Y, Z\}$ $= F\{y_1, y_2 \ldots; z_1, z_2 \ldots\}$, the free superalgebra

170   $I \lhd_2 A$ I is a superideal of A

170   $\mathrm{id}_2(A)$ the superidentities of A

172   $\mathrm{Odd}(b)$ $= \{i \mid b_i \text{ is odd}\}$

172   $\varepsilon(\sigma, I)$ the sign function

173   $\tilde{p}_I^*$ the Grassman involute of a polynomial p

175   $\sigma \bullet (x_1 \cdots x_n)$ $= \varepsilon(\sigma, I)(x_{\sigma(1)} \cdots x_{\sigma(n)})$, the $\bullet$-action of S_n on $T^n(V)$

184   $\mathcal{M}_n$ the superidentities of $M_n(F)$

184   $\mathcal{M}_{k,\ell}$ the superidentities of $M_{k,\ell}(F)$

184   $\mathcal{M}'_n$ the superidentities of $M_n(F[c])$ of Example 6.27

184   $\tilde{\mathcal{M}}_n$ the superidentities common to all three of these superalgebras

186   $\mathrm{index}_2 A$ the Kemer superindex of A

204   $F[\varepsilon]$ $= F[\varepsilon_i : i \in \mathbb{N}]$, the multi-dual numbers

204   G^+ the extended Grassman algebra

207   $\bar{\varepsilon}_t$ $= \sum_{j=0}^{[t/2]} \sum_{i_1 < i_2 \cdots < i_{2j}} \varepsilon_{i_1} \cdots \varepsilon_{i_{2j}}$

213   P_n $= x_1[x_1, x_2]x_2 x_3[x_3, x_4]x_4 \ldots x_{2n-1}[x_{2n-1}, x_{2n}]x_{2n}$, for characteristic 2

216   $\mathcal{G}$ the relatively free superalgebra of the Grassman algebra

216   $P(x_1, x_2)$ $= x_1^{p-1}[x_1, x_2]x_2^{p-1}$

218   P_n $= \prod_{i=1}^n P(x_{2i-1}, x_{2i})$, for odd characteristic

219   $\hat{P}'$ $= \sum_{\sigma \in S_p} x_{1,\sigma(2)} \cdots x_{1,\sigma(p)} z_{1,\sigma(1)}$, half of the multilinearization of P

220   $\hat{P}$ $= \hat{P}'$

228   e, f, t special elements in the test algebra $\widehat{A}$

229   Q_n the polynomials which generate the non-finitely based T-ideal in odd characteristic

240   $\Delta = \Delta(G)$ the elements of a group G having only a finite number of conjugates

240   $C_G(g)$ the centralizer of g in the group G

Page Symbol

Page	Symbol	Meaning
242	$\Delta = \Delta(L)$	the elements of a Lie algebra L whose centralizer has finite index
245	$\tilde{A}_n$	the filtering subspaces of A
245	H_A	the Hilbert series of A (with respect to the expected set of generators)
245	d_n	$= \dim_F(\tilde{A}_n/\tilde{A}_{n-1})$
248	GKdim	Gelfand-Kirillov dimension
248	$\tilde{d}_n$	$= \dim_F \tilde{A}_n$
260	$H_{A;V}$	the Hilbert series of A with respect to subspace V
267	$\chi_n(A)$	the n-th cocharacter of A
268	$m_\lambda(A)$	the multiplicity of the partition λ in a cocharacter
271	$B(n,k)$	$= \langle\psi_{n,k}(GL(V))\rangle$, the centralizer of $A(n,k)$ in $E_{n,k}$
273	V_k^λ	$= e_{T_\lambda}T^n(V)$, the Weyl module
276	$\chi_{n,k}(A)$	$= \chi_{GL(V)}(T^n(V)/\widehat{\Gamma}_n)$, the homogeneous cocharacter
282	$U_{\mathcal{V}}$	the relatively free algebra of the variety $\mathcal{V}$
293	Tr	formal trace operator
294	$\mathrm{Tr}(x_{i_1}\ldots x_{i_t})$	simple trace monomial
295	g_k	the Hamilton-Cayley mixed trace identity
296	$(e_i \otimes \theta_j)$	identified with the matrix unit e_{ij}
302	$f_{e_{\bar\mu}}$	the fundamental pure trace identity
305	$\mathrm{Tid}_{k,n}$	the pure trace identities of $M_k(F)$, a submodule of T_n
306	$t_n(M_k(F))$	the pure trace codimension

Chapter 1

Basic Results

In this chapter, we introduce PI-algebras and review some well-known results and techniques, most of which are associated with the structure theory. In this way, the tenor of the exposition is different from that of the subsequent chapters. The emphasis is on matrix algebras and their subalgebras (called *representable* PI-algebras).

1.1   Definitions and Examples

Given any algebra A, $M_n(A)$ denotes the algebra of $n \times n$ matrices, and e_{ij} denotes the *matrix unit* having 1 in the i, j position and 0 elsewhere. The set of $n \times n$ *matrix units* $\{e_{ij} : 1 \leq i, j \leq n\}$ stisfies the properties:

$$\sum_{i=1}^{n} e_{ii} = 1,$$

$$e_{ij}e_{k\ell} = \delta_{jk}e_{i\ell},$$

where δ_{jk} denotes the *Kronecker delta* (which is 1 if $j = k$, 0 otherwise).

The ring $\mathbb{Z}/n\mathbb{Z}$ of integers modulo n is also denoted $\mathbb{Z}/n$.

Given elements a, b of an algebra A, we define $[a, b] = ab - ba$.

S_n denotes the symmetric group, i.e., the permutations on $\{1, \ldots, n\}$. We write $\text{sgn}(\pi)$ for the sign of a permutation π.

Throughout, C will denote a commutative ring (often a field). Unless otherwise indicated, an algebra over C is assumed to be associative with a unit element 1.

We need the noncommutative analog of polynomials. Recall the free (associative) monoid $\mathcal{M}\{X\}$ in $X = \{x_i : i \in I\}$ is the monoid of *words* $\{x_{i_1} x_{i_2} \ldots x_{i_t} : t \in \mathbb{N}\}$ permitting duplication of subscripts, and whose unit

1

element is the blank word $\emptyset$; the monoid operation is given in terms of juxtaposition of words.

$C\{X\}$, often denoted $C\langle X\rangle$ in the literature, denotes the free algebra in the set $X = \{x_i : i \in I\}$ of noncommuting indeterminates. (Usually $I = \mathbb{N}$, but often I is taken to be finite.) In other words, $C\{X\}$ is the monoid algebra of $\mathcal{M}\{X\}$. The elements of $C\{X\}$ are called *polynomials*. We recall that $C\{X\}$ is free as a C-module, with base consisting of $\mathcal{M}\{X\}$; thus, any $f \in C\{X\}$ is written uniquely as $\sum c_j h_j$ where $h_j \in \mathcal{M}(X)$. We call these $c_j h_j$ the *monomials* of f.

Given $f \in C\{X\}$, we write $f(x_1, \ldots, x_m)$ to denote that $x_1, \ldots, x_m$ are the only indeterminates occurring in f. Later, when the notation becomes more cumbersome, we shall have occasion to use y or y_j (and at times z_k) to denote extra indeterminates that do not enter the computations as actively as the x_i.

The main feature of $C\{X\}$ is the following.

Remark 1.1. Given a C-algebra A and elements $\{a_i : i \in I\} \subseteq A$, there is a unique algebra homomorphism $\phi : C\{X\} \to A$, called the *substitution homomorphism*, such that $\phi(x_i) = a_i$, $\forall i \in I$. Indeed, one defines

$$\phi(x_{i_1} \ldots x_{i_m}) = a_{i_1} \ldots a_{i_m}$$

and extends this linearly to all of $C\{X\}$.

The *evaluation* $f(a_1, \ldots, a_m)$ denotes the image of f under the homomorphism of Remark 1.1. We also say f *specializes* to $f(a_1, \ldots, a_m)$, and $a_1, \ldots, a_m$ are *substitutions* in f. Let us write $f(A)$ for the set of evaluations $\{f(a_1, \ldots, a_m) : a_i \in A\}$.

Definition 1.2. An element $f \in C\{X\}$ is an *identity* of a C-algebra A if $f(A) = 0$, i.e., $f \in \ker \phi$ for every homomorphism $\phi : C\{X\} \to A$.

Identities pass to related algebras as follows.

Remark 1.3. If f is an identity of A, then f is an identity of any homomorphic image of A and also of any subalgebra of A. Furthermore, if f is an identity of each C-algebra A_i, $i \in I$, then f is an identity of $\prod_{i \in I} A_i$.

Remark 1.3 provides an alternate approach to identities, cf. Chapter 11.

Definition 1.4. For a monomial h we define $\deg_i h$ to be the number of occurrences of x_i in h, and the *degree* $\deg h = \sum_i \deg_i h$, which is the length of the word; for a polynomial f, we define $\deg f$ to be the maximum degree of a monomial of f. For example, $\deg(x_1 x_2 + x_3 x_4) = 2$.

One needs some way of excluding the identity px_1, which only says A has characteristic p. Towards this end, we formulate:

Definition 1.5. An identity f is a *PI* for A if at least one of its coefficients is 1. An algebra A is a *PI-algebra* of PI-*degree* d if A satisfies a PI of degree d.

This definition might seem restrictive, but in fact is enough to yield the entire PI-theory, cf. [Am71].

Since PI-algebras are the subject of our study, let us address a subtle distinction in terminology. A ring R is a PI-*ring* when it is a PI-algebra for $C = \mathbb{Z}$. Although most of the general structure theory holds for PI-rings in general, our focus in this book is usually on a particular base ring C, often a field, in which case we denote it as F instead of C; often we require $\text{char}(F) = 0$, for reasons to be discussed shortly. Thus, we shall always assume A is a PI-algebra over a given commutative ring C.

When viewing an algebra A as a module over C, we often consider the submodule *spanned* or *generated* by a given subset of A. We say a module M over a ring R is *finitely generated*, denoted by *f.g.*, if $M = \sum_{i=1}^{t} Rw_i$ for suitable $w_i \in M$, $t \in \mathbb{N}$.

Here is a notion closely related to PI.

Definition 1.6 (Central polynomials). A polynomial $f(x_1, \ldots, x_n)$ is *A-central* if $0 \neq f(A) \subseteq \text{Cent}(A)$, the center of A. In other words, $f(x_1, \ldots, x_n)$ is A-central iff $[y, f]$ (but not f) is an identity of A.

The most important examples of PI-algebras are the matrix algebra $M_n(C)$ for arbitrary n, finite dimensional algebras over a field, and the Grassman algebra G. Since these examples require a bit more theory, we whet the reader's appetite with some easily verified examples.

Example 1.7.

(i) x_1 is central for any commutative algebra.

(ii) Any product of n strictly upper triangular $n \times n$ matrices is 0. Since $[a, b]$ is strictly upper triangular, for any upper triangular matrices a, b, we conclude that the algebra of upper triangular matrices satisfies the identity

$$[x_1, x_2][x_3, x_4] \ldots [x_{2n-1}, x_{2n}].$$

(iii) (Hall's identity) If F is a field, then $M_2(F)$ satisfies the identity $[[x, y]^2, z]$ or, equivalently, the central polynomial $[x, y]^2$, cf. Exercise 1.

(iv) Fermat's Little Theorem translates to the fact that any field F of $n = p^t$ elements satisfies $a^n = a$ for all $a \in F$, i.e., $x_1^n - x_1$ is an identity. (See Exercise 4 for a generalization.)

(v) Any Boolean algebra satisfies the identity $x^2 - x$.

When dealing with arbitrary PIs, it is convenient to work with certain kinds of polynomials. We say a polynomial $f(x_1, \ldots, x_m)$ is *homogeneous* in x_i if x_i has the same degree in each monomial of f. We say f is *homogeneous* if f is homogeneous in every indeterminate. (Sometimes this is called "completely homogeneous" or "multi-homogeneous" in the literature.) If x_i has degree d_i in f_i for $1 \leq i \leq m$, we say f has *multi-degree* $(d_1, \ldots, d_m)$, or *degree* $d = d_1 + \cdots + d_m$. There is a very important special case.

Definition 1.8. A monomial h is *linear* in x_i if $\deg_i h = 1$; the polynomial f is *linear* in x_i if each monomial of f is linear in x_i; f is *t-linear* if f is linear in each of $x_1, \ldots, x_t$.

$f(x_1, \ldots, x_n)$ is *multilinear* if f is linear in each of $x_1, \ldots, x_n$, i.e., is n-linear.

Thus, $x_1 x_2 - x_2 x_1$ is multilinear. However, $x_1 x_2 x_3 - x_2 x_1$ is not multilinear, since x_3 does not appear in the second monomial.

Given a multilinear polynomial $f(x_1, \ldots, x_t)$, we pick any nonzero monomial h, and renaming the variables appropriately, we may assume $h = \alpha x_1 x_2 \ldots x_t$ for some $\alpha \in C$. Thus, the general form for a multilinear polynomial is

$$f(x_1, \ldots, x_t) = c x_1 x_2 \ldots x_t + \sum_{1 \neq \sigma \in S_t} \alpha_\sigma x_{\sigma(1)} x_{\sigma(2)} \ldots x_{\sigma(t)}. \qquad (1.1)$$

Furthermore, if C is a field, then we can divide by c and assume $c = 1$. The main reason we focus on multilinear identities is because of Proposition 1.18 below. However, the linearity property already is quite useful:

Remark 1.9. If f is linear in x_i, then

$$f(a_1, \ldots, \sum_j c_j a_{ij}, \ldots, a_m) = \sum_j c_j f(a_1, \ldots, a_{ij}, \ldots, a_m)$$

for all $c_j \in C$, $a_{ij} \in A$.

Lemma 1.10. *Suppose A is spanned over C by a set B. Then a multilinear polynomial f is an identity of A iff f vanishes on all substitutions to elements of B; f is A-central iff every substitution of f on B is in $\mathrm{Cent}(A)$ but some substitution on B is nonzero.*

Proof.

$$f\left(\sum c_{i1}b_{i1}, \ldots, c_{im}b_{im}\right) = \sum c_{i1} \ldots c_{im} f(b_{i1}, \ldots, b_{im}),$$

in view of Remark 1.9.   □

These observations raise the question of how to go back and forth from arbitrary identities (or central polynomials) to multilinear ones. The answer is in the process of *multilinearization*, or *polarization*. This will be treated in depth in Section 5.5, (also cf. Exercise 2), but can be described briefly as follows:

Definition 1.11 (Multilinearization). Suppose the polynomial $f(x_1, \ldots, x_m)$ has degree $n_i > 1$ in x_i. We focus on one of the variables, x_i, and define the *partial linearization*

$$\Delta_i f(x_1, \ldots, x_{i-1}, x_i, x'_i, x_{i+1}, \ldots, x_m)$$
$$= f(\ldots, x_i + x'_i, \ldots) - f(\ldots, x_i, \ldots) - f(\ldots, x'_i, \ldots) \tag{1.2}$$

where x'_i is a new indeterminate. Clearly $\Delta_i f$ remains an identity for A if f is an identity, but all monomials of degree n_i in x_i cancel out in $\Delta_i f$. The remaining monomials have x'_i replacing x_i in some (but not all) instances, and thus, have degree $< n_i$ in x_i, the maximum degree among them being $n_i - 1$.

Remark 1.12. Since this procedure is so important, let us rename the variables more conveniently, writing x_1 for x_i and y_j for the other variables, and write n for n_i.

(i) Now our polynomial is $f(x_1; y_1, \ldots, y_m)$ and our partial linearization may be written as

$$\Delta_1 f(x_1, x_2; y_1, \ldots, y_m) =$$
$$f(x_1 + x_2; y_1, \ldots, y_m) - f(x_1; y_1, \ldots, y_m) - f(x_2; y_1, \ldots, y_m), \tag{1.3}$$

where x_2 is the new indeterminate.

(ii) Iterating this procedure $n - 1$ times (each time introducing a new indeterminate x_i) yields a polynomial $\bar{f}(x_1, \ldots, x_n; y_1, \ldots, y_m)$ linear in $x_1, \ldots, x_n$, which preserves only those monomials h originally of degree n in x_1. For each such monomial h in f, we now have $n!$ monomials in $\bar{f}$ (according to the order in which $x_1, \ldots, x_n$ appears), each of which specializes back to h when we substitute x_1 for each x_i. Thus, when f is homogeneous in x_1, we have

$$\bar{f}(x_1, \ldots, x_1; y_1, \ldots, y_m) = n! f \tag{1.4}$$

In characteristic 0 this is about all we need, since $n!$ is invertible and we have recovered f from $\bar{f}$. We call $\bar{f}$ the *linearization* of f in x_1.

(iii) Repeating the linearization process for each indeterminate appearing in f yields a multilinear polynomial, called the *multilinearization*, or *total multilinearization*, of f.

(iv) There is a refinement of (ii) that helps us in nonzero characteristic. Write

$$\Delta_1 f = \sum_{j=1}^{n-1} f_j(x_1, x_2; y_1, \ldots, y_m)$$

where $\deg_1 f_j = j$ (and thus, $\deg_2 f_j = n-j$). Then, for any j, we get $\bar{f} = \bar{f}_j$, i.e., we get $\bar{f}$ by performing the multilinearization procedure on any f_j, first on x_1 and then on x_2, and (1.4) becomes

$$\bar{f}(x_1, \ldots, x_1, x_2, \ldots, x_2; y_1, \ldots, y_m) = j!(n-j)!\Delta_1 f. \qquad (1.5)$$

Example 1.13. The multilinearizations of Example 1.7:
(i), (ii) are already multilinear.
(iii) multilinearizes to the central polynomial

$$[x_1, x_2][x_3, x_4] + [x_3, x_4][x_1, x_2] + [x_1, x_4][x_3, x_2] + [x_3, x_2][x_1, x_4];$$

The multilinearization of (iv) is called the *symmetric polynomial*

$$\tilde{s}_n = \sum_{\pi \in \mathrm{Sym}_n} x_{\pi(1)} \cdots x_{\pi(n)}. \qquad (1.6)$$

The multilinearization of (v) is $x_1 x_2 + x_2 x_1$. Specializing $x_2 \mapsto x_1$ yields $2x_1^2$, which is weaker than the original identity and only says A has characteristic 2.

1.1.1 PI-equivalence

Definition 1.14. We write $\mathrm{id}(A)$ for the set of identities of A.

Often we shall be interested in encoding information into $\mathrm{id}(A)$. This leads us to

Definition 1.15. Algebras A_1, A_2 are called PI-*equivalent* if $\mathrm{id}(A_1) = \mathrm{id}(A_2)$; in this case we write $A_1 \sim_{\mathrm{PI}} A_2$.

It is convenient to start with the multilinear identities.

Proposition 1.16. *If f is a multilinear identity of A and H is a commutative C-algebra, then f is an identity of $A \otimes_C H$.*

Proof. $A \otimes 1$ spans $A \otimes H$ over H. $\qquad\qquad\qquad\qquad\qquad$ □

Thus, the class of PI-algebras is closed under tensor extensions.

Proposition 1.17. *If F is an infinite field, then every identity of an algebra A is a consequence of homogeneous identities.*

Proof. This is a well-known application of the Vandermonde argument, cf. Exercise 8. $\qquad\qquad\qquad\qquad\qquad\qquad\qquad\qquad\qquad\qquad$ □

When F is finite, the proposition fails miserably, cf. Exercise 4.

Proposition 1.18. *If A is an algebra over a field F of characteristic zero, then* id(A) *is determined by the multilinear identities.*

Proof. We look a bit closer at the linearization procedure Δ_i, writing $\mathcal{I}$ for the multilinear identities of A. We want to recover f from $\mathcal{I}$. By Proposition 1.17 we may assume f is homogeneous, so we conclude with Remark 1.12(ii). $\qquad\qquad\qquad\qquad\qquad\qquad\qquad\qquad\qquad\qquad$ □

Thus, PI-equivalence in characteristic 0 is established by checking the multilinear identities. For this reason, some of our main theorems only hold for an algebra over a field of characteristic 0. The situation for characteristic $p \neq 0$ is more delicate, because an identity need not be the consequence of its multilinearizations; the simplest example is the Boolean algebra (Example 1.7(v) and Example 1.13). Nevertheless, let us record the following result.

Proposition 1.19. *If F is an infinite field and K is a commutative F-algebra, then* id$(A \otimes_F K) =$ id(A), *for any F-algebra A.*

Proof. We need only consider homogeneous polynomials, say of degree d_i, in x_i.

$(\subseteq)$ $0 = f(r_1 \otimes k_1, \ldots, r_m \otimes k_m) = f(r_1, \ldots, r_m) \otimes k_1^{d_1} \ldots k_m^{d_m}$ implies $f(r_1, \ldots, r_m) = 0$;

$(\supseteq)$ A modification of the linearization argument; see [Row80, Theorem 23.29], for details. $\qquad\qquad\qquad\qquad\qquad\qquad\qquad\qquad\qquad$ □

1.2   Capelli Identities and
Finite Dimensional Algebras

Finite dimensional algebras are so important that we often use the abbreviation *f.d.* for them. Lemma 1.10 also gives us a way of constructing identities for f.d. algebras.

Definition 1.20. A t-linear polynomial $f(x_1, \ldots, x_m)$ is *t-alternating* if f becomes 0 when we substitute x_i for x_j, any $1 \le i < j \le t$.

For example, $[x_1, x_2]$ is 2-alternating. More generally the *standard polynomial*

$$s_t = \sum_{\pi \in S_t} \mathrm{sgn}(\pi) x_{\pi(1)} \cdots x_{\pi(t)}$$

is t-multilinear and t-alternating, since the terms cancel out when any two x are the same. Our main example will be the *Capelli polynomial*

$$c_t = \sum_{\pi \in S_t} \mathrm{sgn}(\pi) x_{\pi(1)} x_{t+1} x_{\pi(2)} x_{t+2} \cdots x_{2t-1} x_{\pi(t)} x_{2t},$$

which is $2t$-multilinear and t-alternating for the analogous reason. In order to distinguish the permuted indeterminates from the fixed ones, we sometimes rewrite this as

$$c_t(x; y) = c_t(x_1, \ldots, x_t, y_1, \ldots, y_t) = \sum_{\pi \in S_t} \mathrm{sgn}(\pi) x_{\pi(1)} y_1 x_{\pi(2)} y_2 \cdots x_{\pi(t)} y_t.$$

For example, for $t = 2$, we have

$$c_4 = x_1 y_1 x_2 y_2 - x_2 y_1 x_1 y_2.$$

Proposition 1.21. *Any t-alternating polynomial f is an identity for every algebra A spanned as C-module by $< t$ elements.*

Proof. It is enough to check that f vanishes on a spanning set B of A. But by choosing B to have $< t$ elements, every evaluation of f on B is 0, by definition.   □

Note. This proposition has obvious analogs both in the nonassociative case and even for algebraic structures involving multiplication that is not necessarily binary. Zubrilin's work on c_t is formulated in this more general setting, although our presentation in Chapters 2 and 4 will only involve the associative case.

Corollary 1.22. $M_n(C)$ *is a PI-algebra, since any* $(n^2 + 1)$-*alternating polynomial is an identity.*

In particular, s_{n^2+1} and c_{n^2+1} are PIs of $M_n(C)$. The standard polynomial upstaged the Capelli polynomial throughout the early history of PI-theory, in view of the Amitsur-Levitzki Theorem 1.49, which shows that s_{2n} is the unique PI of minimal degree of $M_n(C)$. However, as the alternating property of polynomials became more important over time, the Capelli polynomial assumed its proper leading role. Today it is impossible to delve deeply into the PI-theory without constant recourse to the Capelli polynomial, and most of this book is devoted to exposing its more subtle properties. Let us see now why Capelli polynomials arise so naturally in the theory of alternating polynomials.

Definition 1.23. Anticipating Chapter 5 a bit, given a polynomial $f(x_1, \ldots, x_m)$ and $\pi \in S_t$, we write πf for $f(x_{\pi(1)}, \ldots, x_{\pi(t)}, x_{t+1}, \ldots, x_m)$.

This defines an action of the symmetric group S_t on the free algebra $C\{X\}$. In this notation,

$$s_t = \sum_{\pi \in S_t} \mathrm{sgn}(\pi)\pi h$$

where $h = x_1 \ldots x_t$;

$$c_t = \sum_{\pi \in S_t} \mathrm{sgn}(\pi)\pi h$$

where $h = x_1 x_{t+1} \ldots x_t x_{2t}$.

Proposition 1.24. *If* f *is* t-*alternating, then* $\tau f = -f$ *for each transposition* $\tau \in S_t$. *The converse holds in characteristic* $\neq 2$.

Proof. Suppose $\tau = (i, j)$. Then

$$f + \tau f = f(\ldots, x_i, \ldots, x_j, \ldots) + f(\ldots, x_i, \ldots, x_j, \ldots)$$

which, as in Definition 1.11(i), can be rewritten as

$$f(\ldots, x_i + x_j, \ldots, x_i + x_j, \ldots) - f(\ldots, x_i, \ldots, x_i, \ldots) - f(\ldots, x_j, \ldots, x_j, \ldots),$$

all of whose terms are 0 by definition. □

Corollary 1.25. *Assume* $2a = 0$ *implies* $a = 0$. *The following are equivalent for a* t-*linear polynomial* $f(x_1, \ldots, x_t; y_1, \ldots, y_m)$, *where* $f_{(t)}$ *denotes the sum of monomials of* f *in which* $x_1, \ldots, x_t$ *appear in ascending order:*

(i) f is t-alternating.

(ii) $\pi f = \mathrm{sgn}(\pi) f$ for every $\pi \in S_t$.

(iii) $f = \sum_{\pi \in S_t} \mathrm{sgn}(\pi) \pi f_{(t)}$.

(iv) $f = \sum_i g_{0,i} c_t(x_1, x_2, \ldots x_t, g_{1,i}, g_{2,i}, \ldots, g_{t,i})$ for suitable polynomials $g_{0,i}, g_{1,i}, g_{2,i}, \ldots, g_{t,i}$.

Proof. $(i) \Leftrightarrow (ii)$ by Proposition 1.24, since every permutation is a product of transpositions.

$(ii) \Rightarrow (iii)$ by comparing the different monomials of f.

$(iii) \Rightarrow (iv)$ is immediate.

$(iv) \Rightarrow (i)$ is clear, since c_t is t-alternating. $\square$

Remark 1.26. Thus, by (iv), each t-alternating polynomial can be written in terms of c_t. If $c_t \in \mathrm{id}(A)$, then $\mathrm{id}(A)$ contains every t-alternating polynomial.

Definition 1.27. Define the *alternator* of a polynomial f *in indeterminates* $x_{i_1}, \ldots, x_{i_t}$ to be

$$f_{\mathcal{A}(i_1, \ldots, i_t; X)} = \sum \mathrm{sgn}(\pi) f_\pi,$$

summed over all permutations π of $i_1, \ldots, i_t$. Up to sign, this is independent of the order of $i_1, \ldots, i_t$. When $i_1 = 1, \ldots, i_t = t$, we call this the *t-alternator*

$$\sum_{\pi \in S_t} \mathrm{sgn}(\pi) \pi f.$$

For example, the t-alternator of $x_1 \cdots x_t$ is s_t, and the t-alternator of $x_1 y_1 \cdots x_t y_t$ is c_t.

Remark 1.28. If f already is alternating, then its t-alternator is $t! f$. On the other hand, if $f = \tau f$ for some odd permutation τ (for example, for a transposition), then its t-alternator is 0. Thus, before applying the alternator, we should make sure that f lacks such symmetry.

See Exercise 14 for some examples. There is an analog to this that is needed in Chapter 6. We say a polynomial f is *symmetric* in $x_1, \ldots, x_t$ if $f = \tau f$ for any transposition $\tau = (i_1, i_2)$.

Definition 1.29. Define the *t-symmetrizer* $\tilde{f}$ of a t-linear polynomial $f(x_1, \ldots, x_t, y_1, \ldots, y_m)$ to be

$$\sum_{\pi \in S_t} \pi f.$$

For example, the n-symmetrizer of $x_1 \cdots x_n$ is $\tilde{s}_n$ of (1.6). (Note the slight descrepancy of notation; there is no ambiguity since the symmetrizer of s_n is 0.)

Remark 1.30.

(i) If $f(x_1, \ldots, x_t; y_1, \ldots, y_m)$ is t-linear and symmetric in $x_1, \ldots, x_t$, then f is the linearization in x_1 of $f(x_1, \ldots, x_1; y_1, \ldots, y_m)$, cf. Remark 1.12(ii).

(ii) The symmetrizers of πf and of f are the same, for any $\pi \in S_t$.

(iii) If $f(x_1, \ldots, x_t; y_1, \ldots, y_m)$ is the linearization in x_1 of a polynomial $g(x_1; y_1, \ldots, y_m)$, each of whose monomials has degree n_1 in x_1, then

$$f(x_1, \ldots, x_1; y_1, \ldots, y_m) = n_1! g,$$

cf. Remark 1.12.

Thus, there is a natural correspondence between symmetric polynomials and linearizations with respect to a single indeterminate.

1.3   Graded Algebras and Grassman Algebras

We turn to another example. An algebra A is *graded* by an Abelian monoid $(M, +)$ if one can write $A = \{\oplus A_m : m \in M\}$, a direct sum of modules, such that $A_m A_n \subseteq A_{m+n}$. Note that A_0 is always a subalgebra of A, and each A_m is an A_0-module. (One can also do this just as well for nonabelian monoids, but the additive notation is consistent with our examples.) An element in some A_m is called *homogeneous*. Given any $a \in A$, we can write $a = \sum a_m$ (uniquely) where $a_m \in A_m$, i.e., any element can be written uniquely as a sum of homogeneous elements. In particular, $(a + b)_m = a_m + b_m$ for all $a, b \in A$ and all $m \in M$.

One grading monoid of great interest for us here is $\mathbb{Z}/2$, for which we call A a *superalgebra*. Thus, $A = A_0 \oplus A_1$ where $A_0^2, A_1^2 \subseteq A_0$, and $A_0 A_1, A_1 A_0 \subseteq A_1$. A_0 (respectively A_1) is called the *even* (respectively *odd*) part of A.

For example, the commutative polynomial algebra $F[\Lambda] = F[\lambda_1, \ldots, \lambda_n]$ can be graded by $\mathbb{N}^{(n)}$, by associating to any monomial f the "multi-degree" $(d_1, \ldots, d_n)$, where d_i is the degree of λ_i in f; then $F[\Lambda]_{(d_1, \ldots, d_n)}$ is spanned by the monomials of multi-degree $(d_1, \ldots, d_n)$. On the other hand, $F[\Lambda]$ is $\mathbb{N}$-graded when we take $F[\Lambda]_d$ to be spanned by all monomials of total degree d.

The obvious example in the PI-theory is the noncommutative analogue, motivated by Definition 1.4.

Example 1.31. The free algebra $C\{X\}$ is $\mathbb{N}$-graded, by taking $C\{X\}_n$ to be the homogeneous polynomials of degree n.

An ideal I of A is a *graded ideal* if $I = \sum(I \cap A_m)$, i.e., each element of I is a sum of homogeneous elements of I. It is easy to see that an ideal is graded iff it is generated by homogeneous elements.

Remark 1.32. If I is a graded ideal of A, then $A/I = \bigoplus_m (A_m / I \cap A_m)$ is also graded by the same monoid, i.e., we put $(a + I)_m = a_m + I$. This is easily seen to be well-defined.

Remark 1.33. If N is a submonoid of the Abelian monoid M, then any M-graded algebra also is M/N-graded, defining $A_{m+N} = \sum_{n \in N} A_{m+n}$. In particular, any $\mathbb{Z}$-graded or $\mathbb{N}$-graded algebra A becomes a superalgebra where $\sum\{A_m : m \text{ even}\}$ becomes the even part and $\sum\{A_m : m \text{ odd}\}$ becomes the odd part.

Here are some basic examples of this procedure.

Example 1.34.

(i) The free algebra $C\{X\}$ becomes a superalgebra, by taking the even (respectively odd) part to be generated by all homogeneous polynomials by even (respectively odd) degree. This is the grade obtained by applying Remark 1.33 to Example 1.31.

(ii) We can grade the matrix algebra $A = M_n(C)$ by $\mathbb{Z}$, setting $A_m = \sum\{Ce_{ij} : i - j = m\}$; this gives rise naturally to a $\mathbb{Z}/m$-grading, taking the components modulo m. The most important special case is for $m = 2$, yielding the superalgebra with respect to the *checkerboard grading*. (Here e_{ij} is even iff $i - j$ is even.)

(iii) Fixing any $k \leq n$, $M_n(C)$ becomes a superalgebra where

$$M_n(C)_0 = \left\{ \sum \alpha_{ij} e_{ij} : 1 \leq i, j \leq k \text{ or } k < i, j \leq n \right\}; \qquad (1.7)$$

$$M_n(C)_1 = \left\{ \sum \alpha_{ij} e_{ij} : 1 \leq i \leq k < j \leq n \text{ or } 1 \leq j \leq k < i \leq n \right\}. \qquad (1.8)$$

Letting $\ell = n - k$, it is customary to call this superalgebra $M_{k,\ell}(C)$. (This is a generalization of (ii), cf. Exercise 16.)

Our immediate interest is a superalgebra arising from (i).

1.3.1 The Grassman algebra

Definition 1.35. Let F be a field, and $V = \sum F x_i \subset F\{X\}$. The *Grassman algebra*, or *exterior algebra*, G is $G = F\{X\}/I$ where I is the ideal $\langle v^2 : v \in V \rangle$ of $F\{X\}$. (This is defined more generally in Exercise 17.)

Since I is generated by even polynomials, we see $I \cap V = 0$, so we can view $V \subseteq G$; from this point of view (which is the one we shall take), G is the algebra generated by V modulo the relations $v^2 = 0$, $\forall v \in V$.

Remark 1.36. Since we shall use the Grassman algebra so often, let us fix some basic notation. We fix a base $e_1, e_2, \ldots$ of V. By definition, for any α_i in F,

$$0 = \left(\sum \alpha_i e_i \right)^2 = \sum \alpha_i^2 e_i^2 + \sum \alpha_i \alpha_j (e_i e_j + e_j e_i), \tag{1.9}$$

which clearly is equivalent to the following set of relations:

$$e_i^2 = 0; \qquad e_i e_j = -e_j e_i, \quad \forall i \neq j. \tag{1.10}$$

Let

$$B = \{1\} \cup \{ e_{i_1} \cdots e_{i_m} \mid i_1 < \cdots < i_m, \quad m = 1, 2, \ldots \}. \tag{1.11}$$

Clearly, G is spanned by words $e_{i_1} e_{i_2} \ldots e_{i_m}$. We can use (1.10) to rearrange the indices in ascending order, and any repetition of e_i must be 0; hence, G is spanned by elements of B. We claim B is a base for G. Otherwise, given any relation $\sum \alpha_j b_j = 0$ of shortest length (for $b_j \in B$), suppose e_i appears in b_1 but not in b_2. Then $e_i b_1 = 0$, implying $\sum_{j>1} \alpha_j e_i b_j = 0$ is a relation of shorter length, contradiction.

Remark 1.37. Let us now describe the $\mathbb{Z}/2$-grade directly in terms of B. A word $e_{i_1} \ldots e_{i_m} \in B$ is *even* (respectively *odd*) if m is even or odd, respectively. G_0 is spanned by the even words, and G_1 is spanned by the odd words. Furthermore, since $1 \in G_0$, we have $G_1 = G_0 V$.

By definition, G_0 is an algebra generated by all words $e_{i_1} e_{i_2}$. But for any $v \in V$,

$$e_{i_1} e_{i_2} v = -e_{i_1} v e_{i_2} = v e_{i_1} e_{i_2};$$

thus, $e_{i_1} e_{i_2} \in \text{Cent}(G)$. The same argument shows more generally that if $a, b \in G$ are homogeneous, then $ab = \pm ba$, the $-$ sign occurring iff both a, b are odd. In particular, $\text{Cent}(G) = G_0$.

We can also describe G in the following way: We say two elements a, b of an algebra A *strictly anticommute* if $arb = -bra$ for all $r \in A$. (One also defines "strictly commute" analogously, but we shall not need that notion.) Whereas any even element of G is central, we have the following:

Remark 1.38.

(i) Any two odd elements a, b of G strictly anticommute. (Indeed, take any element $r = r_0 + r_1$, the sum of the even and odd parts;

$$arb = r_0 ab - r_1 ab = -r_0 ba + r_1 ba = -b(r_0 + r_1)a = -bra.)$$

(ii) In characteristic $\neq 2$, it follows that $ara = 0$ for any odd a, since $ara = -ara$.

This innocent computation has far-reaching consequences. Its first application was an observation half a century ago by P.M. Cohn that

$$s_m(e_1, \ldots, e_m) = m! e_1 \ldots e_m, \tag{1.12}$$

since

$$e_{\pi(1)} \cdots e_{\pi(m)} = \text{sgn}(\pi) e_1 \ldots e_m \quad \text{in} \quad G,$$

and in fact, Remark 1.38 analogously implies

$$c_m(e_1, \ldots, e_{2m}) = m! e_1 \ldots e_{2m}. \tag{1.13}$$

Consequently, the Grassman algebra of an infinite dimensional vector space in characteristic 0 cannot satisfy any Capelli identity! This observation turns out to have deep significance since, as we shall see, many PI-algebras (including all affine PI-algebras) do satisfy suitable Capelli identities. It was not until a generation later that Kemer fully utilized the fact that G lies on the opposite side of the coin from matrix algebras. See Exercise 18 for a description of the structure of G. Here is the most important PI of G.

Proposition 1.39. G *satisfies the* Grassman *identity* $[[x_1, x_2], x_3]$*; equivalently,* $[x_1, x_2]$ *is* G*-central.*

Proof. It is enough to check for homogeneous elements a, b that $[a, b]$ is even (and thus, central). If a or b is even, then $[a, b] = 0$, so we may assume a, b are odd, in which case $[a, b]$ is even, as desired.   □

In fact, all identities of G are consequences of the Grassman identity, by a fundamental result of Regev, cf. Corollary 3.46.

1.4   Identities and Central Polynomials of Matrices

Now we shall prove two of the most basic PI-theorems of matrix algebras: The Amitsur-Levitzki Theorem and the existence of central polynomials.

Amazingly, neither of these theorems is used much in this book. The standard polynomial has been replaced largely by the Capelli polynomial, and the combinatorics of the Capelli polynomial can be used to replace the transition to commutative theory afforded by central polynomials. Nevertheless, both the results and their proofs are important and instructive, and fully merit inclusion in an in-depth treatment of PIs. Each theorem has a short proof, which also utilizes an extension of the notion of PI.

1.4.1 Identities of matrices

The following result enables us to pass to the special case $C = \mathbb{Z}$, when considering identities with integer coefficients.

Remark 1.40. Any multilinear identity of $M_n(\mathbb{Z})$ also is an identity of $M_n(C)$ for any commutative ring C, the reverse direction holding if C has characteristic 0. (Indeed, $M_n(C) \approx M_n(\mathbb{Z}) \otimes_{\mathbb{Z}} C$, so we appeal to Proposition 1.16. If $\mathrm{char}(C) = 0$, then $\mathbb{Z} \subseteq C$ so $\mathrm{id}(M_n(C)) \subseteq \mathrm{id}(M_n(\mathbb{Z}))$.)

This fact actually holds for all identities, cf. Exercise 3.3. Let us turn to evaluating specific polynomials. Suppose we want to show a multilinear polynomial $f(x_1, \ldots, x_n)$ is *not* an identity of an algebra A. Writing f as in (1.1), it suffices to find $a_1, \ldots, a_t$ with $ca_1 \ldots a_t \neq 0$ but every other rearrangement $a_{\sigma(1)} \ldots a_{\sigma(t)} = 0$; then the summation in the right side of (1.1) is 0, but the first term is nonzero. (This is accomplished most readily when $a_j a_i = 0$ for all $j > i$.) Let us see how this idea works with some examples.

Remark 1.41. If $f(x_1, \ldots, x_t) \in \mathrm{id}(M_n(C))$ of degree d, then $d \geq 2n$. (Indeed, otherwise, we may assume f is multilinear; say

$$f = \alpha x_1 x_2 \ldots x_d + \sum_{\sigma \neq 1} \alpha_\sigma x_{\sigma(1)} x_{\sigma(2)} \ldots x_{\sigma(d)}.$$

Then

$$0 = f(e_{11}, e_{12}, e_{22}, e_{23}, \ldots, e_{k-1,k}, e_{kk}, \ldots) = \alpha e_{1m}$$

where $m = \left[\frac{d}{2}\right] + 1$, a contradiction.)

In particular, the standard polynomial s_{2n-1} is not an identity of $M_n(C)$. For $n = 2$, we already see s_3 is not a PI, but s_5 is a PI. To check s_4 it is enough to evaluate s_4 on the spanning set of matrix units $e_{11}, e_{12}, e_{21}, e_{22}$, and observe $s_4(e_{11}, e_{12}, e_{21}, e_{22}) = 0$. In general, the famous Amitsur-Levitzki Theorem (proved shortly) says s_{2n} is a PI of $M_n(C)$, for any commutative ring C, and in characteristic 0, this is the unique identity of degree $2n$, cf. Exercise 12.

Remark 1.42. The Capelli polynomial, although more complicated to write than the standard polynomial, is easier to analyze on matrices. It is easy to see that c_{n^2} is not an identity of $M_n(C)$. Indeed, we shall find matrix units $\bar{x}_i$ and $\bar{y}_i$ such that

$$c_{n^2}(\bar{x}_1, \ldots, \bar{x}_{n^2}, \bar{y}_1, \ldots, \bar{y}_{n^2}) \neq 0.$$

First, we list a set of matrix units in any order, say

$$\bar{x}_1 = e_{11}, \quad \bar{x}_2 = e_{12}, \ldots, \quad \bar{x}_n = e_{1n}, \quad \bar{x}_{n+1} = e_{21}, \ldots, \quad \bar{x}_{n^2} = e_{nn}.$$

Note for any matrix unit a, we have $e_{i_1 j_1} a e_{i_2 j_2} = 0$ unless $a = e_{j_1 i_2}$, in which case $e_{i_1 j_1} a e_{i_2 j_2} = e_{i_1 j_2}$. Thus, there is a unique specialization of the y_i to matrix units $\bar{y}_i$ such that $\bar{x}_1 \bar{y}_1 \bar{x}_2 \bar{y}_2 \ldots \bar{x}_{n^2} \bar{y}_{n^2} = e_{11}$. But the same argument shows that $\bar{y}_{i-1} \bar{x}_{\pi(i)} \bar{y}_i = 0$ unless $\pi(i) = i$; hence

$$\bar{x}_{\pi(1)} \bar{y}_1 \bar{x}_{\pi(2)} \bar{y}_2 \ldots \bar{x}_{\pi(n^2)} \bar{y}_{n^2} = 0$$

for every $\pi \neq (1)$. We conclude

$$c_{n^2}(\bar{x}_1, \ldots, \bar{x}_{n^2}, \bar{y}_1, \ldots, \bar{y}_{n^2}) = \bar{x}_1 \bar{y}_1 \ldots \bar{x}_{n^2} \bar{y}_{n^2} = e_{11}.$$

(Note that by multiplying $\bar{y}_{n^2}$ on the right by e_{1k}, we see c_{n^2} takes on the value e_{1k}, and hence, by symmetry, any matrix unit is a value of c_{n^2}.)

Let us formalize this idea, for use in Chapter 4: Let us call a set of matrix units $u_1, \ldots, u_t; v_1, \ldots, v_t$ *compatible with* the monomial $x_1 y_1 x_2 y_2 \ldots x_t y_t$ if

$$u_1 v_1 u_2 v_2 \ldots u_t v_t = e_{11}$$

but

$$u_{\sigma(1)} v_1 u_{\sigma(2)} v_2 \ldots u_{\sigma(t)} v_t = 0$$

for every permutation $\sigma \neq (1)$. Such a set exists iff $t \leq n^2$; when $t = n^2$ the $u_1, \ldots, u_t$ must be a base, i.e., must run over all the $n \times n$ matrix units. We just saw examples of this phenomenon. Then in this case

$$c_t(u_1, \ldots, u_t; v_1, \ldots, v_t) = u_1 v_1 u_2 v_2 \ldots u_t v_t = e_{11} \neq 0.$$

Example 1.43. Here is a related result: If H is an algebra such that $c_{n^2+1} \in \mathrm{id}(M_n(H))$, then H is commutative. Indeed, take any $a, b \in H$ and, as in Remark 1.42, we list the set of matrix units

$$\bar{x}_1 = e_{11}, \quad \bar{x}_2 = e_{12}, \quad \ldots, \quad \bar{x}_{n^2} = e_{nn},$$

and take the unique matrix units $\bar{y}_i$ such that $\bar{x}_1\bar{y}_1\bar{x}_2\bar{y}_2\ldots\bar{x}_{n^2}\bar{y}_{n^2} = e_{11}$. Note $\bar{y}_1 = e_{11}$. Then (repeating $\bar{y}_1 = e_{11}$ in the evaluation)

$$
\begin{aligned}
0 &= e_{11}c_{n^2+1}(a\bar{x}_1, b\bar{x}_1, \bar{x}_2 \ldots, \bar{x}_{n^2}, \bar{y}_1, \bar{y}_1, \bar{y}_2 \ldots, \bar{y}_{n^2}) \\
&= (abe_{11} - bae_{11})\bar{y}_1 \ldots \bar{x}_{n^2}\bar{y}_{n^2} \\
&= (ab - ba)e_{11},
\end{aligned}
\tag{1.14}
$$

proving $ab = ba$, $\forall a, b \in H$, as desired.

One of the main themes in PI-theory is to utilize the (monic) characteristic polynomial

$$ f_a = \det(\lambda I - a) $$

for any $a \in M_n(C)$. The Hamilton-Cayley theorem says $f_a(a) = 0$.

When C is a $\mathbb{Q}$-algebra, the coefficients can be expressed in terms of $\{\operatorname{tr}(a^m) : m \leq n\}$, via *Newton's Formulas*, cf. [MacD95, p. 23], or [Jac80, p. 140], which are best formulated in terms of symmetric polynomials in commuting indeterminates $\lambda_1, \ldots, \lambda_n$ (not to be confused with the noncommutative symmetric polynomials defined earlier):

Remark 1.44 (Newton's Formulas). Let

$$ e_m = e_m(\lambda_1, \ldots, \lambda_n) = \sum_{1 \leq i_1 < \cdots < i_m \leq n} \lambda_{i_1} \cdots \lambda_{i_m} $$

be the *elementary symmetric polynomials*, and let

$$ p_k = p_k(\lambda_1, \ldots, \lambda_n) = \lambda_1^k + \cdots + \lambda_n^k $$

be the *power symmetric polynomials* [MacD95]. Then for $m = 1, 2, \ldots$

$$ me_m = \sum_{k=1}^{m} (-1)^{k-1} p_k e_{m-k}. $$

Recursively, Newton's formulas allow us to derive formulas of the form

$$ e_m = q_m(p_1, \ldots, p_m), $$

where q_m is a suitable polynomial with rational coefficients and with constant term 0. (This is where we use characteristic 0; actually, we need only divide by $m!$.)

Thus, $m = 1$ implies $e_1 = p_1$; together with the case $m = 2$, we see $e_2 = \frac{1}{2}(p_1^2 - p_2)$. Similarly, one deduces that $e_3 = \frac{1}{6}(p_1^3 - 3p_1p_2 + 2p_3)$, etc.

Remark 1.45. If $p_k = 0$ for $1 \leq k \leq m$, then $me_m = 0$, implying $e_m = 0$.

Let us translate this fact into matrix theory. Suppose a is an $n \times n$ matrix with characteristic values $\gamma_1, \ldots, \gamma_n$ in a field K. Clearly $\text{tr}(a) = \sum_{i=1}^{n} \gamma_i$; hence

$$p_k(\gamma_1, \ldots, \gamma_n) = \text{tr}(a^k).$$

Furthermore, let $f_a = \det(\lambda I - a)$ be the characteristic polynomial of a and write

$$f_a = \lambda^n + \alpha_1 \lambda^{n-1} + \cdots + \alpha_n.$$

Then $\alpha_m = (-1)^m e_m(\gamma_1, \ldots, \gamma_n)$.

Proposition 1.46. *Suppose $a \in M_n(C)$, where C is a commutative $\mathbb{Q}$-algebra, and denote $t_k = \text{tr}(a^k)$. Then each coefficient α_m can be written as a polynomial in $t_1, \ldots, t_m$ with rational coefficients and with constant term 0. In particular, if $t_k = 0$ for $1 \leq k \leq n$, then $f_a = \lambda^n$.*

Proof. First we assume C is an integral domain. Write $a = (c_{i,j})$ $(c_{i,j} \in C)$. Let K be the algebraic closure of the field of fractions of C. Then f_a factors completely over K:

$$f_a = \lambda^n + \alpha_1 \lambda^{n-1} + \cdots + \alpha_n = \prod_{i=1}^{n} (\lambda - \gamma_i).$$

As noted above, $\alpha_m = e_m(\gamma_1, \ldots, \gamma_n)$, so the first assertion follows from Newton's formulas. The assumption for the second assertion is

$$0 = t_k = p_k(\gamma_1, \ldots, \gamma_n)$$

for $1 \leq k \leq n$, so Remark 1.45 shows $\alpha_m = 0$ for $1 \leq m \leq n$, as desired.

To prove the proposition in general, we turn to a trick that comes up rather frequently in PI-theory. Let $\xi_{i,j}$ be commuting variables, and form the $n \times n$ *generic* matrix $\overline{x} = (\xi_{i,j}) \in M_n(D)$ where D is the integral domain $D = F[\xi_{i,j} : 1 \leq i, j \leq n]$. The homomorphism $\varphi : D \to C$, given by $\varphi : \xi_{i,j} \mapsto c_{i,j}$, extends to $\varphi : M_n(D) \to M_n(C)$ with $\varphi : \overline{x} \mapsto a$, and obviously, $\varphi(\text{tr}(\overline{x}^j)) = \text{tr}(a^j)$. Thus, it suffices to prove the proposition for the matrix $\overline{x}$, or, in other words, we may replace C by the integral domain D, for which case the proposition has just been proved. $\square$

Corollary 1.47. *Suppose C is a commutative $\mathbb{Q}$-algebra. If $a \in M_n(C)$ satisfies $\text{tr}(a^r) = 0$, $r = 1, \ldots, n$, then $a^n = 0$.*

These considerations lead us to consider identities involving a formal linear map tr satisfying $\text{tr}(1) = n$ and $\text{tr}(ab) = \text{tr}(ba)$, thereby giving rise to *trace identities* which will be studied in detail in Chapter 12; Newton's formulas enable us to translate the Hamilton-Cayley theorem into a trace identity for $M_n(C)$. On the other hand, we can find trace identities which hold for *all* n. By definition, $M_n(C)$ satisfies the trace identity $\mathrm{tr}[x_1, x_2]$. More generally, Kostant observed the following fact:

Remark 1.48. $\mathrm{tr}\, s_{2k}(x_1, \ldots, x_{2k})$ is a trace identity of $M_n(C)$ for any k, n. (Indeed for any matrices $a_1, \ldots, a_{2k}$ and any permutation π, we see

$$\mathrm{tr}(a_{\pi(1)} a_{\pi(2)} \cdots a_{\pi(2k)}) = \mathrm{tr}(a_{\pi(2)} \cdots a_{\pi(2k)} a_{\pi(1)});$$

but the cycle $(\pi(1)\, \pi(2)\, \ldots\, \pi(2k))$ is odd, so the terms $a_{\pi(1)} a_{\pi(2)} \cdots a_{\pi(2k)}$ and $a_{\pi(2)} \cdots a_{\pi(2k)} a_{\pi(1)}$ appear in $s_{2k}(a_1, \ldots, a_{2k})$ with opposite sign, and their contributions cancel.)

Combining these basic remarks about trace identities with the Grassman algebra yields the following basic result:

Theorem 1.49 (Amitsur-Levitzki). s_{2n} *is an identity of* $M_n(C)$, *for every commutative ring* C.

Proof. (Rosset) We need only check $s_{2n} \in \mathrm{id}(M_n(\mathbb{Q}))$, since then clearly $s_{2n} \in \mathrm{id}(M_n(\mathbb{Z}))$, and we would conclude with Remark 1.40. Let us work in $M_n(G) \approx M_n(\mathbb{Q}) \otimes_{\mathbb{Q}} G$. Take arbitrary matrices $a_1, \ldots, a_n$ in $M_n(\mathbb{Q})$, and consider $a_i \otimes e_i$ in $M_n(G)$. Recalling $e_{\pi(1)} \cdots e_{\pi(2k)} = \mathrm{sgn}(\pi) e_1 \cdots e_{2k}$, we have

$$s_{2k}(a_1, \ldots, a_{2k}) \otimes e_1 \ldots e_{2k} = \sum_{\pi \in S_{2k}} (a_{\pi(1)} \otimes e_{\pi(1)}) \cdots (a_{\pi(2k)} \otimes e_{\pi(2k)})$$

$$= (a_1 \otimes e_1 + \cdots + a_{2k} \otimes e_{2k})^{2k}.$$

Thus, letting $c = (a_1 \otimes e_1 + \cdots + a_{2n} \otimes e_{2n})^2$, it suffices to show $c^n = 0$.

Note that $c \in M_n(G_0)$; since G_0 is commutative, using Corollary 1.47, it is enough to show that $\mathrm{tr}(c^k) = 0$ for each $k \le n$. But arguing in reverse, in conjunction with Remark 1.38(ii),

$$\mathrm{tr}((a_1 \otimes e_1 + \cdots + a_{2n} \otimes e_{2n})^{2k})$$

$$= \mathrm{tr}\left(\sum_{1 \le i_1, \ldots, i_{2k} \le 2n} (a_{i_1} \otimes e_{i_1}) \ldots (a_{i_{2k}} \otimes e_{i_{2k}}) \right)$$

$$= \sum_{i_1 < \cdots < i_k} \mathrm{tr}(s_{2k}(a_{i_1}, \ldots, a_{i_{2k}})) \otimes e_{i_1} \ldots e_{i_{2k}} = 0,$$

by Remark 1.48. $\qquad\square$

This proof anticipates a key idea we shall need in Chapter 6. To check s_{2n} on the algebra $A = M_n(\mathbb{Z})$, we looked at the symmetric polynomial evaluated on the odd part of $A \otimes G$. This interplay between $\mathrm{id}(A)$ and certain graded identities of the superalgebra $A \otimes G$ lies at the heart of Kemer's solution of Specht's problem.

1.4.2 Central polynomials for matrices

We turn to central polynomials for matrices. The easiest verification of a central polynomial is obtained by extracting the main idea from Razmyslov's construction, and trivializing the remainder of his proof.

Definition 1.50. A *k-weak identity* of $M_n(F)$ is a polynomial $f(x_1, \ldots, x_m)$ that vanishes on all substitutions to matrices $a_1, \ldots, a_m$ for which $\mathrm{tr}(a_1) = \cdots = \mathrm{tr}(a_k) = 0$.

For example, the Capelli polynomial c_{n^2} is a multilinear n^2-weak identity of $M_n(F)$ (since it is n^2-alternating but the matrices of trace 0 have dimension only $n^2 - 1$), but c_{n^2} is not an identity of $M_n(F)$, by Remark 1.42. Here is an example of lower degree, inspired by an example of Amitsur given in Exercise 11.

Lemma 1.51. $s_{n-1}([x, y], [x^2, y], \ldots, [x^{n-2}, y], [x^n, y])$ is a 1-weak identity of $M_n(F)$.

Proof. Suppose $a \in M_n(F)$ has trace 0. Then the characteristic polynomial of a has coefficient 0 for λ^{n-1}, thereby yielding a linear dependence for $1, a, \ldots, a^{n-2}, a^n$. Commuting with b, noting $[1, b] = 0$, we see $[a, b], \ldots, [a^{n-2}, b], [a^n, b]$ are linearly dependent, so can be spanned by $n-2$ elements, implying $s_{n-1}([a, b], [a^2, b], \ldots, [a^{n-2}, b], [a^n, b]) = 0$, i.e., $s_{n-1}([x, y], [x^2, y], \ldots, [x^{n-2}, y], [x^n, y])$ is a 1-weak identity. $\square$

This 1-weak identity has degree $n + \frac{1}{2}(n-2)(n-1) = \frac{1}{2}(n^2 - n + 2)$ in x and $n - 1$ in y. Let $m = \frac{1}{2}(n^2 - n + 2)$. Multilinearizing at x_1 yields an m-weak identity $f(x_1, \ldots, x_m, y_1, \ldots, y_{n-1})$. To get a 1-weak multilinear identity, we need to substitute $[x_{i1}, x_{i2}]$ for x_i, $2 \le i \le m$, since these will specialize to matrices of trace 0. This multilinear 1-weak identity has total degree

$$m + (m - 1) + n - 1 = n^2 - n + 1 + n - 1 = n^2,$$

and is not an identity of $M_n(C)$, cf. Exercise 19.

The easiest way to understand the connection between central polynomials and 1-weak identities is in terms of another generalization of identity.

Definition 1.52. A *linear generalized identity* (LGI) for A is an expression $f = \sum_{j=1}^{k} a_j x b_j$ with a_j, b_j in A, for which $\sum a_j r b_j = 0$ for all r in A. Likewise f is *linear generalized central* (LGC) if $\sum a_j r b_j \in \text{Cent}(A)$ for all r in A.

(More generally, a *generalized polynomial* is a noncommutative polynomial with coefficients from A, interspersed throughout the x_i, and there is a well-developed theory of generalized polynomial identities, cf. [BeMM96] or [Row80, Chapter 7]. However, for our purposes, it suffices to consider this very special case.) Let A^{op} denote the opposite algebra of A; i.e., the same addition, but multiplication in the reverse order.

Remark 1.53. If $\sum a_j x b_j$ is an LGI of A, then $\sum b_j x a_j$ is an LGI of A^{op}.

Lemma 1.54. *Suppose* $f = \sum_j a_j x b_j$. *Then* $\sum b_j x a_j$ *is LGC iff* $f(r) = 0$ *for every matrix* r *of trace 0.*

Proof. The condition is that

$$0 = f([r, s]) = \sum_j a_j r s b_j - \sum_j a_j s r b_j, \qquad \forall r, s \in A,$$

i.e., $\sum a_j x s b_j - \sum a_j s x b_j$ is an LGI, for all s in A. By Remark 1.53, this is equivalent to $\sum s b_j x a_j - \sum b_j x a_j s = [s, \sum b_j x a_j]$ being an LGI of A^{op} for all s in A, which means $\sum b_j x a_j$ is A^{op}-central. But $A^{\text{op}} \approx A$ since the transpose is an anti-automorphism of $M_n(C)$. $\qquad \square$

We are ready for a major result.

Theorem 1.55 (Razmyslov). *There is a 1:1 correspondence between the multilinear central polynomials of* $M_n(C)$ *and the multilinear 1-weak identities that are not identities.*

Proof. Write

$$f(x, x_1, \ldots, x_n) = \sum_{i=1}^{t} f_i(x_1, \ldots, x_n) x g_i(x_1, \ldots, x_n),$$

and let

$$\hat{f}(x, x_1, \ldots, x_n) = \sum_{i=1}^{t} g_i(x_1, \ldots, x_n) x f_i(x_1, \ldots, x_n).$$

Picking matrices $r_1, \ldots, r_n$ arbitrarily, and letting $a_i = f_i(r_1, \ldots, r_n)$ and $b_i = g_i(r_1, \ldots, r_n)$, we see by Remark 1.53 (noting $M_n(C) \cong M_n(C)^{\text{op}}$ via the transpose) that $f \in \mathrm{id}(M_n(C))$ iff $\hat{f} \in \mathrm{id}(M_n(C))$. Likewise, by Lemma 1.54, $\hat{f}$ is central or an identity for $M_n(C)$, iff f is a 1-weak identity. Combining these two observations gives the result. $\square$

Since we have already constructed 1-weak identities which are not PIs for $M_n(C)$, we have the existence of central polynomials. Surprisingly, the minimal possible degree of a 1-weak identity (and thus, central polynomial) for $M_n(F)$ is not known. Formanek's polynomial [For72] has degree n^2, as does the central polynomial constructed via Theorem 1.55 from the Amitsur-Halpin 1-weak identity. However, for $n = 3$, a computer search uncovered a central polynomial of degree 8. The best result known, in [Dr95], is a central polynomial of degree $(n-1)^2 + 4$ for $M_n(\mathbb{Q})$.

1.5   The Identity of Algebraicity

Let us derive an amusing sort of identity, known by Latyshev, for an arbitrary PI-algebra satisfying a multilinear identity f of degree d. At this stage, this identity $(D(x_1 \cdots x_d)$ defined below) may seem just a rather technical curiosity, but we need it for the important task of establishing Capelli identities in Chapter 2. (We can manage without it in characteristic 0.)

Definition 1.56. Given an indeterminate y and a set of indeterminates $x_1, x_2, \ldots$, define the left and right *replacement operators*

$\ell_i : x_i \mapsto y x_i$, and

$\mathfrak{r}_i : x_i \mapsto x_i y$,

fixing all other x_j. For example,

$$\ell_j(x_i y^k x_j) = x_i y^k (y x_j) = x_i y^{k+1} x_j = (x_i y) y^k x_j = \mathfrak{r}_i(x_i y^k x_j). \quad (1.15)$$

Now define $D_{ij} = \ell_j - \mathfrak{r}_i$ for $i \neq j$.

Remark 1.57. $[\ell_i, \ell_j] = [\ell_i, \mathfrak{r}_j] = [\mathfrak{r}_i, \mathfrak{r}_j] = 0$ for all i, j. Indeed, this is clear since for $i \neq j$ the replacements are on different letters and it does not matter which is performed first. For $i = j$,

$$\ell_i \mathfrak{r}_i(\ldots x_i \ldots) = \ldots y x_i y \ldots = \mathfrak{r}_i \ell_i(\ldots x_i \ldots).$$

Hence, all the D_{ij} commute, and we may define

$$D = \prod_{i \neq j-1,\ i \neq j,\ 1 \leq i, j \leq d} D_{ij}.$$

Equation (1.15) implies $D_{ij}(\ldots x_i y^k x_j \ldots) = 0$ for any k, where x_i, x_j does not appear elsewhere in the monomial. Hence, for any monomial h linear in $x_1, \ldots, x_d$ (and in which y appears in arbitrary positions), $D(h) = 0$ unless the x_i appear in h precisely in the order $x_1, x_2, \ldots, x_d$.

Proposition 1.58. *If A satisfies a multilinear identity*

$$f = x_1 \ldots x_d + \sum_{(1) \neq \pi \in S_d} \alpha_\pi x_{\pi(1)} \cdots x_{\pi(d)},$$

then $D(x_1 \cdots x_d) \in \mathrm{id}(A)$.

Proof. Clearly $D(f) \in \mathrm{id}(A)$, since we get sums of different evaluations of f. But $D(x_{\pi(1)} \ldots x_{\pi(d)}) = 0$ for all $\pi \neq (1)$, by Remark 1.57. Hence

$$D(x_1 \cdots x_d) = D(f) \in \mathrm{id}(A).$$

$\square$

Remark 1.59. Let us calculate $D(x_1 \cdots x_d)$. There are $d(d-1)$ ways of choosing (i, j) with $i \neq j$, from which we deduct the $d - 1$ possibilities for $i = j - 1$, yielding $(d-1)^2$ factors. From each factor, we may choose ℓ or $\mathfrak{r}$, so $D(x_1 \cdots x_d)$ is a sum of $2^{(d-1)^2}$ monomials, precisely one of which is

$$x_1 y^{d-1} x_2 y^{d-2} \ldots x_{d-1} y^{d-2} x_d y^{d-2},$$

obtained when we choose $\mathfrak{r}_i$ from D_{ij} at each opportunity. (This is the unique monomial in which each occurence of y is as far to the right as possible.) Likewise, choosing ℓ_i from D_{ij} at each opportunity yields the monomial

$$y^{d-1} x_1 y^{d-2} x_2 y^{d-2} \ldots x_{d-1} y^{d-2} x_d.$$

Thus, $D(x_1 \cdots x_d)$ is a nontrivial polynomial which is a sum of monomials of the form

$$\pm y^{k_0} x_1 y^{k_1} x_2 \ldots x_d y^{k_d},$$

where $k_0 + \cdots + k_d = (d-1)^2$; the "largest" $(d+1)$-tuple $(k_0, \ldots, k_d)$ which appears (under the lexicographic order) is

$$(d-1, d-2, d-2, \ldots, d-2, 0).$$

Given any polynomial linear in $x_1, \ldots, x_d$, we could then use $D(x_1 \cdots x_d)$ to "reduce" the $(d+1)$-tuple of powers of y. Accordingly, $D(x_1 \cdots x_d)$ is called the *identity of algebraicity*. ([DrFor04, page 41] illustrates this for $A = M_n(K)$.)

Kemer [Kem81] first proved that a PI-algebra of characterstic 0 satisfies an identity of algebraicity; the short proof given above holds for all characteristics.

1.6 Review of Major Structure Theorems in PI-Theory

This section lists those basic theorems that we shall need from structure theory, which can be found in [Row80] or [Row88]. They hold for PI-algebras over an arbitrary commutative ring, and are so basic for further reference in this book that we refer to them with letters rather than numbers, for special distinction.

Theorem A (Kaplansky's Theorem). *Any primitive algebra A satisfying a PI of degree d has the form $M_t(D)$ where D is a division algebra of dimension m^2 over $\operatorname{Cent}(A)$, and $mt \leq \left[\frac{d}{2}\right]$. Thus, $[A : \operatorname{Cent}(A)] = (mt)^2$.*

Corollary 1.60. *Every primitive ideal of a PI-algebra A is maximal. In particular, the Jacobson radical $J(A)$ is the intersection of the maximal ideals of A.*

Theorem B (Amitsur). *Any semiprime PI-algebra A has no nonzero left or right nil ideals, and $J(A[\lambda]) = 0$, where $A[\lambda]$ is the polynomial algebra in the commuting indeterminate λ.*

Theorem C (Rowen). *Every ideal of a semiprime PI-algebra A intersects $\operatorname{Cent}(A)$ nontrivially.*

Corollary D (Posner). *If A is prime PI with center C, then for $S = C \setminus \{0\}$, the localization $S^{-1}A$ is simple and f.d. over its center $F = S^{-1}C$, the field of fractions of C.*

Corollary 1.61. *If A is prime PI with center C, then $A \otimes_C \bar{F} \approx M_n(\bar{F})$, for a suitable algebraically closed field $\bar{F}$ and suitable n.*

Proof. By Corollary D the algebra of central fractions of $A \otimes_C F$ is central simple over the field of fractions F of C. If C is finite, then $F = C$, and $A = M_n(F)$ by a well-known theorem of Wedderburn. Thus, we may assume C is infinite, in which case $A \subset A \otimes \bar{F}$ where $\bar{F}$ is the algebraic closure of A. $A \otimes \bar{F}$ is simple and f.d. over $\bar{F}$, so is isomorphic to $M_n(\bar{F})$ since $\bar{F}$ is algebraically closed. $\qquad\square$

Remark 1.62. Under the above assumptions, $A \sim_{\mathrm{PI}} M_n(F)$, by Proposition 1.19.

This leads us to study algebras that are PI-equivalent to $M_n(F)$. The following definition focuses on the key properties.

Definition 1.63. Write $\mathcal{M}_n$ for $\operatorname{id}(M_n(\mathbb{Q}))$. An algebra A has PI-*class* n if

$$\mathcal{M}_n = \operatorname{id}(A) \cap \mathcal{M}_{n-1};$$

in other words, if A satisfies all identities of $M_n(\mathbb{Q})$, but no extra identities of $M_{n-1}(\mathbb{Q})$.

Thus, $M_n(C)$ has PI-class n, whereas it has PI-degree $2n$ by the Amitsur-Levitzki Theorem. Together with the structure theory, this shows

Corollary E. *Any semiprime PI-algebra has suitable PI-class n (and PI-degree $2n$) for suitable n.*

Remark F. If $g(x_1, \ldots, x_d)$ is $M_n(\mathbb{Q})$-central and A has PI-class n, then

$$h_n = g\big(c_{n^2}(x_1, \ldots, x_{n^2}, y_1, \ldots, y_{n^2}), x_{n^2+1}, \ldots, x_{n^2+d-1}\big) \qquad (1.16)$$

is multilinear, A-central and also n^2-alternating. (Indeed, the multilinear and alternating properties are immediate from those of c_{n^2}. Furthermore, $h_n(A) \subseteq g(A)$ and g is either an identity of A or is A-central, so it remains to show $h_n \notin \mathrm{id}(A)$. Since $c_{n^2} \in \mathcal{M}_{n-1}$, clearly $h_n \in \mathcal{M}_{n-1}$, so, in view of Definition 1.63, it suffices to note $h_n \notin \mathcal{M}_n$, which is true by Remark 1.42 frem3.15 since c_{n^2} evaluated on $M_n(\mathbb{Q})$ produces all the matrix units.)

For $i > n^2$ we rewrite the indeterminate x_i as y_i, in order to differentiate between the alternating indeterminates (the x_i for $i \leq n^2$) and the other indeterminates.

Let us reiterate that we have an n^2-alternating central polynomial, despite the fact that c_{n^2+1} is an identity of A since A has PI-class n. Thus, the next remark is relevant, taking $t = n^2$.

Remark G. Suppose $h(x_1, \ldots, x_t, x_{t+1}; y)$ is any t-alternating polynomial. Then the polynomial

$$\tilde{h}(x_1, \ldots, x_{t+1}; y) = \sum_{i=1}^{t+1} (-1)^i h(x_1, \ldots, x_{i-1}, x_{i+1}, \ldots, x_{t+1}, x_i; y)$$

is $t+1$-alternating. Indeed, suppose we specialize $x_j \mapsto x_i$. Then all terms on the right side are 0 except

$$(-1)^i h(x_1, \ldots, x_{i-1}, x_{i+1}, \ldots, x_{t+1}, x_i; y)$$

and

$$(-1)^j h(x_1, \ldots, x_{j-1}, x_{j+1}, \ldots, x_{t+1}, x_i; y)$$
$$= (-1)^j (-1)^{j-i-1} h(x_1, \ldots, x_{i-1}, x_{i+1}, \ldots, x_{j-1}, x_i, x_{j+1}, \ldots, x_{t+1}, x_i; y)$$
$$= (-1)^{i-1} h(x_1, \ldots, x_{i-1}, x_{i+1}, \ldots, x_{j-1}, x_i, x_{j+1}, \ldots, x_{t+1}, x_i; y),$$

which cancel.

In particular, if A satisfies c_{t+1}, then $\tilde{h} \in \mathrm{id}(A)$. Thus, for all $a_1, \ldots,$ $a_{t+1}, r_1, \ldots, r_m$ in A, we have

$$(-1)^t h(a_1, \ldots, a_{t+1}; r_1, \ldots, r_m)$$
$$= \sum_{i=1}^{t} (-1)^i h(a_1, \ldots, a_{i-1}, a_{i+1}, \ldots, a_{t+1}, a_i; r_1, \ldots, r_m). \tag{1.17}$$

In particular, we can apply $\tilde{h}$ when A has PI-class n. (We shall need the more general formulation for the Razmyslov-Zubrilin theory of traces in Chapters 2 and 4.)

Theorem H. *Suppose A has PI-class n and $h_n(x_1, \ldots, x_{n^2}; y)$ is the n^2-alternating central polynomial of Remark F. If*

$$h_n(a_1, \ldots, a_{n^2}, r_1, \ldots, r_m) = 1,$$

then A is a free $\mathrm{Cent}(A)$-module with base $a_1, \ldots, a_{n^2}$.

More generally, if $h_n(a_1, \ldots, a_{n^2}, r_1, \ldots, r_m) = c$, then Ac is contained in a f.g. free C-submodule of A. (For the proof, take

$$h = h_n(x_1, \ldots, x_{n^2}; y) x_{n^2+1}$$

and apply (1.17).)

This result contains the essence of the Artin-Procesi theorem, as shown in [Row80, Chapter 1]. In fact, A of Theorem H is an Azumaya algebra, thereby making it possible to localize any prime PI-algebra by a single element (namely, any nonzero evaluation of h_n) to obtain an Azumaya algebra. Although this technique is fundamental in the structure of PI-theory (and for example is used repeatedly in Braun's proof of the Razmyslov-Kemer-Braun theorem), we hardly use Azumaya algebras in this book.

1.6.1   Hamilton-Cayley properties of alternating polynomials

We now turn to one of the most basic tools in the PI-theory, enabling us to encode the Hamilton-Cayley theorem from linear algebra into alternating polynomials, specifically the Capelli polynomial. The underlying philosophy of the subject is contained in the following observation:

Remark 1.64. Suppose f is a multilinear, alternating form on t variables on a vector space $V = F^{(t)}$ of A, and $T : V \to V$ is a linear transformation of V. Then

$$f(T(x_1), \ldots, T(x_t)) = \det(T) f(x_1, \ldots, x_t).$$

We apply this, viewing a t-alternating polynomial (such as the Capelli polynomial) as a multilinear, alternating A-valued form, taking $V = A$ and

T to be left multiplication by an element of A. Let us first record a special case, given in [Row80, Theorem 1.4.12].

Theorem I. *If A spans $M_n(K)$ over K, so that $t = n^2$, and if we let I denote the additive subgroup of A generated by $h_n(A)$, then not only is $0 \neq I \lhd \text{Cent}(A)$, but $\alpha I \subseteq I$ for every characteristic coefficient α of every matrix over A.*

The same proof word for word gives:

Theorem J. *Suppose $A \subseteq M_n(K)$, and let V be a t-dimensional K-subspace of $M_n(K)$ with a base $a_1, \ldots, a_t$ of elements of A. Given a C-linear map $T : V \to V$, let*

$$\lambda^t + \sum_{i=1}^{t} (-1)^i \alpha_i \lambda^{t-i}$$

denote the characteristic polynomial of T (as a $t \times t$ matrix over K). Then, for any t-alternating polynomial $f(x_1, \ldots, x_t; y_1, \ldots, y_m)$, and any $1 \leq k \leq t$,

$$\alpha_k f(a_1, \ldots, a_t, r_1, \ldots, r_m) = \sum f(T^{k_1} a_1, \ldots, T^{k_t} a_t, r_1, \ldots, r_m), \quad (1.18)$$

summed over all vectors $(k_1, \ldots, k_t)$ for which each

$$k_i \in \{0, 1\}, \quad k_1 + \cdots + k_t = k.$$

In particular,

$$\text{tr}(T) f(a_1, \ldots, a_t, r_1, \ldots, r_m)$$
$$= \sum_{k=1}^{t} f(a_1, \ldots, a_{k-1}, T a_k, a_{k+1}, \ldots, a_t, r_1, \ldots, r_m). \qquad (1.19)$$

Because of its importance, let us recall the main idea of the proof. First we verify (1.18) for $k = t$. The form

$$f(T x_1, \ldots, T x_t, t_1, \ldots, y_m)$$

is linear and alternating in the $x_1, \ldots, x_t$, so by Remark 1.64 is

$$\det(T) f(x_1, \ldots, x_t; y_1, \ldots, y_m),$$

which is (1.18) for the case $k = t$. But now replace T by the transformation $\lambda I_t - T$, whose determinant gives the Hamilton-Cayley polynomial. Matching coefficients yields (1.18) in general, and in particular, (1.19).

When interpreting (1.19), we must exercise care even when $t = n^2$, since we are interpreting T as an $n^2 \times n^2$ matrix and not an $n \times n$ matrix. See Exercises 21 and 20 for a related result.

Theorem J has profound implications in the realization of the "Shir-shov program" described in Chapter 2, especially in connection with the "characteristic closure," since Corollary 1.61 shows that the hypotheses of Theorem I are satisfied by prime PI-algebras. It also is the key to the Razmyslov-Zubrilin theory of adjoining traces, and as such is crucial in Kemer's solution of Specht's problem.

PIs have another fundamental connection to representation theory.

Remark 1.65. Suppose A is a PI-algebra over an algebraically closed field F. Taking all the irreducible representations $\rho_i : A \to M_{n_i}(F)$, for $i \in I$, we note that each ρ_i is onto, by the Jacobson Density Theorem, so PI-$\deg(A) \geq n_i$ for each $i \in I$. On the other hand, by Kaplanksy's theorem, each primitive image of A is simple Artinian, and thus, of the form $M_{n_i}(F)$, so we have a homomorphism $\rho : A \to \prod_{i \in I} M_{n_i}(F)$, whose kernel is $J(A)$. In this way, we see that PI-$\deg(A/J(A))$ is the maximal degree of the irreducible representations of A over the algebraic closure of F.

1.7   Representable Algebras

The previous two results lead us to the most important class of PI-algebras:

Definition 1.66. An algebra A is *representable* if A can be embedded into $M_n(K)$ for a suitable field K.

Representable algebras are more malleable than arbitrary PI-algebras, and provide intuition for the general PI-theory. One basic property of representable algebras is:

Remark 1.67. If A is an algebra over an infinite field and is embedded in $M_n(K)$ for some field K, then AK is a f.d. K-subalgebra of $M_n(K)$, which is PI-equivalent to A. This leads us to a definition:

Definition 1.68. An algebra R is *PI-representable* if $R \sim_{\text{PI}} A$ for some algebra A that is finite dimensional over a field.

PI-representability is a main feature of Kemer's proof of Specht's problem, cf. Chapter 4. Having made the preliminary case for representable algebras, let us give some basic examples.

Corollary 1.61 shows that prime PI-algebras are representable. On the other hand, by Exercise 22, any representable algebra satisfies the ACC (ascending chain condition) on left annihilator ideals, which we write as ACC(ann). Since it is well-known that any semiprime algebra satisfying ACC(ann) is a subdirect product of a finite number of prime algebras, it follows at once that any semiprime PI-algebra satisfying ACC(ann) is representable.

In general, we have

Remark 1.69. Any semiprime PI-algebra A is embeddable in $M_n(H)$ where H is some commutative algebra. (This is seen by embedding A into the direct product of its prime images, each of which can be embedded into matrix algebras $M_{n_i}(H_i)$ for suitable commutative rings H_i and suitable n_i bounded by some m, so A is embedded in $M_n(\prod H_i)$ where $n = m!$)

Remark 1.70. Remark 1.69 has led PI-theorists to a mild generalization, in which an algebra is called representable if it can be embedded in $M_n(H)$ for some commutative ring H. This can be weakened even further to the condition that

$$A \hookrightarrow \operatorname{End}_C(M),$$

for a suitable f.g. module M over a commutative ring C, cf. Exercise 24. The ambiguity in terminology has led to confusion; fortunately, by Proposition 1.92 (of Anan'in) given below, all the definitions coincide when C is a commutative affine algebra (to be defined below) over a field F. Also, the ambiguity can be resolved in certain important cases by means of Exercise 25. We choose the more restrictive definition in our text, since our primary concern here will be for algebras over a field.

What if our PI-algebra is not semiprime? Lewin [Lew74] proved an interesting theorem, which plays a crucial role in the sequel. For convenience, we take $C = F$, a field.

Definition 1.71. A $(n_1, \ldots, n_t)$-*block triangular algebra* is a subalgebra of $M_n(F)$ in which all nonzero entries are in or above the diagonal $n_k \times n_k$ blocks, for $1 \leq k \leq t$.

For example, for $t = 2$, the (m, n)-block triangular algebra has entry 0 in the (i, j) position whenever $i > m$ and $j \leq n$. This example can be thought of in terms of bimodules over algebras A_1, A_2. Recall an A_1, A_2 *bimodule* is a (left) A_1-module M which is also a right A_2-module and module over C satisfying the extra associativity condition

$$(a_1 y)a_2 = a_1(y a_2), \quad \forall r_i \in A_i, \, y \in M,$$

as well as the algebra condition

$$cy = (c1)y = y(c1), \quad \forall c \in C, \, y \in M.$$

Theorem 1.72 (Lewin).

(i) *Suppose A is an algebra and $I_1, I_2 \lhd A$. Then there is an $A/I_1, A/I_2$ bimodule M, such that $A/I_1 I_2$ can be embedded in*

$$\begin{pmatrix} A/I_1 & M \\ 0 & A/I_2 \end{pmatrix} \tag{1.20}$$

via $a \mapsto \begin{pmatrix} a + I_1 & \delta(a) \\ 0 & a + I_2 \end{pmatrix}$, where $\delta : A \to M$ is a derivation.

(ii) *If the algebras $F\{X\}/I_i$ are embeddable in $M_{n_i}(C)$, then $F\{X\}/I_1 I_2$ is embeddable in an (n_1, n_2)-block triangular subalgebra of $M_{n_1+n_2}(C)$.*

A full proof, including an elegant proof of (i) by Bergman-Dicks using universal derivations, is given in [Row88, pp. 136–140].

Corollary 1.73. *If $F\{X\}/I$ is representable, then so is $F\{X\}/I^m$ for any m.*

As we shall see presently, this corollary is false if we try to replace $F\{X\}$ by an arbitrary algebra.

Of course, any representable algebra A satisfies all identities $\mathcal{M}_n$, so one might ask whether $\mathcal{M}_n \subseteq \mathrm{id}(A)$ suffices to make A representable, but examples were found with A nonrepresentable. Other necessary conditions were thrown in (cf. [Row88, Remark 6.3.6]), but people continued to find counterexamples satisfying these extra conditions. Interest focused on the following important class of algebras:

Definition 1.74. An algebra A is *affine* over the commutative ring C if A is generated as an algebra over C by a finite number of elements $a_1, \dots, a_\ell$; in this case we write $A = C\{a_1, \dots, a_\ell\}$.

In most cases, we shall be considering affine algebras over a field F, so unless C is specified otherwise, "affine" will mean "affine over a field."

Since the PI-theory often devolves to the center, let us start by noting the following key result; the easy proof can be found in [Row88, Proposition 6.2.5].

Proposition 1.75 (Artin-Tate). *If A is affine over a Noetherian ring C and is f.g. over its center Z, then Z is also affine over C.*

Proposition 1.76. *If C is Noetherian and $A = C\{a_1, \dots, a_\ell\}$ is affine over C and representable, then $A \subseteq M_n(H)$ for a suitable commutative affine (thus, Noetherian) C-algebra H.*

Proof. For each $1 \leq k \leq \ell$, write each a_k as an $n \times n$ matrix $(a_{ij}^{(k)})$, for $a_{ij}^{(k)} \in K$, and let K_0 be the commutative subalgebra of K generated by these $a_{ij}^{(k)}$. Now the assertion is obvious. $\qquad\square$

This innocent-looking assertion has some far-reaching consequences, as we shall see soon.

Corollary 1.77. *Any semiprime PI-algebra A affine over a commutative Noetherian ring is representable.*

Proof. By Proposition 1.76, $A \subseteq M_n(H)$, which is Noetherian, and thus, satisfies ACC(ann), so this condition descends to the subalgebra A, which thus, is representable. $\qquad\square$

1.7.1   Nil subalgebras of a representable algebra

Let us lay the foundation to prove the Jacobson radical $J(A)$ of an affine PI-algebra A (over a field) is nilpotent.

Remark 1.78. We recall some standard facts.

(i) For a commutative affine algebra H, the "weak Nullstellensatz" says $J(H)$ is nil, but any commutative affine algebra is Noetherian by Hilbert's Basis Theorem; thus, any nil subalgebra (including $J(H)$) is nilpotent.

(ii) For any field F, any nil subalgebra N of $M_n(F)$ is nilpotent; in fact, $N^n = 0$, cf. [Row88].

(iii) The Jacobson radical $J(A)$ of a f.d. algebra is nil, since every prime ideal is maximal, and thus, $J(A)$ is nilpotent by (ii).

Theorem 1.79. *Suppose A is a representable algebra which is affine over a commutative Noetherian ring C. Then any nil subalgebra N of A is nilpotent, of bounded nilpotence index.*

Proof. In view of Proposition 1.76, $A \subseteq M_n(H)$ where H is commutative and C-affine, so we may assume $A = M_n(H)$. For any maximal ideal P of H, $(N + M_n(P))^n = 0$ in $M_n(H)/M_n(P) = M_n(H/P)$ since H/P is a field; hence $N^n \subseteq \cap M_n(P) = M_n(J(H))$. But $J(H)^m = 0$ for some m, so $N^{nm} \subseteq M_n(J(H)^m) = 0$. $\qquad\square$

Theorem 1.80. *If A is a representable affine algebra over a field F, then $J = J(A)$ is nilpotent.*

Proof. Embed A in $M_n(H)$, with H affine over F. Then $J(H)$ is nilpotent, so $M_n(J(H)) = J(M_n(H))$ is nilpotent, and passing to

$$A/(A \cap M_n(J(H))) \subseteq M_n(H/J(H)),$$

we may assume $J(H) = 0$. We shall prove $J^n = 0$.

For any maximal ideal P of H, we see that H/P is an affine field extension of F, and thus, is f.d. over F by Hilbert's Nullstellensatz. But then the image of A in $M_n(H/P)$ is f.d. over F, so the image $\bar{J}$ of J is nilpotent, implying $\bar{J}^n = 0$ by Remark 1.78(ii).

Hence J^n is contained in $\cap M_n(P) = M_n(\cap P) = 0$, where P runs over the maximal ideals of H. □

Note the ease with which these theorems were proved for representable algebras. The proofs for PI-algebras in general are much more difficult, and occupy much of our efforts in Chapter 2.

1.7.2 Representability of affine Noetherian algebras

Example 1.81 (Nonrepresentable affine algebras). Lewin discovered an easy elegant cardinality argument to explain why affine algebras usually are not representable. Suppose the base field F has infinite cardinality γ. (For convenience, one may assume F is countable.) Then the following facts hold:

(i) Any affine F-algebra has cardinality at most γ, since its elements are linear combinations of elements of the form $ca_1^{i_1} \ldots a_\ell^{i_\ell}$.

(ii) Any affine F-algebra A has at most γ affine subalgebras. Indeed, any such subalgebra is specified by its generators, which we pick a finite number of times from the set A of cardinality γ.

(iii) The number of commutative F-affine algebras (up to isomorphism) is $\leq \gamma$. Indeed any such algebra has the form

$$F[a_1, \ldots, a_\ell] = F[\lambda_1, \ldots, \lambda_\ell]/I,$$

where, for suitable ℓ, I is an ideal of the commutative polynomial algebra $F[\lambda_1, \ldots, \lambda_\ell]$, and thus, is f.g. as an ideal. For any given ℓ, there are clearly at most γ choices of the finite number of generators, so at most γ possibilities for the algebra; since there are countably many natural numbers ℓ, we get γ choices altogether.

(iv) The number of representable F-affine algebras (up to isomorphism) is $\leq \gamma$. Indeed, by Proposition 1.76, any such algebra $A \subseteq M_n(H)$ where H is commutative and F-affine.

$$M_n(H) \approx M_n(F[\lambda_1, \ldots, \lambda_\ell])/M_n(I) \approx M_n(F[\lambda_1, \ldots, \lambda_\ell]/I)$$

where $I \triangleleft F[\lambda_1, \ldots, \lambda_\ell]$ (since any ideal of $M_n(F[\lambda_1, \ldots, \lambda_\ell])$ has the form $M_n(I)$), so we conclude with the same argument as in (iii).

(v) Any homomorphism between affine algebras is determined by its action on the finite set of generators, so there are at most γ homomorphisms between F-affine algebras. (This generalizes the three earlier statements.)

Thus, in view of (iv) and (v), in order to prove there are more than γ nonisomorphic, non-representable images of a given affine algebra A, it suffices to show A has more than γ ideals.

The free algebra $\mathbb{Q}\{x_1, x_2\}$ is an affine algebra with an uncountable number of ideals; hence uncountably many of these are nonrepresentable. But we can improve this observation to get a PI-example.

Example 1.82. Suppose $T = F\{a_1, \ldots, a_\ell\}$ is an affine F-algebra and $L = \sum_{j=1}^m Tb_j$ for b_i in T. Then

$$R = \begin{pmatrix} F+L & T \\ L & T \end{pmatrix}$$

is affine. Indeed, one checks easily

$$R = F\{e_{11}, e_{12}, a_i e_{22}, b_j e_{21} : 1 \leq i \leq \ell, \ 1 \leq j \leq m\}.$$

For example, $b_j e_{11} = e_{12} b_j e_{21}$.

Example 1.83. In Example 1.82 let $F = \mathbb{Q}$ and $T = \mathbb{Q}[\lambda, \mu]$ for commuting indeterminates λ, μ and let $L = F\lambda$. In this case T is commutative, so $R \subseteq M_2(T)$ is representable. But for any subset I of $\mathbb{N}$, let

$$J_I = L^2 + \sum_{i \in I} T\lambda\mu^i \triangleleft F + L.$$

Then

$$\mathcal{I}_I = \begin{pmatrix} J_I & J_I \\ L^2 & L^2 \end{pmatrix} \triangleleft R,$$

so we see the affine PI-algebra R has an uncountable number of ideals $\mathcal{I}_I$. We conclude that an uncountable number of the $R/\mathcal{I}_I$ are mutually nonisomorphic.

But the algebraic closure K of $\mathbb{Q}(\lambda, \mu)$ is countable, implying $M_n(K)$ is countable, and thus, has only a countable number of affine subalgebras (since this involves choosing a finite subset), so therefore an uncountable number of the $R/\mathcal{I}_I$ are nonrepresentable affine PI-algebras.

We shall see that the Jacobson radical $J = J(A)$ of an affine PI-algebra always is nilpotent. Thus, when $J^m = 0$, A/J^m could be nonrepresentable although A/J is representable.

Although this argument does not actually display such an algebra, we shall see explicit examples in our study of GK dimension of monomial algebras, cf. Corollary 9.29. Nonetheless, there are certain important classes of affine algebras that are representable.

Recall the three noncommutative generalizations of "Noetherian":

(i) A ring R is *left Noetherian* if it satisfies the ACC (ascending chain condition) on left ideals;

(ii) R is *Noetherian* if it is left and right Noetherian, i.e., satisfies the ACC on left ideals and also satisfies the ACC on right ideals;

(iii) R is *weakly Noetherian* if satisfies the ACC on two-sided ideals.

Lewin's argument fails for Noetherian affine algebras; recently Anan'in proved that all left Noetherian PI-algebras are representable.

Remark 1.84. We recall the important technique of "Noetherian induction": To prove a theorem about weakly Noetherian rings, we suppose on the contrary that we have a counterexample R, and take an ideal I maximal with respect to the theorem failing for R/I. Replacing R by R/I, we may assume that R is a counterexample, but R/J is not a counterexample for every $0 \neq J \triangleleft R$.

Noetherian induction can also be used for proving theorems about Noetherian modules, in an analogous fashion.

One idea of Anan'in's proof is to translate representability from algebras to bimodules. Let us start with a modification of Lewin's theorem (1.72):

Proposition 1.85. *Suppose $I_1, I_2 \triangleleft A$, and write $A_j = A/I_j$, $j = 1, 2$. If A is left Noetherian, with $I_1 I_2 = 0$, then the A_1, A_2 bimodule M of Theorem 1.72 can be taken to be Noetherian as an A_1-module.*

Proof. We identify A as a subring of $\begin{pmatrix} A_1 & M \\ 0 & A_2 \end{pmatrix}$ by the identification

$$a \mapsto \begin{pmatrix} a + I_1 & \delta(a) \\ 0 & a + I_2 \end{pmatrix}.$$

Viewing

$$\begin{pmatrix} A_1 & M \\ 0 & A_2 \end{pmatrix} = A_1 \oplus \begin{pmatrix} M \\ A_2 \end{pmatrix}$$

naturally as an A-module, we consider the projection

$$\pi : A \to \begin{pmatrix} M \\ A_2 \end{pmatrix} \text{ given by } a \mapsto \begin{pmatrix} \delta(a) \\ a + I_2 \end{pmatrix}.$$

$N = \pi(A)$ is clearly Noetherian as an A-module, and by restriction, the further projection N into M is Noetherian as an A_1-module, and N remains a right A_2-module and thus, a bimodule. Replacing M by N yields the desired result. $\qquad\square$

Definition 1.86. Suppose A_1, A_2 are F-algebras, and M is an A_1, A_2 bimodule. For a field extension K of F, letting $\hat{A}_i$ be a K-algebra containing A_i such that $\hat{A}_i = A_i K$, we say a $\hat{A}_1, \hat{A}_2$ bimodule $\widehat{M}$ is a K-*bimodule extension* of M if M is an A_1, A_2 sub-bimodule of $\widehat{M}$ such that $\widehat{M} = KM$. In particular,

$$\begin{pmatrix} A_1 & M \\ 0 & A_2 \end{pmatrix} \hookrightarrow \begin{pmatrix} \hat{A}_1 & \widehat{M} \\ 0 & \hat{A}_2 \end{pmatrix}.$$

We say the A_1, A_2 bimodule M is *representable* if F has a field extension K such that M has a K-bimodule extension that is f.d. over K.

The following observation enables us to pass from representable algebras to representable bimodules.

Remark 1.87. If A_1, A_2 are algebras over a field F and M is an A_1, A_2-bimodule which is f.d. over K and faithful both as A_1-module and right A_2-module, then A_1, A_2 are f.d., so the algebra

$$\begin{pmatrix} A_1 & M \\ 0 & A_2 \end{pmatrix}$$

also is f.d. (Indeed, A_1 can be viewed via the left regular representation as a subalgebra of $\operatorname{End}_F M$, so it has dimension $\leq n^2$, and likewise A_2 is f.d.)

Lemma 1.88. *If A_1, A_2 are representable algebras and M is a representable A_1, A_2-bimodule, then the algebra*

$$\begin{pmatrix} A_1 & M \\ 0 & A_2 \end{pmatrix}$$

is representable.

Proof. One can embed

$$\begin{pmatrix} A_1 & M \\ 0 & A_2 \end{pmatrix}$$

componentwise in

$$\begin{pmatrix} \hat{A}_1 & \hat{M} \\ 0 & \hat{A}_2 \end{pmatrix},$$

where $\hat{M}$ is f.d. over a field K and the $\hat{A}_i$ are K-algebras. Letting $\bar{A}_i = \hat{A}_i / \operatorname{Ann}(M)$ (taking the annihilator from the appropriate side), we see that

$$\begin{pmatrix} \bar{A}_1 & \hat{M} \\ 0 & \bar{A}_2 \end{pmatrix}$$

is f.d., by Remark 1.87. Clearly the map

$$\varphi_1 : \begin{pmatrix} A_1 & M \\ 0 & A_2 \end{pmatrix} \rightarrow \begin{pmatrix} \bar{A}_1 & \hat{M} \\ 0 & \bar{A}_2 \end{pmatrix}$$

restricts to an injection of M. But by hypothesis we can embed A_i into algebras $\tilde{A}$, which are f.d. over a field that contains K (by taking the compositum, if necessary), so the map

$$\varphi_2 : \begin{pmatrix} A_1 & M \\ 0 & A_2 \end{pmatrix} \rightarrow \tilde{A}_1 \times \tilde{A}_2$$

restricts to an injection of the diagonal $A_1 \times A_2$. Putting these together thus yields an injection

$$\varphi_1 \times \varphi_2 : \begin{pmatrix} A_1 & M \\ 0 & A_2 \end{pmatrix} \rightarrow \begin{pmatrix} \bar{A}_1 & \hat{M} \\ 0 & \bar{A}_2 \end{pmatrix} \times \tilde{A}_1 \times \tilde{A}_2,$$

i.e., $\varphi_1 \times \varphi_2$ is an injection into a f.d. algebra. $\qquad\square$

Clearly any sub-bimodule of a representable bimodule is representable. Also if a K-bimodule extension $\widehat{M}$ of M is representable, then M is representable. We can squeeze out more information. First note that the hypothesis implies that $\hat{A}_i$ is a homomorphic image of $K \otimes_F A_i$, given by

$$\sum_j k_j \otimes r_j \mapsto \sum k_j r_j.$$

But then any $\hat{A}_i$-module can be viewed naturally as an $K \otimes_F A_i$-module, and so we could replace $\hat{A}_i$ by $K \otimes_F A_i$ in the definition (and Anan'in actually does that).

Remark 1.89. A K-bimodule extension $\widehat{M}$ of M is finite dimensional over K iff $K \otimes_F M$ has an A_1, A_2 sub-bimodule V of finite codimension (as vector space over K) such that $V \cap M = 0$, where M is identified with $F \otimes_F M \subset K \otimes_F M$. (Indeed, compare V with the kernel of the natural homomorphism $K \otimes_F M \to \widehat{M}$.)

It follows that if $\widehat{M}$ has a representable sub-bimodule of finite codimension over K, then $\widehat{M}$ (and thus, also M) is representable.

Lemma 1.90 (Anan'in). *Suppose A_1, A_2 are algebras, with A_1 Noetherian PI. Then every A_1, A_2-bimodule that is Noetherian as A_1-module is representable.*

Proof. By Noetherian induction, we may assume the assertion is true for A_1/I, for every ideal $I \neq 0$ of A_1. Suppose on the contrary we have a non-representable bimodule which is Noetherian as A_1-module. Given such a non-representable bimodule M, take $I_M \triangleleft A_1$ maximal with respect to $I_M M$ being representable; there is such I_M by virtue of ACC on ideals of A_1, since $0 = 0M$ is certainly representable. Now take I be maximal among all I_M; we choose M so that $I = I_M$.

Thus, we have some field K such that $K \otimes IM$ has a sub-bimodule V of finite codimension over K, with $V \cap IM = 0$. Checking bases, viewing M as $F \otimes_F M$, we have

$$M \cap (K \otimes IM) = IM,$$

so $V \cap M = 0$. Let $\widehat{M} = (K \otimes M)/V$, which by hypothesis is infinite dimensional over K, whereas $I\widehat{M}$ is finite dimensional over K.

Writing $I = \sum_{i=1}^{t} A_1 a_i$ for $a_i \in I$ (since A_1 is left Noetherian), we see that

$$I\widehat{M} = \sum_{i=1}^{t} A_1 a_i \widehat{M}.$$

Letting $\mathcal{A}_{\widehat{M}}$ denote the right annihilator in $\widehat{M}$, we have a right A_2-module isomorphism given by left multiplication by a_i:

$$a_i \widehat{M} \cong \widehat{M}/\mathcal{A}_{\widehat{M}}(a_i).$$

But $a_i \widehat{M} \subseteq I\widehat{M}$ is f.d. over K, so $\mathcal{A}_M(a_i)$ has finite codimension in $\widehat{M}$ (as vector space over K). Hence

$$\mathcal{A}_{\widehat{M}}(I) = \bigcap_{i=1}^{t} \mathcal{A}_{\widehat{M}}(a_i)$$

has finite codimension in $\widehat{M}$ and thus, is infinite dimensional over K. Since this argument also holds for any field containing K and any module extending $\widehat{M}$, we see $\mathcal{A}_{\widehat{M}}(I)$ is not representable, whereas $I\mathcal{A}_{\widehat{M}}(I) = 0$. But $\mathcal{A}_{\widehat{M}}(I)$ has the same structure viewed as an $A_1/I, A_2$ bimodule, and by our Noetherian induction hypothesis $\mathcal{A}_{\widehat{M}}(I)$ is representable (contradiction) unless $I = 0$.

We have thereby proved $I_M = 0$ for *every* non-representable bimodule M that is Noetherian as A_1-module; this means BM is not representable for any $0 \neq B \triangleleft A_1$.

We see at once that A_1 is prime. Indeed, if $0 \neq B_1, B_2 \triangleleft A_1$, then $B_2 M$ is not representable, by the previous paragraph, implying $B_1(B_2 M)$ is not representable, so certainly $B_1 B_2 M \neq 0$; hence $B_1 B_2 \neq 0$.

Let M_0 be the torsion submodule of M over A_1, clearly a bi-submodule of M and thus, Noetherian over A_1. Write $M_0 = \sum_{i=1}^{k} A_1 w_i$. Each w_i is annihilated by some large left ideal of A_1, which contains some $c_i \neq 0$ in C since A_1 is prime PI. Let $c = c_1 \ldots c_k$. Then $cM_0 = 0$, and cM is torsion-free over A_1. (Indeed if cw is torsion for some $w \in M$, then $c'cw = 0$ for some c' in C, implying $w \in M_0$.) But $cM = (cA_1)M$ is not representable, since $0 \neq cA_1 \triangleleft A$, so we may replace M by cM and assume M is torsion-free over A_1.

Furthermore, replacing A_2 by A_2/B where B is the right annihilator of M, we may assume $B = 0$, i.e., M is faithful as right A_2-module.

Now take L to be the field of fractions of C, $\tilde{A}_1 = L \otimes_C A_1$, and $\widetilde{M} = L \otimes_C M$. Since M is torsion-free over C, we can embed

$$\begin{pmatrix} A_1 & M \\ 0 & A_2 \end{pmatrix} \hookrightarrow \begin{pmatrix} \tilde{A}_1 & \widetilde{M} \\ 0 & A_2 \end{pmatrix}.$$

The elements of A_1 can be thought of as transformations of $\widetilde{M}$, which is f.d. over L, so

$$\begin{pmatrix} A_1 & \widetilde{M} \\ 0 & \tilde{A}_2 \end{pmatrix} \hookrightarrow \begin{pmatrix} \tilde{A}_1 & \widetilde{M} \\ 0 & \mathrm{End}_L \widetilde{M} \end{pmatrix},$$

a f.d. L-algebra, contrary to M not being representable. $\square$

We finally can prove the following theorem of Anan'in:

Theorem 1.91 (Anan'in). *Every affine left Noetherian PI-algebra A is representable.*

Proof. By Noetherian induction, one may assume A/I is representable for every $0 \neq I \lhd A$. In particular, since the upper nilradical N is nilpotent, i.e., satisfies $N^n = 0$ for some n, we may assume $\bar{A} = A/N^{n-1}$ is representable. But $(N^{n-1})^2 = 0$, so by Lewin's theorem together with Proposition 1.85, we can embed A in

$$\begin{pmatrix} \bar{A} & M \\ 0 & \bar{A} \end{pmatrix}$$

for some bimodule M that is Noetherian as left $\bar{A}$-module and thus representable, by Lemma 1.88. Hence A is representable, in view of Lemma 1.88. □

One immediate consequence is a result from the folklore which we shall need repeatedly:

Proposition 1.92. *Suppose C is a commutative affine algebra over a field F, and $A \supset C$ is a PI-algebra which is f.g. as a C-module. Then A is representable.*

1.8   ACC in Ring Theory

This section contains basic material about chain conditions on ideals in noncommutative algebras, which is mostly a direct generalization from the commutative theory. The reason we include it is that Kemer's solution to Specht's problem has thrust open the door to a new application of this material, and as long as we shall need the material, we might as well present it here to have it available for other purposes (such as for the structure of affine PI-algebras). Throughout, A is a noncommutative algebra, and we fix a class $\mathcal{I}$ of two-sided ideals. We skip most proofs, since they are in direct analogy to the well-known proofs in commutative algebra.

Definition 1.93. Given $S \subseteq A$, the *ideal generated in $\mathcal{I}$ by S* is defined as $\cap\{I \in \mathcal{I} : S \subseteq I\}$. $I \in \mathcal{I}$ is *finitely generated* if I is generated by some finite set S.

(This contrasts with the notion of f.g. module, although both come from the same philosophy.)

Definition 1.94. $\mathcal{I}$ satisfies the *ACC* if any ascending chain in $\mathcal{I}$ stabilizes, i.e., if

$$I_1 \subseteq I_2 \subseteq \dots$$

for $I_j \in \mathcal{I}$, then there is some k such that $I_j = I_{j+1}$ for all $j \geq k$.

Remark 1.95. The following are equivalent:

(i) $\mathcal{I}$ satisfies the ACC.

(ii) Every member of $\mathcal{I}$ is finitely generated in $\mathcal{I}$.

(iii) Every subset of $\mathcal{I}$ has a maximal member.

Definition 1.96. A member P of $\mathcal{I}$ is *prime* if, for all $I, J \in \mathcal{I}$ properly containing P, we have $IJ \not\subseteq P$. For any $S \subseteq A$, a prime P of $\mathcal{I}$ containing S is *minimal* over S if P does not properly contain a prime of $\mathcal{I}$ containing S.

Remark 1.97. If P is prime and P is minimal over IJ for $I, J \in \mathcal{I}$, then P is minimal over I or P is minimal over J.

Theorem 1.98. *Suppose $\mathcal{I}$ satisfies the ACC. Then for any $I \in \mathcal{I}$, there are only a finite number of primes $P_1, \ldots, P_t$ in $\mathcal{I}$ minimal over I, and some finite product of the P_i is contained in I.*

Proof. Otherwise, there is $I \in \mathcal{I}$ maximal with respect to being a counterexample. Certainly I is not itself prime, so take $J_1, J_2 \supset I$ such that $J_1 J_2 \subseteq I$. By hypothesis, the conclusion of the theorem holds for J_1 and J_2, i.e., there are P_{ik} minimal over J_k with some finite product contained in J_k. But then the product together is in $J_1 J_2$ and thus, in I. Any prime P containing $J_1 J_2$ then must contain some P_{ik}, and thus, if minimal over $J_1 J_2$, is certainly minimal over J_k and thus, must equal P_{ik}. $\square$

Definition 1.99. The *radical* $\sqrt{S}$ of $S \subseteq A$ is the intersection of all primes of $\mathcal{I}$ containing S.

Corollary 1.100. *Suppose $\mathcal{I}$ satisfies the ACC. If $I \in \mathcal{I}$, then $\sqrt{I}$ is a finite intersection of primes, and $\sqrt{I}^n \subseteq I$ for some n.*

Corollary 1.101. *Suppose $\mathcal{I}$ satisfies the ACC, and $0 \in \mathcal{I}$. If $I \in \mathcal{I}$ is contained in every prime, then I is nilpotent.*

Proof. $I \subseteq \sqrt{0}$, so apply the previous corollary. $\square$

Chapter 2

A Few Words Concerning Affine PI-Algebras

Affine algebras were defined in Definition 1.74. Although much of the structure theory of affine PI-algebras has been well understood for some time and exposed in [KrLe00] and [Row88, Chapter 6.3], many of these results can be streamlined and improved when viewed combinatorically in terms of words. Perhaps the first major triumph of PI-theory was Kaplansky's proof [Kap50] of A.G. Kurosh's conjecture for PI-algebras, that every algebraic affine PI-algebra over a field is finite dimensional. (Later Kurosh's conjecture was disproved in general by Golod and Shafarevich [Gol64].) Kaplansky's proof was structural, based on Jacobson's radical and Levitzki's locally finite radical. The first combinatorial proof, obtained by A.I. Shirshov, was seen later to be a consequence of *Shirshov's Height Theorem*, which does not require any assumption on algebraicity, and shows that any affine PI-algebra bears a certain resemblance to a commutative polynomial algebra, thereby yielding an immediate solution to Kurosh's conjecture for PI-algebras. Another consequence of Shirshov's Height Theorem, proved by Berele, is the finiteness of the Gelfand-Kirillov dimension of any affine PI-algebra. A crucial dividend of Shirshov's approach is that it generalizes at once to an affine algebra over an arbitrary commutative ring:

If $A = C\{a_1, \ldots, a_\ell\}$ is an affine PI-algebra over a commutative ring C and a certain finite set of elements of A (e.g., the words in the a_i of degree up to some given number to be specified) is integral over C, then A is f.g. (finitely generated) as a C-module.

Since the theory of commutative Noetherian rings is so well developed, one can sometimes extend the theory to affine algebras over commutative Noetherian rings, but for the most part, we are content here to investigate the theory of affine algebras over a field, which is our focus. Our main goal in this chapter is to present the combinatoric theory behind Shirshov's Height Theorem.

In the process, we give three proofs of the key Shirshov Lemma 2.10. We start with a novel, nonconstructive proof that suffices for the most important applications, to be described presently. Later, we present second and third proofs of Shirshov's Lemma; the second being a modification of Shirshov's original proof, which gives a recursive bound on the height. Our last proof yields a much better bound, following Belov, leading to an independence theorem on hyperwords that implies Ufnarovski'i's indepedence theorem and also solves a conjecture of Amitsur and Shestakov.

Our other main goal is to prove the Razmyslov-Kemer-Braun Theorem on the nilpotence of the Jacobson radical $J(A)$ of an affine PI-algebra A (over a field). The fact that $J(A)^k = 0$ for some k enables one to build proofs of facts about affine PI-algebras by induction on k, building up from the case $k = 1$.

The Razmyslov-Kemer-Braun Theorem has an interesting history. The nilpotence of $J(A)$ is well known for finite dimensional algebras, and is used in the proof of Wedderburn's Principle Theorem, quoted in Theorem 4.3. On the other hand, it is well known that the nil radical of any Noetherian ring is nilpotent. Amitsur[Am57] had proved $J(A)$ is nil for any affine PI-algebra A, and Amitsur [Am70] proved $J(A)$ is nilpotent for A representable, cf. Theorem 1.80. Latyshev [Lat77] conjectured $J(A)$ is nilpotent for all affine algebras, in his doctoral dissertation. However, proving $J(A)$ nilpotent for arbitrary PI-algebras turned out to be very challenging. Razmyslov[Raz74a] (also cf. Schelter [Sch78]) proved that $J(A)$ is nilpotent for any affine PI-algebra which satsifies the identities of $n \times n$ matrices for some n. In the same paper, using Shirshov's Height Theorem, Razmyslov proved in characteristic 0 that the nilpotence of $J(A)$ is equivalent to A satisfying some Capelli identity. Kemer [Kem80] verified that all affine algebras satisfy Capelli identities, thereby completing the proof that $J(A)$ is nilpotent in characteristic 0. Then Braun[Br82], [Br84] put forward a characteristic-free proof based on the structure theory of Azumaya algebras.

In this exposition, we give a characteristic-free proof based on the approach of Razmyslov and Kemer. Thus, after proving Shirshov's Height Theorem, we continue with the "trace ring," a key tool in passing from affine algebras to f.g. modules over their centers, and its generalization as developed by Razmyslov and Zubrilin. This proves that $J(A) = 0$ for any affine algebra A satisfying a Capelli identity, and leads us to Kemer's Capelli Theorem, which we prove twice. First we introduce an elementary but tricky argument, and then we present the more historic argument of

Kemer, which introduces two important techniques: the use of "sparse" identities and the "pumping procedure" developed by Belov.

The theorems and techniques described in this chapter are critical for the theory developed in the remainder of this book.

2.1 Words Applied to Affine Algebras

We work formally with a finite *alphabet* $X = \{x_1, \ldots, x_\ell\}$ of *letters* and the free monoid $\mathcal{M}\{X\}$ described in §1.1, whose elements are now called *words*. If $w = w_1 w_2 w_3$ in $\mathcal{M}\{X\}$, we call the w_i *subwords* of w; w_1 is *initial* and w_3 is *terminal*. A subword v of w has *multiplicity* n if there are n nonoverlapping occurrences of v in w. If the occurrences are consecutive, we write v^k to denote $vv \ldots v$, i.e., v repeating k times. It is useful to introduce the *blank word*, denoted $\emptyset$, which is the unit element in $\mathcal{M}\{X\}$.

One of our main tools is the *lexicographic (partial) order* on words, in which we give each letter x_i the value i, and order words by the first letter in which the values differ. In other words, for two words $w_1 = x_{i_1} v_1$ and $w_2 = x_{i_2} v_2$, we have $w_1 \succ w_2$ if $i_1 > i_2$, or if $i_1 = i_2$ and inductively $v_1 \succ v_2$.

Note that we do not compare a word with the blank word $\emptyset$. Thus, two words $v \neq w$ with $|v| \leq |w|$ are *(lexicographically) comparable* iff v is *not* an initial subword of w. (In particular, any two words of the same length are comparable.)

A word $w = x_{i_1} \ldots x_{i_d}$ will be said to have *length* d, written $|w| = d$.

Our basic set-up: $A = C\{a_1, \ldots, a_\ell\}$ is an affine algebra over a commutative ring C. The homomorphism $\Phi : C\{X\} \to A$, given by $x_i \mapsto a_i \in A$ is fixed throughout this discussion. Given a word $w \in \mathcal{M}\{X\}$, we write $\bar{w}$ for its image under Φ, and shall call $\bar{w}$ a *word in A*.

For a set W of words, we write $\bar{W}$ for $\{\bar{w} : w \in W\}$. A subset S of A is *spanned* by $\bar{W}$ if each element of S can be written in the form $\{\sum_{w \in W} c_w \bar{w} : c_w \in C\}$.

We say $\bar{w}$ is *A-minimal* if $\bar{w}$ is not spanned by the images of words in W less than w (under the lexicographic order).

Remark 2.1. $\{\bar{w} : w$ is A-minimal $\}$ is a base of $\bar{w}$.

The point is that if A satisfies an identity

$$f = x_1 \ldots x_d + \sum_{\pi \neq 1} \alpha_\pi x_{\pi(1)} \ldots x_{\pi(d)},$$

then given any words $w_1 \succ w_2 \succ \cdots \succ w_d$, we can replace $\bar{w}_1 \ldots \bar{w}_d$

by $-\sum_{\pi\neq1}\alpha_\pi\bar{w}_{\pi(1)}\cdots\bar{w}_{\pi(d)}$, a linear combination of words in A of lower order, cf. Proposition 2.8.

2.2   Shirshov's Height Theorem

Definition 2.2.

(i) A word s is a *Shirshov word* of *height* μ over W if $s = u_{i_1}^{k_1}\ldots u_{i_\mu}^{k_\mu}$ where each $u_i \in W$. Note: One could have $i_j = i_{j'}$ for some $j \neq j'$. For example, the word $x_1^7 x_2^5 x_1^3$ has height 3 over $\{x_1, x_2\}$.

Given a set of words W, we write $\widehat{W}_\mu$ for the Shirshov words of height $\leq \mu$ over W.

(ii) A has *height* μ over a set of words W if A is spanned by the set $\{\bar{w} : w \in \widehat{W}_\mu\}$. In this case we say $\bar{W}$ is a *Shirshov base* of A. (This is a misnomer, since it is the evaluations of $\widehat{W}_\mu$ that span A, and not $\bar{W}$, but the terminology has become standard.)

We are ready to formulate the main result of this chapter.

Theorem 2.3 (Shirshov's Height Theorem).

(i) *Suppose* $A = C\{a_1,\ldots,a_\ell\}$ *is an affine PI-algebra of PI-degree* d, *and let* W *be the set of words of length* $\leq d$. *Then* $\bar{W}$ *is a Shirshov base for* A; A *has height* $\leq \mu$ *over* W, *where* $\mu = \mu(\ell, d)$ *is a function of* d *and* ℓ *(to be computed as we develop the proof). In other words,* $\{\bar{w}_1^{k_1}\ldots\bar{w}_\mu^{k_\mu} : w_i \in W, k_i \geq 0\}$ *span* A.

(ii) *If in addition* $\bar{w}$ *is integral over* C *for each word* w *in* W, *then* A *is f.g. as* C-module.

Remark 2.4. A trivial but crucial observation, which we shall need at least half a dozen times, is that the number of words in $X = \{x_1,\ldots,x_\ell\}$ of length m is ℓ^m, since we can choose among ℓ letters in each of the m positions. Thus, the number of words of length $\leq m$ is $\ell^m + \ell^{m-1} + \cdots + 1 = (\ell^{m+1} - 1)/(\ell - 1)$.

Remark 2.5. It is easy to conclude (ii) of Theorem 2.3 as a consequence of (i). Indeed, if $\bar{w}$ satisfies a monic equation of degree k_w, then we can replace each $\bar{w}^{k_w}$ by smaller powers, thereby seeing that A is spanned by a finite number of terms. Of course, this number is rather large.

$$|W| = \frac{\ell^{d+1} - 1}{\ell - 1},$$

and the number of words of height μ over W, with each exponent less than or equal to k, is $\big((k+1)|W|\big)^{\mu}$.

This motivates us to prove (ii) separately with a better bound because of its importance in its own right, and its proof also turns out to be more direct than that of (i).

Our first objective is to prove Shirshov's Height Theorem and give its host of applications to affine PI-algebras. Afterwards, since Shirshov's theorems are so important combinatorically, we shall take a hard look at the Shirshov height function μ, and also use our techniques to answer various problems of Burnside type.

The role of the PI is to provide a reduction procedure whereby we can replace words by linear combinations of words of lower lexicographic order.

Definition 2.6 (Shirshov). A word w is called *d-decomposable* (with respect to $\succ$) if it contains a subword $w_1 w_2 \ldots w_d$ such that for any permutation $1 \neq \sigma \in S_d$,

$$w_1 w_2 \ldots w_d \succ w_{\sigma(1)} w_{\sigma(2)} \cdots w_{\sigma(d)}.$$

There is one main way of proving d-decomposability.

Remark 2.7.

(i) The word w is d-decomposable if it contains a (lexicographically) descending chain of d disjoint subwords, i.e., is of the form

$$s_0 v_1 s_1 v_2 \cdots s_{d-1} v_d s_d,$$

where $v_1 \succ v_2 \succ \cdots \succ v_d$.

(ii) Since comparability of words depends on the initial subwords, we could put $w_i = v_i s_i$ and note that (i) implies w contains a subword comprised of d decreasing subwords

$$w_1 w_2 \ldots w_d, \quad \text{where} \quad w_1 \succ w_2 \succ \cdots \succ w_d.$$

The following proposition relates the PI to d-decomposability.

Proposition 2.8 (Shirshov). *Suppose A has PI-degree d. Then any d-decomposable word w can be written in A as a linear combination of permuted words (i.e., the letters are rearranged) that are lexicographically smaller than w.*

Proof. A satisfies a multilinear identity of degree d, which can be written as

$$x_1 \cdots x_d - \sum_{\eta \neq 1, \, \eta \in S_d} \alpha_\eta x_{\eta(1)} \cdots x_{\eta(d)}.$$

But then

$$a_1 \cdots a_d - \sum_{\eta \neq 1, \, \eta \in S_d} \alpha_\eta a_{\eta(1)} \cdots a_{\eta(d)},$$

so we can replace $\bar{w}_1 \cdots \bar{w}_d$ by

$$\sum_{\eta \neq 1, \, \eta \in S_d} \alpha_\eta \bar{w}_{\eta(1)} \cdots \bar{w}_{\eta(d)}$$

in A. By hypothesis, all the words on the right are smaller than $w_1 \ldots w_d$, and the proof follows. $\square$

In other words, any A-minimal word is d-indecomposable.

Remark 2.9. The use of a polynomial identity to lower the lexicographic order of a d-reducible word does not change the letters used in the word, but merely rearranges them. This somewhat trivial remark is crucial in considerations of growth in Chapter 9.

Shirshov's strategy to prove (ii) of Theorem 2.3 was to show:

Lemma 2.10 (Shirshov's Lemma). *For any ℓ, k, d, there is $\beta = \beta(\ell, k, d)$ such that any d-indecomposable word w of length $\geq \beta$ in ℓ letters must contain a nonempty subword of the form u^k, with $|u| \leq d$.*

Note that this formulation does not mention any PI, but is a purely combinatorical fact about words. To prove it, we shall introduce some standard techniques from combinatorics.

2.2.1   Hyperwords, and the hyperword u^∞

Although words have finite length, by definition, we can keep track of repetitions of words by broadening our horizon.

Definition 2.11. A *hyperword* is a right infinite sequence of letters.

The juxtaposition of a word followed by a hyperword produces a hyperword in the natural way; likewise, for any n, we can factor any hyperword $h = wh'$ where $|w| = n$ and h' is the remaining hyperword; w is called the *initial subword* of length n.

A word is called *periodic*, with *period u*, if it has the form u^k for $k > 1$. A hyperword h is called *periodic* with *period u*, if it has the form u^∞, i.e.,

u repeats indefinitely. $|u|$ is called the *periodicity* of h. (We always choose u such that $|u|$ is minimal possible; in other words, u itself is not periodic.) A hyperword h is *preperiodic* if $h = vh'$ where v is finite and h' is periodic. Such v with $|v|$ minimal possible is called the *preperiod* of h.

Remark 2.12. Analgously, one could say a hyperword h' is initial in a hyperword h if every initial subword of h' is initial in h. But then one has $h' = h$.

Proposition 2.13. *If h is initial in uh, where h is a word or a hyperword, and u is a word, then h is initial in u^∞. (In particular, if h is a hyperword, then $h = u^\infty$.)*

Proof. For all k, $u^k h$ is initial in $u^{k+1}h$, implying h is initial in $u^{k+1}h$, by transitivity; now let $k \to \infty$. $\qquad\qquad\qquad\qquad\qquad\qquad\qquad\qquad\quad\square$

Of course in an alphabet of ℓ letters there are ℓ^n possible words of length n, and one would expect many of them to appear in a given hyperword h. Let us see what happens if this number is restricted. We write $\nu_n(h)$ to be the number of distinct subwords of h of length n. For example, $\nu_1(h)$ is the number of distinct letters appearing in h. Of course $\nu_n(h) \le \ell^n$ is finite, for any n. The following trivial observation is the key to the discussion.

Remark 2.14.

(i) Given any subword v of h of length n, we can get a subword of length $n+1$ by adding on the next letter. Hence $\nu_{n+1}(h) \ge \nu_n(h)$ for any n.

(ii) If $\nu_{n+1}(h) = \nu_n(h)$, then given a subword v of h of length n, h has a unique subword of length $n+1$ starting with v. (Indeed, suppose vx_i and vx_j both appeared as subwords of h. By (i), each other subword of length n starts a subword of length $n+1$, implying $\nu_{n+1}(h) > \nu_n(h)$.)

Proposition 2.15 (Basic combinatoric lemma). *If $\nu_n(h) = \nu_{n+1}(h) = m$ for suitable m, n, then h is preperiodic of periodicity $\le m$.*

Proof. Write $h = x_{i_1} x_{i_2} x_{i_3} \dots$, and, for $1 \le j \le m+1$, let

$$v_j = x_{i_j} x_{i_{j+1}} \cdots x_{i_{j+n-1}}$$

denote the subword of h of length n starting in the j position. Then $v_1, \dots, v_{m+1}$ are $m+1$ words of length n so, by hypothesis, two of these are equal, say $v_j = v_k$ with $j < k \le m+1$, i.e., $j \le m$.

Let h_j denote the hyperword starting in the j position. By Remark 2.14 (ii), any sequence of n letters of h determines the next letter uniquely, so h_j is determined by its initial subword v_j. Hence

$$h_j = h_k = x_{i_j} \ldots x_{i_{k-1}} h_j,$$

so Proposition 2.13 implies $h_j = u^\infty$ where $u = x_{i_j} \ldots x_{i_{k-1}}$. Thus $|u| = k - j \leq m$; furthermore,

$$h = x_{i_1} \ldots x_{i_{j-1}} h_j$$

is preperiodic.   □

Definition 2.16. In analogy with Remark 2.5, a hyperword w is called *d-decomposable* if it contains a (lexicographically) descending chain of d disjoint subwords.

We are ready for our main result about hyperwords.

Theorem 2.17. *Any hyperword h is either d-decomposable or is preperiodic of periodicity $< d$.*

Proof. Say a subword of length d is *unconfined* in h if it occurs infinitely many times as a subword, and is *confined* in h if it occurs only finitely many times.

First, assume h has d unconfined subwords $v_1, \ldots, v_d$ of length d, which we write in decreasing lexicographic order. (This is possible since they have the same length.) Let us take the first occurrence of v_1, and then take the first occurrence of v_2 that starts after this occurrence of v_1, and proceeding inductively, we obtain a subword of h in which the subwords $v_1, \ldots, v_d$ appear disjointly in that order. Hence, by definition h is d-decomposable.

Thus, we may assume that h has $< d$ unconfined subwords of length d. Obviously, h has a (finite) initial subword w containing all the confined subwords of length d. Writing $h = wh'$, we see by assumption that the only subwords of length d that survive in h' are the unconfined subwords. Thus h' has fewer than d words of length d, i.e., $\nu_d(h') < d$. But obviously $\nu_1(h') \geq 1$, so in view of Remark 2.14 (i), there is some $n \leq d$ for which $\nu_n(h') = \nu_{n+1}(h') < d$, and we conclude by Proposition 2.15 that h', and thus h, is preperiodic of periodicity $< d$.   □

This dichotomy of d-decomposability versus periodicity lies at the heart of the Shirshov theory.

Given a d-indecomposable hyperword h, we write $\beta(\ell, k, d, h)$ for the smallest number β for which h contains an initial subword of the form vu^k

with $|v| \leq \beta$ and $|u| \leq d$. In view of Corollary 2.17, such $\beta(\ell, k, d, h)$ exists, being at most the length of the preperiod of h. We want to show this number is bounded by a function independent of the choice of h.

Proposition 2.18. *For any word ℓ, k, d there is a number $\beta(\ell, k, d)$ such that $\beta(\ell, k, d, h) \leq \beta(\ell, k, d)$ for all d-indecomposable hyperwords h.*

Proof. A variant of the König graph theorem, which we spell out. Otherwise, for any given number β let $P_{j\beta} = \{v : v$ is the initial subword of length j of an indecomposable hyperword h with $\beta(\ell, k, d, h) > \beta\}$, and for each word v of length j, let $Q_j = \bigcap \{P_{j\beta} : \beta \in \mathbb{N}\}$.

Since each $P_{j\beta}$ is finite nonempty, this means $Q_j \neq \emptyset$. If $v_j \in Q_j$, then we can find v_{j+1} in Q_{j+1} starting with v_j, so we continue this process by induction to obtain an indecomposable hyperword h. But the initial subword of h having length $\beta(\ell, k, d, h) + dk$ already contains a subword u^k, with $|u| \leq d$, contrary to hypothesis on the v_j. $\square$

Proposition 2.18 constitutes Shirshov's Lemma.

This proof is quite nice conceptually, relying on well-known combinatoric techniques and illustrating the power of the hyperword. On the other hand, it says *nothing* about the function $\beta(\ell, k, d)$. In fact, one could bound $\beta(\ell, k, d)$ by an induction argument based on $\nu_n(h)$, but presently all we need for any of the qualitative applications is the existence of $\beta(\ell, k, d)$. So we shall give the applications first, and turn later to finding a reasonable bound for the Shirshov function. Toward the end of this chapter, we give a second proof (based on Shirshov's original proof) that constructs an explicit bound, but this bound is so astronomical as to be useless. Then we give a third proof that provides a reasonable handle on the Shirshov function. Exercise 2.20 is another good tool for lowering bounds in Shirshov's Lemma.

We could already prove the second part of Shirshov's Height Theorem, but to get the first part, we need another connection between decomposability and periodicity.

Proposition 2.19. *If a word v has multiplicity $\geq d$ in w, and v contains d lexicographically comparable subwords (not necessarily disjoint), then w is d-decomposable.*

Proof. Write $w = s_0 v s_1 v \ldots s_{m-1} v s_d$. We choose the highest subword in the first occurrence of v, the second highest in the second occurrence of v, and so on. Clearly this provides a decreasing chain of non-overlapping subwords of w, so we use Remark 2.7. $\square$

Our goal is then to find d pairwise comparable subwords in a given word. Here is one way.

Remark 2.20. Let u, v be comparable words (i.e., neither is initial in the other). Then $u^{d-1}v$ contains d different comparable subwords:

(i) If $v \prec u$, then $v \prec uv \prec u^2v \prec \cdots \prec u^{d-1}v$.

(ii) If $v \succ u$, then $v \succ uv \succ u^2v \succ \cdots \succ u^{d-1}v$.
So any word in which $u^{d-1}v$ has multiplicity $\geq d$ is d-decomposable, by Proposition 2.19.

Proof of Shirshov's Height Theorem 2.3(ii). Induction on the lexicographic order. Let k be the maximum of d and the maximum degree of integrality of the finite set $\{\bar{w} : w \in W\}$. By Proposition 2.8, we may replace d-decomposable words by linear combinations of smaller words, so we may assume that w is d-indecomposable. Since we can use integrality to reduce the power, we see in Lemma 2.10 that $\bar{w}$ is A-minimal only when $|w| \leq \beta(\ell, k, d)$.   □

Proof of Shirshov's Height Theorem 2.3(i). Let $\beta = \beta(\ell, 2d, d)$ as in Shirshov's Lemma 2.10, let β' be the smallest integer greater than $\frac{\beta}{d}$, and let

$$\mu = \mu(\ell, d) = (d^2\ell^{2d} + 1)(\beta' + 2).$$

Again, we may assume w is d-indecomposable, and we shall prove that the height of w over the words $\{u : |u| < d\}$ is at most μ.

We describe now a process for partitioning w, which eventually will show that the height of w is $\leq \mu$. If $|w| < \beta + d$, stop. Assume $|w| \geq \beta + d$. Write $w = w'w''$ with $|w'| = \beta$, so w' is also d-indecomposable. By Shirshov's Lemma 2.10, w' contains a subword of the form $u_1^{k_1'}$ with $k_1' \geq 2d$, i.e., w' has an initial subword $s_0 u_1^{k_1'}$. Now write $w' = s_0 u_1^{k_1} w^*$ with k_1 maximal; then $k_1 \geq 2d$ and $|s_0| < \beta$. If $|w^*| < |u_1|$ we stop, and if $|w^*| \geq |u_1|$, then we write $w^* = v_1 w_1$ where $|v_1| = |u_1|$. In that case, $w = s_0 u_1^{k_1} v_1 w_1$, and u_1 and v_1 are comparable. This completes the first step.

Repeat the same process with w_1: If $|w_1| < \beta + d$, stop. If $|w_1| \geq \beta + d$, continue the same way: $w_1 = s_1 u_2^{k_2} v_2 w_2$ with $|s_1| < \beta$, $|u_2| = |v_2| \leq d$, $k_2 \geq 2d$ and u_2 and v_2 are comparable. Next apply the same process to w_2. Continue until we get

$$w = s_0 u_1^{k_1} v_1 s_1 u_2^{k_2} v_2 s_2 u_3^{k_3} v_3 \cdots s_{t-1} u_t^{k_t} w_t'$$

with $|w_t'| < \beta + d$. Here all $|s_j| < \beta$, $k_j \geq 2d$, $|u_j| = |v_j| \leq d$ and u_j, v_j are comparable.

We want to view this as a string of subwords of length $\leq d$, so we partition each s_i into β' such subwords. In determining the height of w, we can disregard the powers k_i and thus count β' for each s_i, and 1 for each u_i or v_i. Thus, w has height at most

$$t\beta' + t + (t-1) + \beta' + 1 < (t+1)(\beta' + 2).$$

On the other hand, $u_i^d v_i$ is determined by the last $|u_i| + |v_i| \leq 2d$ letters (since they determine u_i and v_i); since $|u_i| \leq d$, there are fewer than ℓ^{2d} possibilities for $u_i^d v_i$, for any given length of $|u_i|$, or, counting all d possible lengths, fewer than $d\ell^{2d}$ possibilities in all for $u_i^d v_i$.

If $t \geq d^2 \ell^{2d} = d(d\ell^{2d})$, some $u^d v$ must repeat d times in w, contrary to w being d-indecomposable (by Remark 2.20). Hence $t < d^2 \ell^{2d}$, which implies that the height of w is $< (d^2 \ell^{2d} + 1)(\beta' + 2) = \mu$. $\qquad \square$

Remark 2.21. In view of Remark 2.4, the number of elements needed to span A over C in Theorem 2.3 (ii) is $(\ell^{m+1} - 1)/(\ell - 1)$ where $m = \beta(\ell, k, d)$.

Remark 2.22. Since Shirshov's Height Theorem is combinatoric, it is applicable to more general situations of integrality, as indicated in [Row80, Theorem 4.3.5].

Corollary 2.23. *If A is affine PI and algebraic over a field F, then $J(A)$ is nilpotent, and every nil subalgebra of A is nilpotent.*

Proof. A is finite dimensional, by Theorem 2.3, and thus representable. Hence, we apply Theorem 1.79 and Theorem 1.80. $\qquad \square$

Corollary 2.24. *If $A = F\{a_1, \ldots, a_\ell\}$ is an affine PI-algebra without 1 and the words in the Shirshov base are nilpotent, then A is nilpotent.*

Proof. Suppose m is the maximum nilpotence degree of the words of A, and A has PI-degree d. Checking lengths of words, we see that $A^{\ell m \beta(\ell, 2d, d)}$ is a sum of words each of which contains some $\bar{w}^m$ for a suitable $\bar{w}$ in the Shirshov base and thus is 0. $\qquad \square$

We shall strengthen this considerably in Corollary 2.82. These results can be generalized to affine algebras over a commutative Noetherian ring, cf. Exercise 17.

Corollary 2.25. *Any nil subalgebra of a PI-algebra is locally nilpotent (i.e., every finite subset generates a nilpotent subalgebra).*

Proof. Any finite subset generates an affine PI-algebra without 1, so Corollary 2.24 is applicable. $\qquad \square$

We cannot say yet that nil subalgebras of an affine algebra are nilpotent, since they need not be affine (as algebras without 1). This assertion is true, but only as a consequence of the Razmyslov-Kemer-Braun theorem. Let us strengthen this by a direct application of Shirshov's Height Theorem.

2.3   The Shirshov Program

In order to appreciate how Shirshov's Lemma fits into PI-theory as a whole, let us outline a program for proving theorems about PI-algebras. It was really the next generation (including Razmyslov, Schelter, and Kemer) who realized the full implications of Shirshov's approach. The program is:

1. Prove a special case of the theorem for representable algebras.

2. Given an affine algebra $A = F\{a_1, \ldots, a_\ell\}$, adjoin to F the characteristic coefficients of all words of length $\leq d$ in the a_i, thereby making the algebra f.g. over a commutative affine algebra, in view of Shirshov's Height Theorem.

3. Reduce the theorem to an assertion in commutative algebra.

Shortly we shall apply the Shirshov program to prove the Razmyslov-Kemer-Braun Theorem, that the radical of any affine PI-algebra is nilpotent. We shall go through the same sequence of ideas in Chapter 4, in solving Specht's problem for affine PI-algebras.

2.4   The Trace Ring

To proceed further, we need to be able to reduce to the integral (and thus f.g.) case, by Step 2 of Shirshov's program. This was first done for A prime. Although the technique has been described in detail in several books, we review the description briefly here, both for its important role in studying prime affine algebras and also as preparation for the more general situation described below in Section 2.4.3. The main idea is encapsulated in the next definition.

Definition 2.26. Suppose $A = C\{a_1, \ldots, a_\ell\}$ is an affine subalgebra of $M_n(K)$, where K is some commutative C-algebra. Take a finite Shirshov base $\bar{W}$ for A, over which A has height μ, and form $\hat{C}$ by adjoining to C the characteristic coefficients of $\bar{w}$ for every $\bar{w} \in \bar{W}$. Define

$$\hat{A} = A\hat{C} = \hat{C}\{a_1, \ldots, a_\ell\}.$$

Remark 2.27. In characteristic 0, the construction of Definition 2.26 is equivalent to adjoining $\mathrm{tr}(\bar{w}^k)$ for all $k \leq n$. This observation is important since the trace is an additive function.

Remark 2.28. The Hamilton-Cayley theorem shows each $\bar{w}$ is integral over $\hat{C}$ in $\hat{A}$, implying by Theorem 2.3(ii) that $\hat{A}$ is f.g. as a $\hat{C}$-module.

Also $\hat{C}$ is affine as a C-algebra, generated by at most $n|W|$ elements. Hence $\hat{A}$ can be described quite easily in terms of C.

Having constructed this nice algebra $\hat{A}$, our task is to pass information back from $\hat{A}$ to A. Thus, we want to relate A to $\hat{A}$, i.e., what can we say about the characteristic coefficients of these $\bar{w}$?

2.4.1 The trace ring of a prime algebra

We first consider the special case of a prime algebra. Strictly speaking, we do not need to consider prime algebras at all, since the more technical theory to follow encompasses it. Nevertheless, the prime case provides valuable intuition, as well as a quick application, and is one of the most important tools in the study of prime and semiprime affine PI-algebras.

Definition 2.29. The *trace ring*, or *characteristic closure*, of a prime affine PI-algebra $A = C\{a_1, \ldots, a_\ell\}$ is defined as follows:

Let $Z = \text{Cent } A$ and $Q = S^{-1}A$ and $K = S^{-1} Z$, where $S = Z \setminus \{0\}$. Recall by Corollary 1.61 that $[Q : K] = n^2$, so any element a of A, viewed as a linear transformation of Q (given by $q \mapsto aq$), satisfies a characteristic polynomial

$$a^{n^2} + \sum_{i=0}^{n^2-1} \alpha_i a^i = 0, \tag{2.1}$$

where the $\alpha_i \in K$. We work inside Q, which satisfies the standard identity s_{2n}. For each word of length $\le 2n$ in $a_1, \ldots, a_\ell$, adjoin the appropriate $\alpha_0, \ldots, \alpha_{n^2-1}$ to C. Thus, all in all, we adjoin at most $n^2(\ell^{2n+1} - 1)/(\ell - 1)$ elements to C to get a new algebra $\hat{C} \subseteq K$. Although $\hat{C}$ is not a field, when C is Noetherian, K is commutative affine and thus Noetherian. Let $\hat{Z} = Z\hat{C}$, and let $\hat{A} = A\hat{C} \subseteq Q$, called the *characteristic closure* of A.

Note. In the above construction, since Q is central simple, we could have used instead the reduced characteristic polynomial of degree n, and then Equation (2.1) could be replaced by an equation of degree n. This alternate choice does not affect the development of the theory.

Remark 2.30. $\hat{A} = \hat{C}\{a_1, \ldots, a_\ell\}$ is a prime algebra, which is f.g. as a module over $\hat{C}$. In particular, if C is Noetherian, $\hat{A}$ is Noetherian as well as C-affine. (This is a special case of Remark 2.28.)

We need some device to transmit our knowledge back to A. This is provided by the following fact, which is an immediate consequence of Theorem I.

Remark 2.31. $\hat{Z}$ and Z have a nonzero ideal I in common. In other words, $I \subseteq Z$ but $I \triangleleft \hat{Z}$. Consequently, $\hat{A}$ and A have the ideal AI in common.

One application of the trace ring is the following famous result (which we do not need for our proof of the general theorem, but include to show how easily this method works):

Theorem 2.32 (Amitsur). *If A is affine PI over a field, then its Jacobson radical $J(A)$ is locally nilpotent.*

Proof. First note that the nilradical N is locally nilpotent, in view of Corollary 2.25. Modding out N, we may assume $N = 0$, and we want to prove $J(A) = 0$. But A now is a subdirect product of prime algebras, so passing to the prime images, we may assume A is prime. In the notation of Remark 2.31, $J(A)I$ is quasi-invertible, but also an ideal of $\hat{A}$, so $J(A)I \subseteq J(\hat{A})$. Since $I \neq 0$ and A is prime, it suffices to prove $J(\hat{A}) = 0$. Let $J = J(\hat{A})$.

$\hat{C}$ is an affine domain so $J(\hat{C}) = 0$. In view of Corollary 1.60 and "going up" for maximal ideals of an integral extension of a central subalgebra, it follows that $J \cap \hat{C} = 0$. But any element $a \in J$ satisfies an equation $\sum_{i=0}^{t} c_i a^i = 0$ for $c_i \in \hat{C}$; take such an equation with t minimal. Then $c_0 \in J \cap \hat{C} = 0$. If $0 \neq a \in \hat{Z}$, we can cancel out a and lower t, contradiction; hence, $\hat{Z} \cap J = 0$. But then $J = 0$ by Theorem C. $\qquad\square$

Note that Theorem 2.32 does not generalize to affine PI-algebras over commutative Noetherian rings, since any local Noetherian integral domain has non-nilpotent Jacobson radical. It can be generalized to affine PI-algebras over commutative Jacobson rings.

2.4.2 The trace ring of a representable algebra

We would like to backtrack, to show that the hypothesis in Definition 2.29 that A is prime is superfluous, and that we only use the fact that the affine F-algebra A is representable, i.e., $A \subseteq M_n(K)$ for a field K; since K is included in its algebraic closure, we may assume K is algebraically closed. Let $\tilde{A} = AK$, a finite dimensional subalgebra of $M_n(K)$. Since we have little control over the embedding into $M_n(K)$, we look for a more effective way of computing the traces. Anticipating Chapter 4, we rely on a famous classical theorem of Wedderburn:

Theorem 2.33 (Wedderburn's Principal Theorem). *[Row88, p.193]*
If $\tilde{A}$ is a finite dimensional algebra over an algebraically closed field, then there is a vector space isomorphism

$$\tilde{A} = \bar{A} \oplus \tilde{J}$$

where $\tilde{J} = J(\tilde{A})$ is nilpotent and $\bar{A}$ is a semisimple algebra isomorphic to $\tilde{A}/J$.

Take the decomposition into simple algebras

$$\bar{A} = \bar{A}_1 \times \cdots \times \bar{A}_q,$$

where $\bar{A}_k = M_{n_k}(K)$ (since K is algebraically closed), and we arrange the $\bar{A}_k$ so that $n_1 \geq n_2 \geq \cdots \geq n_q$.

For convenience, we assume first $\operatorname{char}(F) = 0$ in order to be able to appeal to Newton's formulas (Remark 1.44), which show that all characteristic coefficients rely on the trace.

We want to define traces in $\bar{A}$ in an effective way. Since any nilpotent element should have trace 0, we ignore $\tilde{J}$ and focus on $\bar{A}$. Since $\bar{A}$ is a direct product of matrix algebras $\bar{A}_1 \times \cdots \times \bar{A}_q$, we can define tr_j to be the trace in the j component and then for any $r = (r_1, \ldots, r_q) \in \bar{A}$, define

$$\operatorname{tr}(r) = \sum_{j=1}^{q} n_j \operatorname{tr}_j(r_j) \in F. \tag{2.2}$$

The reason we weight the traces in the sum is that this gives the trace of r in the regular representation of $\bar{A}$.

Remark 2.34. Notation as above, let $t = n_1^2$, and take j' to be the maximal j such that $n_j^2 = t_1$. In other words, $n_1 = \cdots = n_{j'} > n_{j'+1} \geq \cdots$. Suppose S is any set of t-alternating polynomials, and let I be the ideal of A generated by evaluations of polynomials of S. Thus, any element of I is a sum of specializations that we typically write as

$$f(\bar{x}_1, \ldots, \bar{x}_t; \bar{y}_1, \ldots, \bar{y}_m).$$

Let $\pi_j : \bar{A} \to \bar{A}_j$ be the natural projection. If $j > j'$, then c_t is an identity of $\bar{A}_j$, so $\pi_j(I) = 0$.

If $j \leq j'$, then Theorem J of Chapter 1 implies for any $r \in A$ that $\pi_j(I) \operatorname{tr}(r_j) \subseteq \pi_j(I)$, and in fact, Equation (1.19) of Chapter 1 shows for any $\bar{x}_i, \bar{y}_j \in \bar{A}_j$ that

$$n_1 \operatorname{tr}_1(r) f(\bar{x}_1, \ldots, \bar{x}_t, \bar{y}_1, \ldots, \bar{y}_m)$$

$$= \sum_{j=1}^{t} f(\bar{x}_1, \ldots, r\bar{x}_j, \bar{x}_{j+1}, \ldots, \bar{x}_t, \bar{y}_1, \ldots, \bar{y}_m). \tag{2.3}$$

(We had to multiply by n_1, since left multiplication by r is viewed in (1.19) as a $t \times t$ matrix, whose trace is n_1 times the trace of r as $n_1 \times n_1$ matrix.)

Hence, for any $\bar{x}_i, \bar{y}_j$ in $\tilde{A}$, in view of (2.2), we have

$$\mathrm{tr}(r) f(\bar{x}_1, \ldots, \bar{x}_t, \bar{y}_1, \ldots, \bar{y}_m)$$
$$- \sum_{j=1}^{t} f(\bar{x}_1, \ldots, r\bar{x}_j, \bar{x}_{j+1}, \ldots, \bar{x}_t, \bar{y}_1, \ldots, \bar{y}_m) \in \tilde{J}. \quad (2.4)$$

We conclude that traces are compatible in $\bar{A}$ with t-normal polynomials, where t is the size of the largest matrix component in $\bar{A}$.

In characteristic 0, we now define $\hat{A}$ exactly as in Definition 2.26 and Remark 2.27. We would like to say that $I \lhd \hat{A}$, but unfortunately, we do not know much about how t-normal polynomials interact with $\tilde{J}$. (This is because Wedderburn's principal theorem does not relate the multiplicative structures of $\bar{A}$ and $\tilde{J}$.) Thus, the best we can say is that $I\hat{A} \subseteq A + \tilde{J}$. This is a serious obstacle, and to circumvent it in solving Specht's problem, we need the rather sophisticated analysis of Kemer polynomials in Chapter 4. Of course, since $J(A)$ is locally nilpotent, we see $J(A)K$ is nil so $J(A) \subseteq J(\tilde{A})$. But $J(\tilde{A})$ is nilpotent by Corollary 2.23, so its subset $J(A)$ is nilpotent. (This also is seen at once from the Razmyslov-Kemer-Braun Theorem below.)

Let $J = J(A)$. When $J^s = 0$, we can say $J^{s-1}I\hat{A} \subseteq J^{s-1}I$, so $J^{s-1}I \lhd \hat{A}$. Unfortunately $J^{s-1}I$ could be 0, but this in turn could be valuable information, enabling us to induct on s when passing to A/I, as we shall see in Chapter 9.

With all these difficulties in characteristic 0, one might fear that the characteristic $p > 0$ situation is altogether unwieldy, since we do not have control on the characteristic coefficients (other than trace) of elements of J. But there is a cute little trick to take care of this, which leads to a better final result. First we give the philosophy.

Remark 2.35. If a and b are commuting matrices over an algebraically closed field K of characteristic p, then $(a + b)^{p^k} = a^{p^k} + b^{p^k}$. Hence, if we take the Jordan decomposition $T = T_s + T_r$ of a linear transformation T into its semisimple and radical parts, and if p^k is greater than the degree of T, then $T_r^{p^k} = 0$ and thus $T^{p^k} = T_s^{p^k}$ is semisimple.

In other words, taking high enough powers of an element destroys the radical part. Here is is an abstract formulation.

Lemma 2.36. *Suppose C is a commutative affine algebra of characteristic p, and $q = p^k$ is at least the nilpotence index of the radical of C. Then the map $\varphi : c \mapsto c^q$ is an endomorphism of C, and the algebra $\varphi(C)$ is semiprime and affine.*

Proof. Suppose $a = c^q$ is nilpotent in $\varphi(C)$. Then c is nilpotent, so $0 = c^q = a$. $\square$

Definition 2.37. Suppose $A = F\{a_1, \ldots, a_\ell\}$ is representable.

If $\mathrm{char}(F) = 0$, define $\hat{A}$ as in Definition 2.26 (taking $C = F$), but using the trace as defined above. (By looking inside the individual components, we could replace t by n_1, but that is not significant.)

If $\mathrm{char}(F) = p > 0$, take q greater than the nilpotence index of $J(A)$ (cf. Theorem 1.79). Now take a finite Shirshov base $\bar{W}$ for A over which A has height μ, and for every $\bar{w} \in \bar{W}$, form $\hat{F}$ by adjoining to F the characteristic coefficients of $\bar{w}^q$ to A and let

$$\hat{A} = A\hat{F} = \hat{F}\{a_1, \ldots, a_\ell\}.$$

In either case, $\hat{A} = A\hat{F}$ is called the *characteristic closure* of A.

Let us summarize these results.

Proposition 2.38. *Notation as above, suppose* $\mathrm{char}(F) = 0$, *A is F-affine and representable, and suppose* $J^s = 0$. *Let* $\mathcal{I}$ *be any T-ideal generated by t-alternating polynomials, and let I be their specializations in A. Then*

(i) $J^{s-1}I$ *is a common ideal of $\hat{A}$ and A.*

(ii) $\mathcal{I} \subseteq \mathrm{id}(A/I)$.

(iii) $\hat{A}$ *is algebraic of bounded degree and thus $\hat{A}$ is a f.g. module over its center $\hat{Z}$; in particular, $\hat{Z}$ and $\hat{A}$ are Noetherian algebras.*

Proof. $J^{s-1}I \subseteq A$ by definition, but is closed under multiplication by the traces, so is a common ideal with $\hat{A}$, proving (i); furthermore, (ii) is immediate by definition of I.

(iii) By Shirshov's Height Theorem, the projection of $\hat{A}$ into $\bar{A}$ is algebraic of degree t. But this means for any $a \in \hat{A}$, there a polynomial f of degree t for which $f(a) \in J(\hat{A})$. But $J(\hat{A})^{s'} = 0$ for some s', implying $f(a)^{s'} = 0$ and a is algebraic of degree $\leq ts'$. Thus, we can apply Shirshov's Height Theorem to all of $\hat{A}$. The rest is standard structure theory: $\hat{Z}$ is commutative affine by Proposition 1.75 and so is Noetherian; $\hat{A}$, being finite over $\hat{Z}$, is Noetherian. $\square$

To show that these results subsume the previous case when A is prime, note in that case that $\bar{A}$ is simple so $\bar{J} = 0$ and $s = 1$. The situation is easier in nonzero characteristic, since we bypass J and $\bar{J}$ at the outset.

Proposition 2.39. *Notation as above, suppose* $\mathrm{char}(F) > 0$ *and A is F-affine and representable. Then*

(i) I is a common ideal of $\hat{A}$ and A.

(ii) $\mathcal{I} \subseteq \mathrm{id}(A/I)$.

(iii) $\hat{A}$ is algebraic of bounded degree and thus $\hat{A}$ is a f.g. module over its center $\hat{Z}$; in particular, $\hat{Z}$ and $\hat{A}$ are Noetherian algebras.

Proof. We need a few modifications in the previous proof.

(i) We need to show that I absorbs the characteristic coefficients of $\bar{w}^q$, where q is as in Definition 2.37. But this is true since $\bar{w}^q$ is semisimple, and thus belongs to $\bar{A}$.

(iii) Each $\bar{w}^q$ is algebraic, implying $\bar{w}$ is algebraic, so Shirshov's Height Theorem is applicable again.   □

2.4.3 A Hamilton-Cayley type theorem for algebras satisfying a Capelli identity

Let us present an abstract version of the Hamilton-Cayley theorem, which will put the characteristic closure in a much broader framework. The most direct way would be in the language of generalized identities with operations, to be formulated in Section 11.3.6, but in order to avoid extra formalism let us make a slight detour, which enables us to focus on particular evaluation of polynomials. For notational convenience, we also shall let $y_1, \ldots, y_m$ and z denote extra indeterminates, with $m \geq n$. Our motivation comes from (1.19) from Chapter 1, which we now use as our definition.

Definition 2.40. Suppose $f(x_1, \ldots, x_n, x_{n+1}; y)$ is multilinear in the x_i. Define $\delta_j(f)$ to be the homogeneous component of degree j in a new indeterminate z in the polynomial $f((z+1)x_1, \ldots, (z+1)x_n, x_{n+1}; y)$, or in other words, the sum of all substitutions of zx_i for x_i in j positions of f.

Remark 2.41. We shall assume throughout that f alternates in the indeterminates $x_1, \ldots, x_n$. Then $\delta_j(f)$ also alternates in $x_1, \ldots, x_n$.

When $A = M_n(F)$ and we specialize the x_i, y, and z to matrices $\bar{x}_i, \bar{y}$, and $\bar{z}$, then Theorem J says $\delta_j(f)$ is the characteristic coefficient of $\bar{z}^j$ multiplied by $f(\bar{x}_1, \ldots, \bar{x}_n, \bar{x}_{n+1}; \bar{y})$. Thus, the results that we describe now generalize Theorem J. From this point of view, we can call δ_{n+1} the *Razmyslov-Zubrilin trace function.*

Remark 2.42. $\delta_j\delta_k = \delta_k\delta_j$, since we are working with the same indeterminate z. (However, this is not δ_{j+k}, since the insertions of z do not match even in Remark 2.41.)

The following concept is a bit artificial, but it enables us to avoid introducing free products at this stage.

Definition 2.43. Given a set of indeterminates

$$X = \{x_1, \ldots, x_n; y_1, \ldots, y_m; z\},$$

we fix an arbitrary algebra homomorphism $\psi : C\{X\} \to A$ (which can be defined by specializations $x_i \mapsto \bar{x}_i \in A$, $y_i \mapsto \bar{y}_i \in A$, and $z \mapsto a \in A$), and extend it to $\psi : A \times C\{X\} \to A$ by

$$a'f(x_1, \ldots, x_n; y_1, \ldots, y_m; z) \mapsto a'f(\bar{x}_1, \ldots, \bar{x}_n; \bar{y}_1, \ldots, \bar{y}_m; a) \quad \text{for} \quad a' \in A.$$

Viewed in this way, polynomials in $C\{X\}$ can have (left) coefficients in A. Such $f(x_1, \ldots, x_n; y_1, \ldots, y_m; z)$ is said to A-alternate in $x_1, \ldots, x_n$ if $\psi f = 0$ whenever $\bar{x}_i = \bar{x}_j$ for $1 \le i < j \le n$. (This is analogous to our previous definition of alternating.)

To cut down on the notation, we write y in place of $y_1, \ldots, y_m$ in a polynomial, but it should be understood we are to be allowed as many variables as we wish. The next result is quite technical, and its proof is even more so; before presenting it, let us pause for a minute to see from where we have come and to where we would like to go. The key tool in constructing the trace ring was Theorem I, which enabled us to incorporate the Hamilton-Cayley characteristic coefficients into the evaluations of the Capelli polynomial. A mild generalization, Theorem J, enabled us to construct the trace ring for any representable algebra. However, in the more difficult theorems of PI-theory (especially Kemer's theorem), we need a much more sophisticated trace ring that involves an abstraction of Theorems I and J; the δ_j are to take the role of the characteristic coefficients.

To be sure, Schelter [Sch76] obtained a large chunk of the Razmyslov-Kemer-Braun Theorem by means of a clever application of the trace ring of a prime algebra, and Braun 's proof also manages with this special case (although his proof involves other technical difficulties), but the patient reader should look carefully to see that the proof of this result strongly resembles that of Theorem J, with the same underlying idea.

Proposition 2.44. *Suppose $f(x_1, \ldots, x_n, x_{n+1}; y)$ is a polynomial which A-alternates in $x_1, \ldots, x_n$ and is linear in x_{n+1}. (The symbol y is used shorthand for several other variables.) Also (cf. Remark G) suppose the $(n+1)$-alternating polynomial*

$$\tilde{f}(x_1, \ldots, x_{n+1}; y) = \sum_{k=1}^{n+1} f(x_1, \ldots, x_{k-1}, x_{n+1}, x_{k+1}, x_{k+2} \ldots, x_n, x_k; y)$$

$$(2.5)$$

is an identity on A. Then A also satisfies the identity:

$$\sum_{j=0}^{n}(-1)^j\delta_j\Big(f(x_1,\ldots,x_n,z^{n-j}x_{n+1};y)\Big).\tag{2.6}$$

Proof. Since $\tilde{f}$ is an identity, $f(x_1,\ldots,x_n,z^{n-j}x_{n+1};y)$ can be replaced by

$$-\sum_{k=1}^{n}f(x_1,\ldots,x_{k-1},z^{n-j}x_{n+1},x_{k+1},\ldots,x_n,x_k;y).$$

Hence (2.6) is equivalent to verifying the identity

$$\sum_{j=0}^{n}(-1)^j\sum_{k=1}^{n}\delta_j\Big(f(x_1,\ldots,x_{k-1},z^{n-j}x_{n+1},x_{k+1},\ldots,x_n,x_k;y)\Big)$$

i.e., $\displaystyle\sum_{k=1}^{n}\sum_{j=0}^{n}(-1)^j\delta_j\Big(f(x_1,\ldots,x_{k-1},z^{n-j}x_{n+1},x_{k+1},\ldots x_n,x_k;y)\Big).$

Let us show each

$$\sum_{j=0}^{n}(-1)^j\delta_j\Big(f(x_1,\ldots,x_{k-1},z^{n-j}x_{n+1},x_{k+1},\ldots x_n,x_k;y)\Big)$$

is identically 0. Focusing on k, we define:

$\delta'_{jk}(f)$ is the sum of the monomials of $\delta_j f$ in which zx_k has been substituted for x_k, and

$\delta''_{jk}(f)$ is the sum of the monomials of $\delta_j f$ in which x_k has been left as it is, i.e., z does not appear before x_k. Thus, by definition,

$$\delta_{jk}(f)=\delta'_{jk}(f)+\delta''_{jk}(f).$$

Then for any k, $\delta'_{0k}f(x_1,\ldots,x_{k-1},z^n x_{n+1},x_{k+1},\ldots,x_n,x_k;y)=0$,

$\delta''_{0k}f(x_1,\ldots,x_{k-1},z^n x_{n+1},x_{k+1},\ldots,x_n,x_k;y)$
$$=f(x_1,\ldots,x_{k-1},z^n x_{n+1},x_{k+1},\ldots,x_n,x_k;y),$$

$\delta'_{nk}f(x_1,\ldots,x_{k-1},x_{n+1},x_{k+1},\ldots,x_n,x_k;y)$
$$=f(zx_1,\ldots,zx_{k-1},zx_{n+1},zx_{k+1},\ldots,zx_n,x_k;y),$$

$$\delta''_{nk}f(x_1,\ldots,x_{k-1},x_{n+1},x_{k+1},\ldots,x_n,x_k;y)=0,\tag{2.7}$$

and for $1 < j < n$,

$$\delta'_{jk} f(x_1, \ldots, x_{k-1}, z^{n-j} x_{n+1}, x_{k+1}, \ldots, x_n, x_k; y)$$
$$= \delta''_{j-1,k} f(x_1, \ldots, x_{k-1}, z^{n-j+1} x_{n+1}, x_{k+1}, \ldots, x_n, x_k; y),$$

so

$$\delta_{jk} f(x_1, \ldots, x_{k-1}, z^{n-j} x_{n+1}, x_{k+1}, \ldots, x_n, x_k; y)$$
$$= \delta''_{j-1,k} f(x_1, \ldots, x_{k-1}, z^{n-j+1} x_{n+1}, x_{k+1}, \ldots, x_n, x_k; y)$$
$$+ \delta''_{jk} f(x_1, \ldots, x_{k-1}, z^{n-j} x_{n+1}, x_{k+1}, \ldots, x_n, x_k; y).$$

Summing, we see all

$$\sum_{j=0}^{n} (-1)^j \delta_j (f(x_1, \ldots, x_{k-1}, z^{n-j} x_{n+1}, x_{k+1}, \ldots x_n, x_k; y))$$

$$= \sum_{j=1}^{n} (-1)^j \delta''_{j-1,k} (f(x_1, \ldots, x_{k-1}, z^{n-j+1} x_{n+1}, x_{k+1}, \ldots x_n, x_k; y))$$

$$+ \sum_{j=0}^{n-1} (-1)^j \delta''_{jk} (f(x_1, \ldots, x_{k-1}, z^{n-j} x_{n+1}, x_{k+1}, \ldots x_n, x_k; y))$$

$$\tag{2.8}$$

which cancel out, since the two sums are identical with opposite sign, after changing the index in the second summand. □

(The same result holds, with identical proof, if instead of z we use any linear operator on A; also the result was formulated by Zubrilin much more generally, for nonassociative algebras whose multiplication need not even be binary.)

Corollary 2.45 (Zubrilin). *Suppose a polynomial $h(x_1, \ldots, x_n; y)$ A-alternates in the variables $x_i, 1 \le i \le n$. Then any algebra A satisfying the Capelli identity c_{n+1} also satisfies:*

$$\sum_{j=0}^{n} (-1)^j z^{n-j} \delta_j (h(x_1, \ldots, x_n; y)). \tag{2.9}$$

Proof. Apply Proposition 2.44 where $f = x_{n+1} h$; to apply it, we must show $\tilde{f}$ is $(n+1)$-alternating. But

$$f(x_1, \ldots, x_{k-1}, x_{n+1}, x_{k+1}, \ldots, x_n, x_k)$$
$$= (-1)^{n-(k+1)} f(x_1, \ldots, x_{k-1}, x_{k+1}, \ldots, x_n, x_{n+1}, x_k)$$
$$= (-1)^{n-1} (-1)^k f(x_1, \ldots, x_{k-1}, x_{k+1}, \ldots, x_n, x_{n+1}, x_k),$$

so $\tilde{f}$ is $(n+1)$-alternating by Remark G. We conclude by specializing $x_{n+1} \mapsto 1$. $\square$

2.4.4 Obstruction to integrality

Our next step is to see how far an element of A is from being integral of degree n. The motivation is provided by Shirshov's Height Theorem—If every element of a Shirshov base is integral, then the algebra is f.d. See Exercises 20 and 21 for a converse.

Definition 2.46. Suppose A is an C-algebra. Given $a \in A$, take commuting indeterminates $\xi_i, 1 \le i \le n$, and define

$$A' = A[\xi_1, \ldots, \xi_n]/\langle a^n + \xi_1 a^{n-1} + \cdots + \xi_n \rangle,$$

and consider the natural map $\varphi_a : A \to A'$.

The kernel $\ker \varphi_a = A \cap \langle a^n + \xi_1 a^{n-1} + \cdots + \xi_n \rangle$ is called the *obstruction to integrality* of degree n of the element a.

Remark 2.47.

(i) $\varphi_a(a)$ is integral in A' over the image C' of the commutative ring $C[\xi_1, \ldots, \xi_n]$.

(ii) If a is already integral of degree n satisfying $a^n + \sum \alpha_i a^{n-i} = 0$ for $\alpha_i \in C$, then the composition of φ_a with the natural homomorphism $A' \to A$ given by $\xi_i \mapsto \alpha_i$, $1 \le i \le n$, is the identity, implying $\ker \varphi_a = 0$.

(iii) Since any element integral of degree n is integral of all degrees less than n, it follows from (ii) that the obstruction to integrality of degree n is contained in the obstruction to integrality of any degree $< n$.

Now we are ready for a sleight of hand. We want to use Corollary 2.45 to view elements of A as integral in some sense, but the "coefficients" $\delta_j(h(x_1, \ldots, x_n; y))$ are not in the center! In Chapter 4, we shall see how to arrange for them to be central, but for the time being we can get around this difficulty with the following trick.

Suppose $A = C\{a_1, \ldots, a_\ell\}$ satisfies the Capelli identity c_{n+1}. Given an n-alternating polynomial

$$f = f(x_1, \ldots, x_n, y_1, \ldots, y_m, z),$$

let us view $C(X) = C\{x_1, \ldots, x_n, y_1, \ldots, y_m, z\}$ as an A-module via Definition 2.43, in terms of $\psi : C\{X\} \to A$. Defining $\delta_k^q(f) = \delta_k(\delta_k^{q-1}(f))$

inductively, cf. Definition 2.40, let $M = M_{\psi;f}$ denote the A-submodule of $C\{X\}$ generated by f and the $\delta_k^q(f)$ for each $q \geq 1$. Thus, every element of M could be considered an n-alternating polynomial with coefficients in A. Fixing $\bar{x}_i, \bar{y}_j, \in A$ for $1 \leq i \leq n$ and $1 \leq j \leq m$, we view M as an $A[\xi_1, \ldots, \xi_n]$-module, by putting

$$\xi_j \delta_k^q(f) = (-1)^j \delta_j \delta_k^q(f(\bar{x}_1, \ldots, \bar{x}_n; \bar{y}_1, \ldots \bar{y}_m; a).$$

This indeed defines a module, in view of Remark 2.42.

Proposition 2.48. ker φ_a *annihilates* M.

Proof. We shall show $(a^n + \xi_1 a^{n-1} + \cdots + \xi_n)M = 0$. But any element of M is n-alternating, so for $h \in M$ we have

$$(a^n + \xi_1 a^{n-1} + \cdots + \xi_n)h = \sum (-1)^k a^{n-k} \delta_k h = 0,$$

by Corollary 2.45. $\qquad\square$

We note that if M is an A-module and $I \triangleleft A$ with $IM = 0$, then we can view M naturally as an A/I module. Furthermore, under this action, if $I_1/I \triangleleft A/I$ with $(I_1/I)M = 0$, then $I_1 M = 0$ under the original action.

Thus, we have the following consequence.

Corollary 2.49. M *is also a module over* $A/\ker \varphi_a$, *with the same action.*

Let $A_1 = A/\ker \varphi_a$. We have seen that A_1 has a central extension A_1' in which a is integral (over Cent(A_1)). Let us repeat the process. Since $c_{n+1} \in \text{id}(A)$, A has PI-degree $d \leq 2(n+1)$. Hence, there are a finite number $m < (\ell + 1)^d$ of monomials in the a_i of length $\leq d$, which we denote $\tilde{a}_1, \ldots, \tilde{a}_m$. Inductively, we let

$$\Phi_{j+1} = \Phi_j + \ker \varphi_{\tilde{a}_j},$$

and we see $A_{j+1} = A_j/\ker \varphi_{\tilde{a}_j} = A/\Phi_j$ has a central extension over which $\tilde{a}_1, \ldots, \tilde{a}_{j+1}$ is integral. Letting $\Phi(A) = \Phi_m$ we get a homomorphic image $\tilde{A} = A/\Phi(A) = C\{\tilde{a}_1, \ldots, \tilde{a}_\ell\}$ over which M remains a module.

Remark 2.50.

(i) $\Phi(A)M = 0$ (since the image of $\Phi(A)$ in $\tilde{A} = A/\Phi(A)$ is 0). Recall this was for any $M_{\psi;f}$ as well as any ψ, so in particular, $\Phi(A)$ annihilates every substitution of every n-alternating polynomial.

(ii) By construction, $\tilde{A}$ has a central extension $\tilde{A}' = C'\{\tilde{a}_1, \ldots, \tilde{a}_\ell\}$ for which every word of length $\leq d$ is integral; thus Shirshov's Height Theorem 2.3 implies $\tilde{A}'$ is integral over C', and thus representable. It follows, by Theorem 1.79 and Theorem 1.80, that the image in $\tilde{A}$ of $J(A)$ is nilpotent, and the image of any nil subalgebra N is nilpotent.

2.4.5 t-closure and t-closed ideals

The obstruction to integrality gives rise to a generalization of the characteristic closure. Suppose A is any affine PI-algebra. Let us fix a Shirshov base $\bar{W}$ of A, which by Shirshov's Theorem is finite. Applying Remark 2.47 to each element of $\bar{W}$ in turn, we obtain an algebra homomorphism $\varphi_W : A \to \hat{A} = A\hat{F}$, where $\hat{F}$ is a central affine subalgebra of $\hat{A}$, satisfying the property that $\varphi_W(\bar{w})$ is integral of degree t in $\hat{A}$ over $\hat{F}$, and $\hat{F}$ is generated by the coefficients of the equations of integrality of the elements of $\bar{W}$; furthermore, if $\psi : A \to R$ is any homomorphism of A into an algebra R for which the image of each element of $\bar{W}$ is integral of degree t, then $\ker \psi \supseteq \ker \varphi_W$.

Definition 2.51. We call $\ker \varphi_W$ the *obstruction to integrality for A of degree t* (with respect to W), and $\hat{A}$ the *t-closure of A* (with respect to W). (Although $\hat{A}$ formally depends on the choice of W, this is not important.)

Remark 2.52. Shirshov's Height Theorem implies $\hat{A}$ is f.g. over $\hat{F}$, and in particular, is Noetherian. Thus, $A/\ker \varphi_W$ is a representable algebra. The obstruction to integrality of A of degree t is 0, iff A can be centrally embedded into an algebra $\hat{A}$ integral of bounded degree t over its center.

This leads to a closure operation:

Definition 2.53. Given an ideal $I \lhd A$, define I^{cl} to be the preimage in A of the obstruction to integrality of degree t of A/I. I is *t-closed* if $I = I^{\mathrm{cl}}$, i.e., the obstruction to integrality of degree t of A/I is 0.

Proposition 2.54. *Any affine PI-algebra satisfies ACC on t-closed ideals.*

Proof. Let $\hat{A}$ be as above. Let $\hat{I} = I\hat{F}$. I is t-closed in $\hat{A}$ iff $I = A \cap \hat{I}$. But the $\hat{I}$, being ideals of the Noetherian algebra $\hat{A}$, satisfy the ACC, so their contractions to A also satisfy the ACC. $\square$

2.5   The Razmyslov-Kemer-Braun Theorem

We now turn to proving the nilpotence of $J(A)$ for an affine algebra A. In Chapter 1, we studied nil subalgebras and the Jacobson radical of representable algebras. Of course, if A itself were nil (without 1), then it would be nilpotent by Shirshov's Height Theorem. Nevertheless, the theorem is quite difficult to prove in its entirety.

Braun [Br82] proved this using a clever series of reduction to Azumaya algebras. On the other hand, Razmyslov had proved $J(A)$ is nilpotent if A satisfies a Capelli identity, so when Kemer proved that every affine

PI-algebra satisfies a Capelli identity, one saw that $J(A)$ is nilpotent. We follow this approach, with some streamlining. The first major step is to apply the Razmyslov-Zubrilin theory of abstract traces described above.

Proposition 2.55 (Razmyslov). *If A is affine and satisfies a Capelli identity c_{n+1}, then $J(A)$ is nilpotent, and every nil subalgebra N of A is nilpotent.*

Proof. By induction on n. First assume $n = 1$. Then A is commutative, so the theorem is already known.

Assume the theorem holds if A satisfies c_n. Let I be the ideal of A generated by all evaluations of c_n. Remark 2.50(i) implies $\Phi(A)I = 0$.

Furthermore, Remark 2.50(ii) shows $N^j \subseteq \Phi(A)$ for some j.

On the other hand, by definition, $c_n \in \mathrm{id}(A/I)$, so by induction $N^k \subseteq I$ for some k. Hence, $N^{j+k} \subseteq \Phi(A)I = 0$. □

An unaesthetic aspect of this proof is that we had to use the trick of the evaluations of the module in order to bypass the fact that the coefficients of Corollary 2.45 need not be central. There is a way to ensure a certain commutativity, which we shall give in the appendix to Chapter 4.

2.5.1 Kemer's Capelli Theorem

In order to conclude the proof of the Razmyslov-Kemer-Braun Theorem, we need to show that any affine algebra satisfies a Capelli identity. This was Kemer's contribution.

We shall prove this fact twice. The first proof works in all characteristics, but is quite intricate, relying on a combinatoric argument which requires the identity of algebraicity (Proposition 1.58 and Shirshov's Height Theorem applied in full force). The second proof, closer to Kemer's original proof, works only in characteristic 0 and requires the representation theory developed in Chapter 5, but may be easier to understand and fits in better with the subsequent theory that we shall work with.

Theorem 2.56. *Every affine PI-algebra $A = F\{a_1, \ldots, a_\ell\}$ of PI-degree d satisfies a Capelli identity c_n, where n depends only on ℓ and d.*

Proof. Let $W = \{w_1, \ldots, w_{|W|}\}$ denote the set of words of length $\leq d$ in $a_1, \ldots, a_\ell$; recall $|W| \leq \ell^d + \ell^{d-1} + \cdots + \ell$. By Shirshov's Height Theorem it suffices to find $n = n(\ell, d)$ such that any affine algebra $A = F\{a_1, \ldots, a_\ell\}$ of PI-degree d satisfies c_n, i.e., $c_n(\bar{x}_1, \ldots, \bar{x}_n; \bar{y}_1, \ldots, \bar{y}_n)$ vanishes for all substitutions of "Shirshov words" $\bar{x}_1, \ldots, \bar{x}_n; \bar{y}_1, \ldots, \bar{y}_m$, cf. Definition 2.2, i.e., each

$$\bar{x}_j = w_{j_1}^{k(j_1)} \ldots w_{j_\mu}^{k(j_\mu)}.$$

We call $k(j_i)$ the *exponent* of w_{j_i}. We claim for each $0 \le i \le |W|$, there is $n_i = n_i(\ell, d)$ such that $c_{n_i}(\bar{x}_1, \ldots \bar{x}_{n_i}, \bar{y}_1, \ldots, \bar{y}_{n_i}) = 0$ for all substitutions of Shirshov words such that each exponent for $w_1, \ldots, w_{|W|-i}$ (whenever it occurs in the Shirshov words $\bar{x}, \ldots, \bar{x}_n$) is less than d. Clearly, the claim would yield the theorem, taking $i = \mu$ and $n = n_\mu$, since n_μ involves no restriction on the exponents $k(j_1), \ldots, k(j_\mu)$.

The proof of the claim is by induction on i. For $i = 0$, we see each exponent $\le d$, so the number of distinct $\bar{x}_j$ is less than $((d-1)|W|)^\mu$, which we take to be n_0.

In general, suppose we know n_{i-1}. We need to compute Capelli polynomials on Shirshov words $\bar{x}_j$ for which every exponent of $w_1, \ldots, w_{|W|-i}$ is less than d, wherever it occurs.

Consider $a = c_{n_{i-1}}(\bar{x}_1, \ldots, \bar{x}_{n_{i-1}}, \bar{y}_1, \ldots, \bar{y}_{n_{i-1}})$ for such $\bar{x}_1, \ldots, \bar{x}_{n_{i-1}}$. If the exponent of $w_{|W|-i+1}$ also is less than d, then by hypothesis $a = 0$, so we may assume $w_{|W|-i+1}$ appears in some $\bar{x}_j$ with exponent $\ge d$.

We shall prove that $n_i = d n_{i-1}$. Indeed, take $\bar{x}_1, \ldots, \bar{x}_{n_i}, \bar{y}_1, \ldots, \bar{y}_{n_i}$, arbitrary Shirshov words, where every exponent of $w_1, \ldots, w_{|W|-i}$ appearing in the $\bar{x}_j$ is less than d, wherever it occurs. Recall from Exercise 1.14 that we can rewrite c_{n_i} as a sum of products

$$c_{n_{i-1};\pi} c_{n_{i-1};\pi} \cdots c_{n_{i-1};\pi}, \qquad \pi \in S_{n_i} \tag{2.10}$$

(a product of d factors, each being $c_{n_{i-1}}$, where the indeterminates $x_1, \ldots, x_{n_i}$ are permuted by π). But by hypothesis $c_{n_{i-1}}(\bar{x}_1, \ldots, \bar{x}_{n_i}; \bar{y}_1, \ldots, \bar{y}_{n_i}) = 0$ unless $w_{|W|-i+1}$ has exponent at least d in some $\bar{x}$ in each factor $c_{n_{i-1}\pi}$ in (2.10), and we write these d exponents as $(k_1, \ldots, k_d)$. Thus, $w_{|W|-i+1}^{k_j}$ appears in some $\bar{x}_{t_j}$ in the j-th Capelli polynomial in (2.10). We choose such $\bar{x}_{t_1}, \ldots, \bar{x}_{t_d}$ such that $(k_1, \ldots, k_d)$ is minimal. But now we can use the identity of algebraicity (calculated in Remark 1.59 to lower $(k_1, \ldots, k_d)$). (To be precise, we might have several places that $(k_1, \ldots, k_d)$ occurs, but we could apply the identity of algebraicity for each of them.) This contradicts the minimality of $(k_1, \ldots, k_d)$ and thus

$$c_{n_i}(\bar{x}_1, \ldots, \bar{x}_{n_i}; \bar{y}_1, \ldots, \bar{y}_{n_i}) = 0,$$

as desired. $\square$

Ironically, going over the proof, we can now see that Kemer's Capelli Theorem holds for an affine PI-algebra over any commutative ring C, although the next theorem requires a stronger requirement on C, as mentioned earlier. For convenience, we take C to be a field.

Theorem 2.57 (Razmyslov-Kemer-Braun). *The Jacobson radical of any affine PI-algebra (over a field) is nilpotent.*

Proof. It satisfies a Capelli identity, so apply Proposition 2.55.   □

This proof, although completely self-contained, involved several hard combinatorical techniques, including the identity of algebraicity and the Razmyslov-Zubrilin theory.

2.5.2 Affine algebras satisfying a sparse identity

For our second proof of the existence of a Capelli identity, we need a technical definition.

Definition 2.58. A multilinear polynomial $g = \sum \alpha_\sigma x_{\sigma(1)} \ldots x_{\sigma(d)}$ is a *sparse identity* of A if, for any polynomial $f(x_1, \ldots, x_d; Y)$ (writing Y for extra indeterminates if necessary), we have

$$\sum \alpha_\sigma f(x_{\sigma(1)}, \ldots, x_{\sigma(d)}; Y) \in \text{id}(A).$$

The major example of a sparse identity is the Capelli identity, cf. Exercise 14. Although there are PI-algebras (such as the Grassman algebra in characteristic 0) which do not satisfy a Capelli identity, a key feature of PI-theory is that every PI-algebra over a field satisfies a suitable sparse identity. This will be proved in Theorem 5.58 in characteristic 0, using the theory of Young diagrams, and in the Appendix to Chapter 5, we shall see that the proof also works in characteristic > 0.

Sparse identities play a key role in the representation theory in characteristic 0, since they generate two-sided ideals in the group algebra of S_n, in a way to be made clear in Chapter 5 (especially in the discussion of Amitsur-Capelli polynomials). At this stage, we have a combinatoric reason why sparse identities are important.

Any sparse identity can be viewed as a powerful *sparse reduction procedure*. Namely, we may assume $\alpha_{(1)} = 1$; given $a_1, \ldots, a_d$ in A, we can replace any term $f(a_1, \ldots, a_d)$ by

$$-\sum_{\sigma \neq 1} \alpha_\sigma f(x_{\sigma(1)} \ldots x_{\sigma(d)}, x_{d+1}, \ldots, x_n).$$

(The analogous assertion also holds for c_d.)

Lemma 2.59 (Sparse pumping lemma). *Let A be a PI-algebra satisfying a sparse identity g of degree d. Then for each homogeneous d-linear polynomial $f(x_1, \ldots, x_n; y_1, \ldots, y_m)$, any specializations $\bar{x}_i, \bar{y}$ of the x_i and y, and any words v_i in the $\bar{x}_i$,*

$$f(v_1, \ldots, v_n; \bar{y}_1, \ldots, \bar{y}_m)$$

is a linear combination of elements of the form $f(v'_1, \ldots, v'_n; \bar{y})$ where at most $d - 1$ of the v'_i have length $\geq d$, and the $\bar{y}_i$ remain unchanged.

Proof. By induction on the number of v'_i of length $\geq d$. Assume the lemma is false. Then we have some specialization with at least d of the v_i of length $\geq d$, which cannot be written as a linear combination of specializations with a smaller number of v_i of length $\geq d$. Renumbering the indices if necessary, we may assume $|v_1| \geq |v_2| \geq \cdots \geq |v_d| \geq d$.

Claim. We can rearrange the letters of these v_i (without altering $v_{d+1}, \ldots, v_n$) to reduce v_d to a word of length $< d$, contrary to hypotheses.

The proof of the claim is by induction on $|v_1|, |v_2|, \ldots, |v_d|$ (In other words, we induct first on $|v_1|$, then on $|v_2|$, and so forth). We want to use the sparse identity, which we now write as

$$g = \sum_{\sigma \in S_d} \alpha_\sigma z_{\sigma(1)} \cdots z_{\sigma(d)}$$

for indeterminates $z_1, \ldots, z_d$, to decrease the length of the first component that it alters and thereby conclude by the induction hypothesis. Towards this end, we are done unless $v_1, \ldots, v_d$ all have length at least d.

Write $v_i = u'_i u_i$ where $|u_i| = d - i$, for $1 \leq i \leq d$. Define

$$h(x_1, \ldots, x_d, x_n; y; z_1, \ldots, z_d) = f(z_1 x_1, z_2 x_2, \ldots, z_d x_d, x_{d+1}, \ldots, x_n; y)$$

where we have expanded our set of indeterminates to include the $z_1, \ldots, z_d$. (We have condensed $y_1, \ldots, y_m$ to a single letter y, to simplify notation.) The sparse reduction procedure (for $g(z_1, \ldots, z_d)$) says that we can replace h by

$$-\sum_{\sigma \neq 1} \alpha_\sigma h(x_1, \ldots, x_n; y; z_{\sigma(1)}, \ldots, z_{\sigma(d)})$$

$$= -\sum_{\sigma \neq 1} \alpha_\sigma f(x_1 z_{\sigma(1)}, x_2 z_{\sigma(2)}, \ldots, x_d z_{\sigma(d)}, x_{d+1}, \ldots, x_n; y). \tag{2.11}$$

But then substituting u'_i for x_i and u_i for z_i shows that we could replace $f(v_1, \ldots, v_d, v_{d+1}, \ldots, v_n; \bar{y})$ by linear combinations of

$$f(u'_1 u_{\sigma(1)}, u'_2 u_{\sigma(2)}, \ldots, u'_d u_{\sigma(d)}, v_{d+1}, \ldots, v_n; \bar{y}). \tag{2.12}$$

Letting $v_{i,\sigma} = u'_i u_{\sigma(i)}$, we see that

$$|v_{1,\sigma}| = |v_1| - (d - 1) + d - \sigma(1) = |v_1| + 1 - \sigma(1) \leq |v_1|,$$

with inequality holding iff $\sigma(1) = 1$. Hence, each term in (2.12) with $\sigma(1) \neq 1$ is 0 by induction. Likewise, each term in (2.12) with $\sigma(1) = 1$ but $\sigma(2) \neq 2$ is 0 by induction. Continuing down the line, each term in (2.12) is 0 by induction, implying $f(v_1, \ldots, v_n; \bar{y}) = 0$, as desired.   □

We are ready to utilize sparse identities. Note that this proof, like the previous one, is characteristic-free.

Theorem 2.60. *Every affine algebra $A = F\{a_1, \ldots, a_\ell\}$ satisfying a sparse identity g of degree d satisfies the Capelli identity c_n, where $n = \ell^d + d$.*

Proof. We claim $c_n(A) = 0$. By linearity, we can specialize all x_i to words v_i in $a_1, \ldots, a_\ell$, and need to show $c_n(v_1, \ldots, v_n; \bar{y})$ always vanishes. Applying Lemma 2.59 repeatedly, it suffices to show that $c_n(v'_1, \ldots, v'_n; \bar{y})$ vanishes under this substitution when at most $d - 1$ of the v'_i have length $\geq d$. But this means there are at least

$$n - (d - 1) = \ell^d + 1$$

positions for which $|v'_i| < d$. The number of such words is

$$\ell^{d-1} + \ell^{d-2} + \cdots + 1 < \ell^d.$$

Thus, two of the words must be the same, so by the alternating property, $c_n(v'_1, \ldots, v'_n, \bar{y})$ vanishes.   □

Although the proof here is remarkably easy, the value for n is huge. By a clever substitution and careful analysis of Young tableaux, Kemer [Kem80] actually proved that $n \leq d$. Parts of his argument are given in Exercise 5.4.

To conclude this second proof, we rely on Theorem 5.58 below to show the hypothesis in Theorem 2.60 is superfluous. We shall refer to this method of rearranging entries as *pumping*.

2.6 Shirshov's Lemma Revisited

Having seen how prominently Shirshov's Height Theorem appears in proving the Razmyslov-Kemer-Braun Theorem, we would like methods of computing his function $\beta(\ell, k, d)$. Thus, as promised earlier, we shall now give two other proofs of Shirshov's Lemma 2.10.

2.6.1   Second proof of Shirshov's Lemma

Here we use Shirshov's original idea, which although rather straightforward in principle, is a bit tricky to execute. Although this proof is of mostly historical interest, we present it in full. We start with a weak version of Lemma 2.10, which does not have any stipulation on the length of the repeated word u:

The proof is by double induction, on d and ℓ. If $d = 1$, then the assertion is tautological since every word is 1-decomposable. If $\ell = 1$, then just take $\beta = k$, and $v = x_1$.

Suppose we know $\beta(\ell - 1, k, q)$ and $\beta(p, k, d - 1)$ for any p, q.

Assume w is any word which does not contain any subword of the form v^k. We shall calculate $\beta = \beta(\ell, k, d)$ large enough such that w must be d-decomposable whenever $|w| \geq \beta$.

We list each occurrence of x_ℓ separately in w, i.e.,

$$w = v_0 x_\ell^{t(1)} v_1 x_\ell^{t(2)} v_2 \cdots x_\ell^{t(m)} v_m,$$

where v_i are words in $x_1, \ldots, x_{\ell-1}$ and v_0, v_m could be blank, but the other v_i are not blank. (We wrote the power as $t(i)$ instead of the more customary t_i, in order to make the notation more readable in what follows). By induction on ℓ, we are done unless each $|v_i| < \beta(\ell - 1, k, d)$. Also, by hypothesis on w, $t(i) < k$ for each i.

We form a new alphabet X' as follows:

$$X' = \{v_1 x_\ell^{t(2)}, v_2 x_\ell^{t(3)}, \ldots, v_{m-1} x_\ell^{t(m)}\}.$$

Thus, every new letter $x' \in X'$ starts with some x_j where $1 \leq j \leq \ell - 1$. To estimate $|X'|$, we noted above that $|v_i| < \beta(\ell - 1, k, d)$, and each letter in v_i has $\ell - 1$ possibilities, so there are fewer than $\ell^{\beta(\ell-1,k,d)}$ choices for v_i. Multiplying this by the k possible values for $t(i)$, we see the number of letters in X' is bounded by "only" $k\ell^{\beta(\ell-1,k,d)}$.

We introduce the following new order $\succ'$ on these letters:

$v_{i-1} x_\ell^{t(i)} \succ' v_{j-1} x_\ell^{t(j)}$, iff $v_{i-1} x_\ell^{t(i)} \succ v_{j-1} x_\ell^{t(j)}$ in the old lexicographic order $\succ$, or $v_{j-1} x_\ell^{t(j)}$ is an initial subword of $v_{i-1} x_\ell^{t(i)}$ (in which case, $v_{i-1} = v_{j-1}$ and $t(j) < t(i)$). This is a total order on X' which induces the corresponding lexicographic order on the submonoid $\mathcal{M}(X')$ generated by X', denoted also by $\succ'$.

Claim 1. Let $f, g \in \mathcal{M}(X')$. If $f \succ' g$, then either $f \succ g$ or g is an initial subword of f with respect to the original alphabet X.

Indeed, assume g is not an initial subword of f. Then $f = uxr$ and $g = uys$, with $u, r, s \in \mathcal{M}(X')$, $x, y \in X'$ and $x \succ' y$. Let $x = v_{i-1} x_\ell^{t(i)}$ and $y = v_{j-1} x_\ell^{t(j)}$.

Case 1. $v_{i-1} \succ v_{j-1}$. Then clearly $f \succ g$.

Case 2. $v_{i-1} = v_{j-1}$. Then $t(i) > t(j)$. Rewrite $f = uv_{i-1}x_\ell^{t(i)}r$ and $g = uv_{j-1}x_\ell^{t(j)}s$. Since g is not an initial subword of f, $s \neq 1$. Now s starts with some letter from $\{x_1, \ldots, x_{\ell-1}\}$; hence $x_\ell \succ s$, which implies that $f \succ g$.

Claim 2. Let $w = vw'$ where $w' \in \mathcal{M}(X')$, and assume for some integer q that w' is q-decomposable in the $\succ'$ order. Then w', and therefore also w, is q-decomposable in the original $(\succ)$ order.

Indeed, by assumption, $w' = w_1 \cdots w_q$ satisfies that for any $1 \neq \pi \in S_q$, $w_1 \cdots w_q \succ' w_{\pi(1)} \cdots w_{\pi(q)}$. Since both words have same length and are different, one cannot be an initial subword of the other. By Claim 1 it follows that $w_1 \cdots w_q \succ w_{\pi(1)} \cdots w_{\pi(q)}$, as desired.

We can now complete the proof of Shirshov's Lemma. Write

$$w = v_0 x_\ell^{t(1)} v_1 x_\ell^{t(2)} v_2 \cdots x_\ell^{t(m)} v_m = vw',$$

where $v = v_0 x_\ell^{t(1)}$ and $w' = v_1 x_\ell^{t(2)} v_2 \cdots x_\ell^{t(m)} v_m$. Let

$$\beta(\ell, k, d) = \beta(\ell - 1, k, d) + k + \beta\big(k\ell^{\beta(\ell-1,k,d)}, k, d-1\big).$$

Then we must have at least one of three possibilities:

(i) $|v_0| \geq \beta(\ell - 1, k, d)$ or

(ii) $t(1) \geq k$ or

(iii) $|w'| \geq \beta\big(k\ell^{\beta(\ell-1,k,d)}, k, d-1\big)$.

In case (i), we are done, since by induction (on $\ell - 1$) v_0—hence w—contains a subword that is d-decomposable.

In case (ii), we are done since x_ℓ^k appears in w.

So assume case (iii) holds. By induction (on $d - 1$), w' is $(d-1)$-decomposable with respect to $\succ'$ and therefore also with respect to $\succ$. Thus,

$$w' = w_0 w_1 \cdots w_{d-1}$$

and $w_0 w_1 \cdots w_{d-1} \succ w_0 w_{\pi(1)} \cdots w_{\pi(d-1)}$ for any $1 \neq \pi \in S_{d-1}$. Thus, w contains the subword

$$(x_\ell^{t(1)} w_0) w_1 \cdots w_{d-1}$$

which is $(\succ)$ d-decomposable, since $w_1, \ldots, w_{d-1}$ all start with letters $\prec x_\ell$. This completes the proof of the weak version of Shirshov's Lemma.   □

It remains to bound the size of the repeating subword u. We need a basic fact about periodicity, which could be seen via Proposition 2.13, but for which we give a direct proof:

Lemma 2.61. *If $w = uv = vu$ for u, v nonempty, then w is periodic of periodicity dividing $\gcd(|u|, |v|)$, and u, v are periodic with the same period as w.*

Proof. We may assume $|u| \le |v|$. Then u is an initial subword of v, so we write $v = uv'$. Now $uuv' = w = vu = uv'u$, so $uv' = v'u$ and by induction on $|w|$, we have uv' periodic with period $\hat{u}$ whose length divides $\gcd(|u|, |v'|)$. Writing $u = \hat{u}^i$ and $v' = \hat{u}^j$, we see that $v = \hat{u}^{i+j}$ and $w = \hat{u}^{2i+j}$. $\qquad\square$

We formalize Lemma 2.61 with a definition.

Definition 2.62. Words w and w' are called *cyclically conjugate* if $w = uv$ and $w' = vu$ for suitable subwords v, u of w. Note that w' is a subword of w^2.

The *cyclic shift* δ is defined by $\delta(vx) = xv$, for any word v and letter x.

Remark 2.63. $\{\delta^k(u) : 0 \le k < |u|\}$ is the set of words which are cyclically conjugate to the word u. Lemma 2.61 shows that $\delta^k(u) = u$ for some $0 < k < |u|$, iff u is periodic with periodicity dividing k. Thus, we have the following dichotomy:

(i) u is periodic of periodicity $< d$ (in case $\delta^k(u) = u$ for some $0 < k < d$), or

(ii) $\{\delta^k(u) : 0 \le k < d\}$ are d distinct comparable subwords of u^2. In this case, u^{2d} is d-decomposable by Proposition 2.19.

Applying this dichotomy to the weak version of Shirshov's Lemma yields the stronger result:

Proposition 2.64. *If w is d-indecomposable and $|w| \ge \beta(\ell, k, d)$ with $k \ge 2d$, then w contains a word of the form u^k with $|u| \le d$.*

Proof. We already know w contains a subword u^k. Take such u with $|u|$ minimal; in particular, u is non-periodic. We are done unless $|u| > d$, then (i) of Remark 2.63 fails, so u^{2d} is d-decomposable, implying w is d-decomposable. $\qquad\square$

This completes the second proof of Shirshov's Lemma, since we may replace k by the maximum of k and $2d$.

2.6.2   Third proof of Shirshov's Lemma: quasiperiodicity

Our third proof of Shirshov's Lemma yields a much more reasonable estimate than what we gave above. This study will also have other applications, to Burnside-type problems and, later, to Gelfand-Kirillov dimension.

We start with a nice dichotomy result modifying Remark 2.63.

Lemma 2.65. *Suppose* $|v| \geq \frac{1}{2}dk$. *Then either*

(i) v *has a subword* u^k *with* $|u| \leq d$, *or*

(ii) v *contains* d *comparable subwords* $(\delta^j(v)$ *for* $0 \leq j < d)$.

Proof. (ii) holds unless $\delta^j(v) = v$ for some $0 \leq j < d$, in which case, by Lemma 2.61, v is quasiperiodic with periodicity $\gcd(d, j)$. But $\gcd(d, j) \leq \frac{d}{2}$, so the period u repeats at least k times in v, yielding (i). $\square$

Remark 2.66. We call a word w *Shirshov reducible* if w has either a subword u^k with $|u| \leq d$ or is d-decomposable. Discarding possibility (i) of Lemma 2.65 (working with $\beta > \frac{1}{2}dk$), we search for β such that every word of greater length is Shirshov reducible. Then β would be an upper bound for $\beta(\ell, k, d)$.

Let us focus on two ways for a word to be Shirshov reducible. We call a word w *Shirshov admissible* if either w has a subword of the form u^k for $|u| \geq 1$ or w contains d comparable subwords. By Lemma 2.65, any word of length $\frac{1}{2}dk$ is Shirshov admissible. But then by Proposition 2.19, if w has a Shirshov admissible subword of multiplicity d, then w is Shirshov reducible. So we look for the smallest β to guarantee this condition.

The quickest way to do this is to take all words of length $\frac{1}{2}dk$, namely $\ell^{dk/2}$ such words. If $|w|$ can be partitioned into $d\ell^{dk/2}$ subwords of length $\frac{1}{2}dk$, then at least one of these has multiplicity d and is Shirshov admissible, having length $\frac{1}{2}dk$; thus, w is d-decomposable. So we could take

$$\beta = \frac{1}{2}d^2 k \ell^{dk/2}.$$

To lower this estimate, we need to exclude many of these words. The underlying question is, "What conditions are satisfied by a word, such that condition (ii) of Lemma 2.65 fails?" The key is a more careful analysis of periodicity, so let us record some related conditions.

Definition 2.67. The *normal form* of a word w is $vu^k r$, where r is an initial subword of u, and $|v| + |u|$ is minimal possible. Then v is called the *preperiod* of w, u is the *period* and $|v| + |u|$ is the *preperiodicity* of w. If $v = \emptyset$, i.e., $w = u^k r$, we say w is *quasiperiodic* with *period* u.

Quasiperiodicity is left-right symmetric. Explicitly, if $w = u^k v$, where $u = vv'$, then $w = v(v'v)^k$. In the same way, we see that words with cyclically conjugate periods can have arbitrarily large common subwords.

Remark 2.68. If a word is not quasiperiodic, then deleting its first letter reduces its preperiodicity by 1.

Proof. Write $w = vu^k r$ in normal form, and $v \neq \emptyset$. Then cutting off the first letter of $|v|$ decreases $|v| + |u|$ by 1. But the preperiodicity cannot decrease by more than 1, since then adding back the initial letter would give us a lower preperiodicity for w. $\qquad\square$

We turn again to hyperwords to simplify a proof.

Proposition 2.69.

(i) *Suppose a hyperword* $h = vuh' = vh'$ *for words* u, v. *Then* h' *is periodic with period* u, *so* h *is preperiodic with preperiod an initial subword of* v.

(ii) *If a finite word* $w = vuw'$ *with* $|u| < |w|$, *and* vw' *is an initial subword of* w, *then* w *is has preperiod an initial subword of* v, *and* w' *is periodic with period* u.

Proof. (i) Apply Proposition 2.13 to $uh' = h'$. (ii) is a restatement of (i), in conjunction with Lemma 2.61. $\qquad\square$

Now we can establish a new dichotomy which underpins our estimate.

Lemma 2.70. *Assume* $|w| \geq 2d$, *and write*

$$w = u_0 w' = v_1 u_1 w' = v_2 u_2 w' = \cdots = v_{d-1} u_{d-1} w',$$

where $|v_i| = i$ *and* $|u_i| = d - i$. *Then one of the following two statements holds:*

(i) *the words* $u_i w'$ *are lexicographically comparable and comprise a chain;*

(ii) w' *is periodic of some period* u, *and* w *is preperiodic of preperiodicity* $< d$.

Proof. We are done unless some $u_j w'$ is initial in $u_i w'$ for $i < j$. Write $u_i = u_j u$. Then w' is initial in uw', so Proposition 2.69(ii) shows w' is periodic with period u. But

$$v_j u_j = v_i u_i = v_i u_j u,$$

so we see that the preperiod is contained in $v_i u_j$, and the preperiodicity $\leq |v_i u_j| + |u| < d$. $\qquad\square$

Definition 2.71. A word w is called *d-critical*, if w has preperiodicity d (so is Shirshov admissible), but its initial subword of length $|w| - 1$ has preperiodicity $d - 1$.

Remark 2.72. Any d-critical word w has the form $vu^k x_i$ where

$$|v| + |u| = d - 1.$$

Thus, $|w| \le d - 1 + (d-1)(k-1) + 1 = dk - k + 1$.

Let us use this observation to tighten our estimates on β. We consider the "extreme case" for a word w of length $\ge dk$, i.e., w is Shirshov admissible, but the initial subword w' obtained by omitting the last letter is not Shirshov admissible. By Lemma 2.70, w' has preperiodicity $< d$. On the other hand, assume $|w| > (d-1)k$. If w has preperiodicity $< d$, then its period u has length $\le d - 1$, and its preperiod v has length $\le d - 1 - |u|$, implying vu^k appears in w', implying w' is already Shirshov admissible. Thus we may assume w has preperiodicity $\ge d$. Now Remark 2.68 shows w has preperiodicity d and w' has preperiodicity $d - 1$; i.e., w is critical.

In other words, any word of length $\ge kd$ which is not Shirshov admissible contains a d-critical initial subword of length $< kd$. Let us count such subwords.

Proposition 2.73. Let $|X| = \ell$.

(i) The number of all d-critical words of length t is at most $d(\ell-1)\ell^{d-1}$.

(ii) The number of d-critical words of length $\le t$ is at most $d(\ell-1)\ell^{d-1}t$.

(iii) The number of all d-critical words that do not contain the k-th power of a word of length $\le d$ is at most $d^2 k(\ell-1)\ell^{d-1}$.

Proof. (i) The initial subword of length $d - 1$ can be chosen in ℓ^{d-1} ways. There are d ways to separate it as a period and a preperiod. The last letter can be chosen in $\ell - 1$ ways.

(ii) There are t different possible lengths, and apply (i) to each.

(iii) A d-critical word that does not contain a k-th power has length $< dk$ by Remark 2.72, so we take $t = dk$ in (ii). $\square$

Theorem 2.74 (Improved bound for Shirshov's function).

$$\beta(\ell, k, d) \le d^4 k^2 (\ell-1)\ell^{d-1}.$$

Proof. Partitioning any word w of length $> d^4 k^2 (\ell-1)\ell^{d-1}$ into subwords each of length dk as above, we proved each of these subwords contains a d-critical subword, so we have at least $d^3 k(\ell-1)\ell^{d-1}$ disjoint d-critical subwords. By Proposition 2.73 (iii), at least d of them are the same. Thus, w is Shirshov reducible (cf. Remark 2.20). $\square$

Corollary 2.75. *Suppose $A = C\{a_1, \dots, a_\ell\}$ is an affine PI-algebra (not necessarily with 1), of PI-degree d. If all words of length $\leq d$ in the a_i are nilpotent of index k, then A is nilpotent of index $\leq n = \ell^d d^4 k^2$.*

Proof. In view of Proposition 2.8 , A^n is spanned by those $\bar{w}$ for which w has length n and is not d-decomposable and thus contains a k-th power, so $\bar{w} = 0$ by hypothesis.   □

2.7   Appendix: The Independence Theorem for Hyperwords

Having utilized hyperwords to good effect, we pause for one more application—a version of a theorem of Ufnarovski'ĭ—which also settles a conjecture of Amitsur and Shestakov.

Clearly, if two hyperwords h, h' are unequal, then they have unequal initial subwords of some length n. It follows that hyperwords (unlike finite words) are totally ordered with respect to the lexicographic ordering. Also, each subset of right hyperwords has an infimum and a supremum (obtained by starting with the smallest or largest possible letter, and continuing inductively).

Although we cannot define $\bar{h}$ for a right hyperword, since we do not have infinite products in A, we can define when $\bar{h}$ is 0:

Definition 2.76. $\bar{h} = 0$ iff $\bar{w} = 0$ for some (finite) initial subword w of h. We say h is *nonzero in A* if $\bar{h} \neq 0$.

Remark 2.77. Suppose A is a non-nilpotent algebra without 1. Then the set of all right hyperwords such that $\bar{h} \neq 0$ is nonempty and thus has maximal and the minimal elements.

Let u be a word in A, which is maximal in the set of all nonzero words of length $\leq n$. It is conceivable that u cannot be extended to a longer nonzero word in A. Hence, to use hyperwords we shall apply the following construction.

Example 2.78. Let $A = F\{a_1, \dots, a_\ell\}$. Take an indeterminate $x_{\ell+1}$ and formally declare $x_{\ell+1} \succ x_i$ for all $1 \leq i \leq \ell$. Consider the free product $A' = A * F\{x_{\ell+1}\}$.

Each word u in A is initial in some hyperword in A'. If u is the minimal (nonzero) word in A in the set of all words of length $\leq |u|$, then clearly u is initial in certain hyperwords of A' (for example, $u x_\ell^\infty$), among which the hyperwords in A are minimal (if they exist).

Theorem 2.79 (Independence of hyperwords). *Let h be a minimal nonzero hyperword on A, and w_i denote its initial subword of length i. Then, given a homomorphism $C\{X\} \to A$, one of the following two conditions holds:*

(i) *$\bar{w}_1, \ldots, \bar{w}_n$ are linearly independent.*

(ii) *h is preperiodic of preperiodicity $\leq n$, and its period is a terminal subword of w_n.*

Proof. Let us define the hyperwords h_i by $h = w_i h_i$. As in Lemma 2.70, if (ii) does not hold, then the set $\{h_i\}$ constitutes a chain. Suppose $\sum \alpha_i \bar{w}_i = 0$. Take a minimal hyperword h_j in the set $\{h_i : \alpha_i \neq 0\}$. Then $h = w_i h_i \succ w_i h_j$, for all $i \neq j$ where $\alpha_i \neq 0$. We can truncate the hyperwords $\{h_j\}$ to words $\{\hat{h}_j\}$ which comprise a chain under the same order, but where $h \succ w_i \hat{h}_j$ for each j. If (i) were false, then $\bar{w}_j$ would be a linear combination of the $\bar{w}_i$, so $\hat{h} = w_j \hat{h}_j$ would also be spanned by a linear combination of the $w_i \hat{h}_j$, each of which is smaller. If all of the $w_i \hat{h}_j$ were 0 on A, then $\hat{h}$, being a linear combination, would also be 0 on A. Hence, some $w_i h_j$ is nonzero on A, contradicting the minimality of h. $\qquad\square$

Minimality can be replaced by maximality in the formulation of Theorem 2.79.

Corollary 2.80. *Let $A \subset M_n(F)$. Then any minimal nonzero (hyper)word h is preperiodic of preperiodicity $\leq n^2 + 1$, since (i) cannot hold.*

By the independence theorem for hyperwords and Example 2.78, we also have

Corollary 2.81 (Ufnarovski'i's independence theorem). *Suppose:*

(i) *$w = a_{i_1} \ldots a_{i_d}$ is minimal with respect to the left lexicographic ordering in the set of all nonzero products of length $\leq d$.*

(ii) *The terminal subwords of w are nilpotent.*

Then the initial subwords of w are linearly independent.

This theorem was proved by Ufnarovski'i and Chekanu [Che94b].

Corollary 2.82 (Amitsur and Shestakov's conjecture). *If A is an affine subalgebra without 1 of $M_n(F)$, such that all words of length $\leq n$ in the generators are nilpotent, then $A^n = 0$.*

Proof. Any nilpotent word w satisfies $w^n = 0$. Corollary 2.81 shows any word of length $d > n^3$ contains some word w^n and thus is 0, so A is nilpotent. Then $A^n = 0$ by a theorem of Wedderburn, cf. Remark 1.78. □

Corollary 2.83. *If $A = F\{a_1, \ldots, a_\ell\}$ is semiprime without 1, of PI-class n and if all words of degree $\leq n$ in the a_i are nilpotent, then $A = 0$.*

Proof. First suppose A is prime of PI-class n. Then $A \subset M_n(K)$ for some field K by Corollary 2.82, so $A^n = 0$.

Now, for A semiprime, we have $A^n \subseteq P$ for each prime ideal, implying $A^n \subseteq \cap P = 0$. □

Chapter 3

T-Ideals and Relatively Free Algebras

Until now, we have focused on the properties of a ring satisfying a given PI (whether the PI comes from a central polynomial, or a Capelli identity). At this stage we turn to the set of identities of a given algebra, to see what extra information is available. This turns out to be a powerful multi-purpose tool, which we shall need in several related contexts. Accordingly, we start this chapter with a general description of identities and varieties. The collection of identities, described as a subset of the free algebra, has a special structure, called a T-*ideal*, and enables us to construct the free object of the variety, called the *relatively free*, or *universal* PI-algebra. The relatively free algebra has important structural properties, thereby providing an excellent source of examples as well as enabling us to unify aspects of the theory.

The study of T-ideals leads us to *Specht's problem*, whether every T-ideal is finitely generated (as a T-ideal); in other words, is any set of identities the consequence of a finite number of identities? The answer, discovered in characteristic 0 by Kemer, is "Yes," although Belov later discovered counterexamples in positive characteristic. Nevertheless, Specht's problem has an affirmative answer for all affine algebras (proved by Kemer for algebras over an infinite field, and by Belov for algebras over a commutative Noetherian ring). See Chapter 7 for the discussion of the characteristic p case.

The solution to Specht's problem requires techniques that will be developed in the following three chapters, and only will be completed in Chapter 6. We shall describe the T-ideal of identities of the Grassman algebra as a special case, showing it is generated by a single identity.

3.1   Sets of Identities and T-Ideals

Definition 3.1. Fix a commutative ring C. Given a set S of C-algebras, we let $\mathrm{id}(S)$ denote the set of identities holding in each algebra in S, i.e., $\cap_{A \in S} \mathrm{id}(A)$.

The underlying questions of this chapter are, "Why are we interested in $\mathrm{id}(S)$, and what can we say about it?"

In the other direction, given $\mathcal{I} = \mathrm{id}(S)$, define the *variety* $\mathcal{V}_\mathcal{I}$ to be the class of algebras for which each $f \in \mathcal{I}$ is an identity.

First an easy observation. It is apparent that $\mathrm{id}(\mathcal{V}_{\mathrm{id}(S)}) = \mathrm{id}(S)$, and so we have an analogue with algebraic sets from algebraic geometry which was described in the Preface and motivates much of the theory.

Remark 3.2. Clearly if f, g are identities of an algebra A, then so is $f \pm g$, hf, and fh, for all $h \in C\{X\}$, so $\mathrm{id}(A) \triangleleft C\{X\}$; hence we also have $\mathrm{id}(S) \triangleleft C\{X\}$.

However, there is another crucial property.

Remark 3.3. If $f(x_1, \ldots, x_m) \in \mathrm{id}(A)$ and $h_1, \ldots, h_m \in C\{X\}$, then for any specializations of x_i to $\bar{x}_i$ in A, writing $\overline{h_j}$ for the corresponding value of h_j in A, we clearly have

$$0 = f(\overline{h_1}, \ldots, \overline{h_m}) = \overline{f(h_1, \ldots, h_m)},$$

implying $f(h_1, \ldots, h_m) \in \mathrm{id}(A)$. In other words, an identity remains an identity under any specialization of its indeterminates.

But by Remark 1.1, for any algebra R, any set mapping $\psi : X \to R$ extends uniquely to an algebra homomorphism $\psi : C\{X\} \to R$. Hence, taking $R = C\{X\}$ we can restate Remark 3.3 more concisely as:

Remark 3.4. If $f \in \mathrm{id}(A)$, then $\psi(f) \in \mathrm{id}(A)$ for all algebra homomorphisms $\psi : C\{X\} \to C\{X\}$.

In view of Remark 3.4, we formulate a basic notion.

Definition 3.5. A T-*ideal* of an algebra A is an ideal $\mathcal{I}$ such that $\psi(\mathcal{I}) \subseteq \mathcal{I}$ for all homomorphisms $\psi : A \to A$. In this case, we write $\mathcal{I} \triangleleft_T A$.

Thus $\mathrm{id}(S) \triangleleft_T C\{X\}$, for any set of algebras S. In the same spirit, we have

Remark 3.6. Suppose $\psi : C\{X\} \to A$ is any algebra homomorphism. Then $\mathrm{id}(A) \subseteq \ker \psi$, since

$$\psi(f(x_1, \ldots, x_m)) = f(\psi(x_1), \ldots, \psi(x_m)) \in f(A) = 0.$$

Example 3.7. The most familiar T-ideal is $\mathrm{id}(M_n(F))$, which we denote as $\mathcal{M}_n$.

$\mathcal{M}_n$ gains in prominence through the following observations.

Remark 3.8. $\mathcal{M}_n \leq \mathcal{M}_m$ for all $m \leq n$. (This is because we can embed $M_m(F)$ into $M_n(F)$ as an algebra without 1, by adding a fringe of 0s along the last $n - m$ rows and columns.)

Being interested in T-ideals, i.e., the set of identities of suitable algebras, we are led at once to PI-equivalence, cf. Definition 1.15.

Lemma 3.9. *If R is a semiprime PI-algebra over an infinite field F, then $R \sim_{\mathrm{PI}} M_n(F)$, i.e., $\mathrm{id}(R) = \mathcal{M}_n$, where n is the PI-class of R.*

Proof. When R is prime, this is Corollary 1.61.

In general, each prime image of R has PI-class $\leq n$, and at least one image has PI-class n, so we conclude by Remark 3.8. $\square$

Thus, the $\mathcal{M}_n$ are the only T-ideals which are prime as ideals. However, there are other important T-ideals, such as the identities of the Grassman algebra.

Definition 3.10. Here are two interesting ways of generalizing T-ideals of the free algebra $F\{X\}$ over a field.

(i) A T-*space* of $F\{X\}$ is an F-subspace V closed under substitutions, i.e., if $f(x_1, \ldots, x_m) \in V$, then $f(h_1, \ldots, h_m) \in V$ for all $h_i \in F\{X\}$.

(ii) Suppose A is an F-algebra and V is an F-subspace of A. Then those polynomials vanishing for all substitutions in V are denoted $\mathrm{id}(A; V)$, and this is also an ideal of $F\{X\}$, which Kemer calls an S-*ideal*.

Note that $\mathrm{id}(A; V)$ was used in proving a basic result, the construction of central polynomials, when we took A to be $M_n(F)$ and V the subspace of matrices of trace 0.

3.2 Relatively Free Algebras

It is also enlightening to work backwards.

Definition 3.11. Given a T-ideal $\mathcal{I} \lhd_T C\{X\}$, we form the *relatively free* algebra

$$U_\mathcal{I} = F\{X\}/\mathcal{I}$$

for $\mathcal{I}$. We write $\bar{x}_i$ for the image of x_i in $U_\mathcal{I}$. When $\mathcal{I} = \mathrm{id}(A)$, we write U_A or $U_\mathcal{I}$.

Lemma 3.12. *If* $\mathrm{id}(A) \supseteq \mathcal{I}$, *then for any* $a_1, a_2, \ldots \in A$, *there exists a unique algebra homomorphism* $U_{\mathcal{I}} \to A$ *satisfying* $\bar{x}_i \to a_i$, $i = 1, 2, \ldots$.

Proof. We know there is a unique homomorphism $\psi : C\{X\} \to A$ given by $x_i \mapsto a_i$. The kernel contains $\mathcal{I}$, by Remark 3.6, so we conclude with Noether's isomorphism theorem. □

Proposition 3.13. $\mathrm{id}(U_{\mathcal{I}}) = \mathcal{I}$, *for any* $\mathcal{I} \triangleleft_T C\{X\}$.

Proof. $\mathrm{id}(U_{\mathcal{I}}) \subseteq \mathcal{I}$, by Remark 3.6. Conversely, for any $f(x_1, \ldots, x_m) \in \mathcal{I}$ and $\overline{h_1}, \ldots, \overline{h_m}$ in $U_{\mathcal{I}}$, where $\overline{h_i} = h_i + \mathcal{I}$, we see

$$f(\overline{h_1}, \ldots, \overline{h_m}) = f(h_1, \ldots, h_m) + \mathcal{I} = 0$$

since $\mathcal{I} \triangleleft_T C\{X\}$, proving $f \in \mathrm{id}(U_{\mathcal{I}})$. □

Geometrically, any T-ideal $\mathcal{I} \triangleleft_T C\{X\}$ can be considered as the ideal of polynomials of a given variety $(\mathcal{V}_{\mathcal{I}})$, in analogy to the ideal of the commutative polynomial ring annihilating a given algebraic variety.

Relatively free algebras enable us to encode identities, in the following manner.

Definition 3.14. Elements $y_1, y_2, \cdots \in A$ are *PI-generic* for A if they test identities in the following sense:
$$f(x_1, \ldots, x_m) \in \mathrm{id}(A) \text{ iff } f(y_1, \ldots, y_m) = 0.$$

Remark 3.15. The elements $\bar{x}_i : i = 1, 2, \ldots$ are PI-generic for $U_{\mathcal{I}}$. Indeed $f(\bar{x}_1, \ldots, \bar{x}_m) = 0$ iff $f(x_1, \ldots, x_m) \in \mathcal{I} = \mathrm{id}(U_{\mathcal{I}})$.

We have the following analog of Lemma 3.12.

Lemma 3.16. *Suppose an algebra* $\hat{A} \in \mathcal{V}_{\mathcal{I}}$ *is generated by PI-generic elements* $\{r_i : i \in I\}$. *Then for any* A *in* $\mathcal{V}_{\mathcal{I}}$ *and any* $a_i \in A$, *there is a (unique) homomorphism* $\varphi : \hat{A} \to A$ *obtained by sending* $r_i \mapsto a_i$, $\forall i \in I$.

Proof. Define φ by

$$\varphi(f(r_1, \ldots, r_m)) = f(a_1, \ldots, a_m).$$

This is well-defined since if $f(r_1, \ldots, r_m) = 0$, then $f(x_1, \ldots, x_m) \in \mathrm{id}(A)$, implying $f(a_1, \ldots, a_m) = 0$. □

The next result is almost trivial but turns out to be surprisingly useful.

Proposition 3.17. *Any algebra* $\hat{A}$ *in* $\mathcal{V}_{\mathcal{I}}$ *generated by PI-generic elements* $\{r_i : i \in I\}$ *is relatively free for* $\mathcal{I}$.

Proof. By Lemma 3.16 we can define a homomorphism $\varphi : \hat{A} \to U_{\mathcal{V}}$ by

$$f(r_1, \ldots, r_m) \mapsto f(\bar{x}_1, \ldots, \bar{x}_m),$$

and likewise we define a homomorphism $\psi : U_{\mathcal{V}} \to \hat{A}$ by

$$f(\bar{x}_1, \ldots, \bar{x}_m) \mapsto f(r_1, \ldots, r_m).$$

The composition in either way is the identity, so $\hat{A}$ is isomorphic to $U_{\mathcal{V}}$. $\square$

We shall have occasion to weaken this property to multilinear polynomials:

Definition 3.18. The elements $y_1, \ldots, y_n$ of an algebra are called ml-generic (ml for "multilinear") for A if they satisfy the following property:
  For any multilinear polynomial $f(x_1, \ldots, x_n)$, $f(x) \in \mathrm{id}(A)$ if and only if $f(y_1, \ldots, y_n) = 0$.

Here is an important example.

Example 3.19. Let A, B be two PI-algebras with corresponding relatively free algebras $U_A = F\{\bar{x}_1, \bar{x}_2, \ldots\}$ and $U_B = F\{\bar{x}'_1, \bar{x}'_2, \ldots\}$. Then the elements $\bar{x}_1 \otimes \bar{x}'_1, \bar{x}_2 \otimes \bar{x}'_2, \ldots \in U_A \otimes U_B$ are ml-generic for $A \otimes B$.
  Indeed, assume $f(\bar{x}_1 \otimes \bar{x}'_1, \ldots, \bar{x}_n \otimes \bar{x}'_n) = 0$ for f multilinear. We need to show that $f(x) \in \mathrm{id}(A \otimes B)$. It suffices to show that $f(a_1 \otimes b_1, \ldots, a_n \otimes b_n) = 0$ for any $a_1 \ldots, a_n \in A$ and $b_1 \ldots, b_n \in B$. This follows by extending the homomorphisms given by $\bar{x}_i \to a_i$ and $\bar{x}'_i \to b_i$, to the homomorphism given by $\bar{x}_i \otimes \bar{x}'_i \to a_i \otimes b_i$.

3.3 Relatively Free Algebras Without 1, and Their T-Ideals

Although we emphasize algebras with 1 in this book, there are important results in the theory for associative algebras without 1, especially concerning nil algebras (which obviously cannot contain 1), such as Theorem 3.40. So by a slight abuse of terminology, "nil subalgebra" denotes an algebra without 1, whereas all other algebras are with 1. To salve our consciences, let us take a minute to consider the identities of an algebra A without 1.

Remark 3.20. If $f(x_1, \ldots, x_m) \in \mathrm{id}(A)$, then the constant term of f is $f(0, \ldots, 0) = 0$. Thus, writing $F\{X\}^+$ for the polynomials of constant term 0, so that $F\{X\} = F \oplus F\{X\}^+$, we have $\mathrm{id}(A) \subseteq F\{X\}^+$.

Remark 3.21. There is a well-known construction of adjoining 1 to any algebra without 1, namely by forming $R = F \oplus A$ as a vector space, with multiplication

$$(\alpha_1, a_1)(\alpha_2, a_2) = (\alpha_1 \alpha_2, \alpha_1 a_2 + \alpha_2 a_1 + a_1 a_2).$$

Thus, we could view A as an ideal of R, and $\mathrm{id}(A) \lhd F\{X\}$. Furthermore, given $f \in \mathrm{id}(A)$, we have $0 = f(0, \ldots, 0)$, the constant term of f, so we can view $\mathrm{id}(A) \lhd F\{X\}^+$, and obtain the relatively free algebra $U_A = F\{X\} / \mathrm{id}(A) = F \oplus (F\{X\}^+ / \mathrm{id}(A))$. Then $U_A^+ = F\{X\}^+ / \mathrm{id}(A)$ is the *relatively free algebra without* 1 of A. Thus, we see that the theory of T-ideals of algebras without 1 embeds into the theory of S-ideals (cf. Definition 3.10(ii)) in a natural way.

3.3.1   Examples of relatively free algebras

Let us note some very important examples of relatively free algebras, which play a central role both in the general theory and in applications.

Algebra of generic matrices. We start with a famous example. Fixing a field F, let $\mathcal{M}_n = \mathrm{id}(M_n(F))$. We want to describe its relatively free algebra.

Definition 3.22. Take a set $\Lambda = \{\lambda_{ij}^{(k)} : 1 \le i, j \le n, k \in \mathbb{N}\}$ of commuting indeterminates over a field F, and define the *algebra of generic $n \times n$ matrices* $F\{Y\}_n$ to be the F-subalgebra of $M_n(F[\Lambda])$ generated by $y_k = \left(\lambda_{ij}^{(k)}\right)$ for $k = 1, 2 \ldots$. We call each y_k a "generic $n \times n$ matrix."

Remark 3.23. If $f(y_1, \ldots, y_m) = 0$, then $f(x_1, \ldots, x_m) \in \mathcal{M}_n$. (This is seen by specializing the $\lambda_{ij}^{(k)}$ to arbitrary elements of F, thereby specializing the y_k to arbitrary matrices in $M_n(F)$.)

It follows that the generic matrices are PI-generic elements, so by Proposition 3.17, $F\{Y\}_n$ is relatively free for $\mathcal{M}_n$.

Let $F(\Lambda)$ denote the field of fractions of the integral domain $F[\Lambda]$.

Proposition 3.24. $F\{Y\}_n$ *is prime of PI-class n.*

Proof. Using the notation of Definition 3.22, let A be the $F(\Lambda)$ subalgebra of $M_n(F(\Lambda))$ spanned by $F\{Y\}_n$. Clearly, c_{n^2} is not an identity of $F\{Y\}_n$. Hence, $A = M_n(F(\Lambda))$, by Proposition 1.21, and thus $F\{Y\}_n$ has PI-class n. Furthermore, if $B_1, B_2 \lhd F\{Y\}_n$ with $B_1 B_2 = 0$, then $B_1 F(\Lambda) B_2 F(\Lambda) = B_1 B_2 F(\Lambda) = 0$, implying $B_1 = 0$ or $B_2 = 0$ since $M_n(F(\Lambda))$ is simple. $\square$

One can push this result further.

Theorem 3.25. *For every field F and every n, $F\{Y\}_n$ is a (noncommutative) domain, whose ring of central fractions, denoted $UD(n, F)$, is a division algebra D of dimension n^2 over its center.*

Proof. In view of Corollary D, $UD(n, F)$ is central simple of degree n, so in view of the Wedderburn-Artin theorem, it suffices to show $UD(n, F)$ has no nonzero nilpotent elements, or equivalently $F\{Y\}_n$ has no nonzero nilpotent elements. But in this case, there would be a polynomial $f(x_1, \ldots, x_m)$ not in $\mathcal{M}_n$, such that $f^t \in \mathcal{M}_n$ for some $t > 0$. This is false, because there are examples of division algebras D of PI-class n containing any given field (namely, the "generic" symbol). Taking $a_1, \ldots, a_n \in D$ with $a = f(a_1, \ldots, a_n) \neq 0$, we would have $a^t = 0$, contradiction. $\square$

Although we do not pursue this direction here, $UD(n, F)$ has become a focal example in the theory of division algebras, since it is the test case for many important theorems and conjectures.

Relatively free algebras of finite dimensional (f.d.) algebras. There is another, less well-known example, which nevertheless also plays an important role in the PI-theory.

Example 3.26. Suppose A is a f.d. F-algebra with base $b_1, \ldots, b_t$, and let

$$\Lambda = \{\lambda_{ij} : 1 \leq i \leq \ell, 1 \leq j \leq t\}$$

be commuting indeterminates over A. Define the *generic element*

$$y_i = \sum_{j=1}^{t} b_j \lambda_{ij}, \quad 1 \leq i \leq \ell,$$

and let $\mathcal{A}$ be the subalgebra of $A[\Lambda]$ generated by these generic elements $y_1, \ldots, y_\ell$. Assume that F is infinite.

We claim that $y_1, \ldots, y_\ell$ are PI-generic, cf. Definition 3.14, and therefore $\mathcal{A}$ is isomorphic to the relatively free PI-algebra of A. Indeed, suppose $f(y_1, \ldots, y_\ell) = 0$. Then, given any $a_1, \ldots, a_\ell$ in A, write

$$a_i = \sum_{j=1}^{t} \alpha_{ij} b_j, \quad 1 \leq i \leq \ell,$$

and define a homomorphism $\psi : A[\Lambda] \to A$ given by $\lambda_{ij} \mapsto \alpha_{ij}$. Then ψ restricts to a homomorphism $\hat{\psi} : \hat{A} \to A$ and

$$f(a_1, \ldots, a_\ell) = f(\hat{\psi}(y_1), \ldots, \hat{\psi}(y_\ell)) = \hat{\psi}(f(y_1, \ldots, y_\ell)) = 0,$$

as desired.

Note that the structure of the relatively free algebra $\mathcal{A}$ can be quite different from that of the f.d. algebra A; for example, if $A = M_n(F) \times A_1$ where A_1 is an algebra of upper triangular $n-1 \times n-1$ matrices, then its relatively free algebra $\mathcal{A}$ is also the relatively free algebra of $M_n(F)$, and so "lost" the structure of A_1. Even when $\mathcal{A}$ has nilpotent elements, they might be hard to recover. For example, if A/J has PI-class n, then all values of c_{n^2+1} in $\mathcal{A}$ are nilpotent, but these can be difficult to handle. To get hold of the radical, we offer still another generic construction in Chapter 4, cf. Example 4.50, although it is not quite relatively free.

3.3.2   *T*-ideals of relatively free algebras

T-ideals of relatively free algebras are especially important.

Remark 3.27. The *T*-ideals of an algebra A comprise a sublattice of the lattice of ideals of A; namely, if $\mathcal{I}, \mathcal{J} \lhd_T A$, then $\mathcal{I} + \mathcal{J} \lhd_T A$ and $\mathcal{I} \cap \mathcal{J} \lhd_T A$. (This is immediate since $\psi(\mathcal{I} + \mathcal{J}) \subseteq \psi(\mathcal{I}) + \psi(\mathcal{J}) \subseteq \mathcal{I} + \mathcal{J}$ and $\psi(I \cap J) \subseteq \psi(I) \cap \psi(J) \subseteq \mathcal{I} \cap \mathcal{J}$, for any homomorphism $\psi : A \to A$).

Proposition 3.28. *Given a T-ideal $\mathcal{I}$ of a relatively free algebra U, there is lattice isomorphism between the T-ideals of U containing $\mathcal{I}$ and the T-ideals of $U/\mathcal{I}$, given by $\mathcal{J} \mapsto \mathcal{J}/\mathcal{I}$.*

Proof. Suppose $\mathcal{J} \lhd_T U$ contains $\mathcal{I}$. Then for any homomorphism $\psi : U/\mathcal{I} \to U/\mathcal{I}$ we write $\psi(\bar{x}_i + \mathcal{I}) = y_i + \mathcal{I}$ for suitable $y_i \in U$. By Lemma 3.16, there is a homomorphism $\hat{\psi} : U \to U$ given by $\hat{\psi}(\bar{x}_i) = y_i$. By definition of *T*-ideal, $\hat{\psi}(\mathcal{I}) \subseteq \mathcal{I}$ and $\hat{\psi}(\mathcal{J}) = \mathcal{J}$. Thus, for any $a + \mathcal{I}$ in $\mathcal{J}/\mathcal{I}$,

$$\psi(a + \mathcal{I}) = \hat{\psi}(a) + \mathcal{I} \subseteq \mathcal{J}/\mathcal{I},$$

proving $\psi(\mathcal{J}/\mathcal{I}) \subseteq \mathcal{J}/\mathcal{I}$, as desired.

Conversely, suppose $\mathcal{J}/\mathcal{I} \lhd_T U/\mathcal{I}$. For any $\psi : U \to U$, we have $\psi(\mathcal{I}) \subseteq \mathcal{I}$, so ψ induces a homomorphism of $U/\mathcal{I}$, given by $a + \mathcal{I} \mapsto \psi(a) + \mathcal{I}$. But then, for $a \in \mathcal{J}$, $\psi(a) + \mathcal{I} \subseteq \mathcal{J}/\mathcal{I}$, implying $\psi(a) \in \mathcal{J}$. □

Corollary 3.29. *If $\mathcal{J} \lhd_T U$ and U is relatively free, then the algebra $U/\mathcal{J}$ is relatively free.*

Proof. Write $U = C\{X\}/\mathcal{I}$ for $\mathcal{I} \lhd_T C\{X\}$. Then $\mathcal{J} = \mathcal{J}_0/\mathcal{I}$ for some $\mathcal{J}_0 \lhd_T C\{X\}$ containing $\mathcal{I}$, and $U/\mathcal{J} = (C\{X\}/\mathcal{I})/(\mathcal{J}_0/\mathcal{I}) \approx C\{X\}/\mathcal{J}_0$ is relatively free. □

3.4 Recognizing T-Ideals

A natural question to ask at this stage is, "Which ideals occurring naturally in the structure theory of relatively free algebras are T-ideals?" The theory divides at this point into two directions: Where the algebra is affine or not affine; Kemer calls these the *local* and *global* cases, respectively.

In the affine (local) case, we have the theory of Chapter 2 to draw from, which will be the point of view taken in Chapter 4. This case works in all characteristics, although we shall give full proofs only in characteristic 0.

The infinite (global) case is harder (and parts of the theory fail in characteristic > 0), but it does have one specific advantage.

Remark 3.30. Suppose U is a relatively free algebra generated by the infinite set $\bar{X} = \{\bar{x}_i : i \in I\}$, and $\mathrm{id}(A) \supseteq \mathrm{id}(U)$.

(i) If $r_j = f_j(a_1, \ldots, a_{m_j})$, then there is a homomorphism $\Phi : U \to A$ such that

$$r_1 = \Phi(f_j(\bar{x}_1, \ldots, \bar{x}_{m_1})), \quad r_2 = \Phi(f_j(\bar{x}_{m_1+1}, \ldots, \bar{x}_{m_1+m_2})), \quad \ldots;$$

the point here is that different $\bar{x}_i$ appear in the different positions.

(ii) If A is generated by a set of cardinality $\leq |\bar{X}|$, then Φ can be taken to be a surjection. (Indeed, deleting a subset of letters needed to produce $r_1, r_2, \ldots$, still leaves the same cardinality of letters in $\bar{X}$, enough to produce a surjection on A.)

(iii) It follows from (ii) that to check that $\mathcal{I} \lhd U$ is a T-ideal, it is enough to check that $\psi(\mathcal{I}) \subseteq \mathcal{I}$ for all surjections $\psi : U \to U$.

The point of this is that some of the ideals in structure theory are preserved under surjections but not necessarily under arbitrary homomorphisms, because surjections take ideals to ideals. For example, the Jacobson radical $J = J(U)$ is a quasi-invertible ideal containing all quasi-invertible left ideals. If $\psi : U \to U$ is a surjection, then $\psi(J)$ is an ideal of U, clearly quasi-invertible, and thus contained in J. The same reasoning holds for the nilradical $N = N(U)$: If $\psi : U \to U$ is a surjection, then $\psi(N)$ is a nil ideal of U, implying $\psi(N) \subseteq N$.

Lemma 3.31. *If* $J(A) = 0$ *then* $J(U_A) = 0$.

Proof. Otherwise, take $0 \neq f \in J(U_A)$, and elements $a_1, \ldots, a_m$ in A such that $f(a_1, \ldots, a_m) \neq 0$. Using Remark 3.30, take a surjection $\psi : U \to A$ such that $\bar{x}_i \mapsto a_i$. Then $0 \neq \psi(J(U_A)) \subseteq J(A) = 0$, contradiction. $\square$

Proposition 3.32. *If* U *is relatively free, then* $J(U) = N(U)$ *are T-ideals.*

Proof. We just saw they are T-ideals, so we need to prove equality; we shall do this for F infinite. Since $N(U) \subseteq J(U)$, we can pass to $U/N(U)$ and assume $N(U) = 0$. In particular, U is semiprime and thus has some PI-class n, cf. Corollary E. But $J(U[\lambda]) = 0$ by Theorem B, and $\mathrm{id}(U[\lambda]) = \mathrm{id}(U)$ by Proposition 1.19. Hence, $U = U_{U[\lambda]}$, which has Jacobson radical 0 by Lemma 3.31.   □

Here is another application of Remark 3.30:

Remark 3.33. The relatively free algebra $U = U_A$ is a subdirect product of copies of A. (Indeed, for each surjection $\varphi : U_A \to A$, write A_φ for this copy of A; the homomorphism $U \to \prod A_\varphi$ given by sending $\bar{f}$ to the vector $(\varphi(\bar{f})$ is an injection, since $\varphi(\bar{f}) = 0)$ for all surjections φ iff $f \in \mathrm{id}(A)$, iff $\bar{f} = 0$.)

We shall improve this result in Exercise 6.3.

3.5   Specht's Problem Posed

Recapitulating, we started with a set $\mathcal{S}$ of algebras, passed to $\mathcal{I} = \mathrm{id}(\mathcal{S})$, and then to $U_\mathcal{I}$, noting $\mathrm{id}(U_\mathcal{I}) = \mathcal{I}$. Thus, we have 1:1 correspondences between T-ideals of $F\{X\}$ and relatively free algebras. Although relatively free algebras are very interesting, we shall focus on the T-ideal aspect.

Perhaps the most interesting question about T-ideals is generating them as efficiently as possible, in the sense of Definition 1.93, where $\mathcal{I}$ are the T-ideals. This is because if a set $S \subseteq \mathrm{id}(A)$ and f is in the T-ideal generated by S, then $f \in \mathrm{id}(A)$. In this case, we say f is a *consequence of S*; we also say S *implies* f.

Example 3.34. The standard s_n is a consequence of the Capelli polynomial c_n, since we get s_n from c_n by specializing $x_i \mapsto 1$, for $n < i \leq 2n$. On the other hand, for $m \geq 2$, c_{2m} is not a consequence of s_{2m}, seen by noting that $s_{2m} \in M_m(F)$ by the Amitsur-Levitzki theorem, whereas $c_{2m} \notin M_m(F)$ by Remark 1.42.

The first basic question about generation of T-ideals was asked by Specht. Since we know by Hilbert's Basis Theorem that every algebraic variety can be defined by a finite number of commutative polynomials, we might ask whether any T-ideal can be generated (as a T-ideal) by a finite number of noncommutative polynomials. In order to avoid confusion with f.g. as a module, a finitely generated T-ideal is usually called *finitely based*, so we have:

Problem 3.35 (Specht's problem). [Sp50] Is every T-ideal of $F\{X\}$ finitely based, for a given field F? (This property is called the *finite basis property* for T-ideals.)

If so, then the set of identities of any PI-algebra A is generated as a T-ideal by a finite set of identities $\{f_1, \ldots, f_n\}$, and so every identity of A is a consequence of $\{f_1, \ldots, f_n\}$.

Specht's problem generated a large literature which dealt with various special cases. In a striking breakthrough, Kemer [Kem87] solved Specht's problem affirmatively for char$(F) = 0$. The proof required several important ideas, to be presented in the next few chapters, culminating in the proof of Kemer's theorem in Chapter 6, after extensive preparation. The depth of Kemer's theorem can be measured in part by the very special case that id$(M_2(\mathbb{Q}))$ is generated as a T-ideal by the obvious candidates s_4 and $[[x_1, x_2]^2, x_3]$ but is surprisingly difficult to prove, cf. Exercise 7.

Kemer's theorem can be viewed more structurally—namely, $\mathbb{Q}\{X\}$ satisfies the ACC on T-ideals, by Remark 1.95. In this way, parts of the theory of Noetherian rings become applicable, and we shall describe these in Chapter 6. Let us note here that the applications of Specht's problem should contain an analysis of the "prime" T-ideals in the sense of Definition 1.96.

The naive approach to answering Specht's problem is to try to search for a set of candidates for the generators, usually polynomials having "nice" properties so that we can work with them.

Specht contributed the next result. We define *higher (Lie) commutators* inductively:

$[f, g]$ is a higher commutator if f is a letter or higher commutator and g is a letter or a higher commutator.

For example, $[[x_1, x_2], [x_5, [x_3, x_4]]]$ is a higher commutator. We call a polynomial *Spechtian* if it is a sum of products of higher commutators.

Proposition 3.36. *Every identity of an algebra A can be written*

$$f(x_1, \ldots, x_m) = \sum_{(u_1, \ldots, u_m) \in \mathbb{N}} f_{(u_1, \ldots, u_m)} x_1^{u_1} \ldots x_m^{u_m}$$

where each $f_{(u_1, \ldots, u_m)}$ is a Spechtian polynomial.

Proof. Note $x_{i_1} x_m x_{i_2} = x_{i_1} x_{i_2} x_m + x_{i_1}[x_m, x_{i_2}]$. Iterating yields

$$x_{i_1} \ldots x_{i_{u-1}} x_m x_{i_u} \ldots x_{i_{m-1}}$$
$$= x_{i_1} \ldots x_{i_{m-1}} x_m + \sum_{j=u}^{m-1} x_{i_1} \ldots x_{i_{j-1}}[x_m, x_{i_j}] x_{i_{j+1}} \ldots x_{i_{m-1}},$$

for each monomial in which x_m appears, and thus any polynomial $f(x_1, \ldots, x_m)$ can be written

$$f = p + g(x_1, \ldots, x_{m-1})x_m + h,$$

where x_m does not appear in p, and h can be written in such a way as that x_m appears (at least) once in a Lie commutator. We continue this argument in turn with each appearance of x_m, so we may assume x_m appears in h *only* in Lie commutators. Now we repeat this procedure with x_{m-1}, and so forth, to yield the desired result.   □

Corollary 3.37 (Specht's Lemma). *Every multilinear identity of an algebra A (with 1) is a consequence of Spechtian identities.*

Proof. The proposition shows $f = g(x_1, \ldots, x_{m-1})x_m + h$, where h can be written in such a way as that x_m only appears in a Lie commutator. Thus $h(x_1, \ldots, x_{m-1}, 1) = 0$, implying $g(x_1, \ldots, x_{m-1}) = f(x_1, \ldots, x_{m-1}, 1)$ is an identity of A, so by induction on m is a consequence of Spechtian identities. Applying the same argument to x_{m-1} in h, and iterating, yields the desired result.   □

A question closely related to Specht's problem is, "Does $f \in \mathrm{id}(A)$ imply $g \in \mathrm{id}(A)$?" Alternatively, "When is g in the T-ideal generated by f?" Here is a quick but significant example, due to Amitsur.

Theorem 3.38. *Every algebra A of PI-degree d satisfies a power s_{2n}^k of the standard identity, for $2n \leq d$ and for suitable k.*

Proof. We can replace A by its relatively free algebra. Let N be the nilradical. A/N is a semiprime PI-algebra and thus has some PI-class n, implying s_{2n} is a PI of A/N, i.e., $s_{2n}(\bar{x}_1, \ldots, \bar{x}_{2n}) \in N$. But this means

$$s_{2n}(\bar{x}_1, \ldots, \bar{x}_{2n})^k = 0$$

for some k, and because A is relatively free, $s_{2n}(x_1, \ldots, x_{2n})^k \in \mathrm{id}(A)$.   □

An explicit bound for k in terms of d is computed in Remark 5.49.

Besides providing a canonical PI, Theorem 3.38 shows more generally that any variety of algebras (defined perhaps with extra structure), in which the semiprime members satisfy a standard identity, actually is a variety of PI-algebras satisfying the PI of Theorem 3.38. This argument has been used in the literature to show that classes of algebras satisfying more general sorts of identities described in Section 11.3 are indeed PI-algebras, cf. Exercise 5.3.

This sort of result can be extended using Lewin's theorem:

Proposition 3.39. *If A is a relatively free PI-algebra and its upper nil-radical $N(A)$ is nilpotent, then $\mathcal{M}_n \subseteq \mathrm{id}(A)$ for some n.*

Proof. $\mathcal{M}_m \subset \mathrm{id}(A/N(A))$ for some m, so apply Theorem 1.72; $n = mt$ where $N(A)^t = 0$. $\qquad\square$

For example, the hypothesis holds by the Razmyslov-Kemer-Braun theorem when A is relatively free and affine. Although we shall soon see such A is representable, the method here is much easier, and also leads to the stronger result in Exercise 4.8, that any algebra satisfying a Capelli identity satisfies $\mathcal{M}_n$ for some n.

One celebrated result (dating from 1943) predates most of the PI-theory.

Theorem 3.40 (Dubnov-Ivanov-Nagata-Higman theorem). *Suppose a $\mathbb{Q}$-algebra A without 1 is nil of bounded index. Then A is nilpotent; more precisely, there is a function $\beta : \mathbb{N} \to \mathbb{N}$ such that if A satisfies the identity x^n, then A satisfies the identity $x_1 \ldots x_{\beta(n)}$.*

A quick proof due to Higgens can be found in Jacobson [Ja56, Appendix C], which gives the bound $\beta(n) = 2^{n-1} - 1$, cf. Exercise 4. Later Razmyslov calculated the improved bound $\beta(n) = n^2$, using the theory of trace identities; in Exercise 12.4, we give a quick argument along the lines of his reasoning to show that $\beta(n) \leq 2n^2 - 3n + 2$; Razmyslov's bound is obtained in Exercise 12.8. On the other hand, Higman[Hig56] also gave the lower bound of n^2/e^2, which was improved by Kuznin to $\frac{n(n+1)}{2}$ in [Kuz75] (which is only available in Russian and not listed in MathScinet), cf. Exercise 12.9, and his student Brangsburg proved $\beta(4) = 10$.

To make best use of Theorem 3.40, we recall the symmetric polynomial $\tilde{s}_t$ from Equation 1.6, i.e., the multilinearization of x_1^t.

Corollary 3.41. *Any $\mathbb{Q}$-algebra without 1 satisfying the identity x_1^t satisfies its multilinearization $\tilde{s}_t$, and thus $x_1 \ldots x_{\beta(t)}$. It follows that $x_1 \ldots x_{\beta(t)}$ is contained in the T-ideal generated by $\tilde{s}_t$ in the free algebra without 1.*

3.6 Solution of Specht's Problem for Grassman Algebras

Before starting our study of T-ideals in earnest, let us solve an interesting special case of Specht's problem. Let us see now that Specht's problem has a very strong affirmative answer for the Grassman algebra G of a vector space V; recall G is generated by $\{e_i : i \in I\}$, satisfying the relations $e_i e_j = -e_j e_i$ and $e_i^2 = 0$. In Proposition 1.39 we noted G satisfies the

Grassman identity $[[x_1, x_2], x_3]$, and we shall show this generates id(G) as a T-ideal. Although this was first proved by Regev using the theory to be espoused in Chapter 5, there is a very direct identity-theoretic proof, which we present here.

Remark 3.42. The following identities hold in any associative algebra, seen by inspection:

(i) $[x, yz] = [x, y]z + y[x, z]$.

(ii) $[xy, z] = x[y, z] + [x, z]y$.

(iii) $[x, yz] = [xy, z] + [zx, y]$.

Lemma 3.43 (Latyshev). *Every algebra A satisfying the Grassman identity satisfies the identity $[x_1, x_2][x_3, x_4] + [x_1, x_3][x_2, x_4]$.*

Proof. For any $a, b, c, d \in A$, we have (using (ii) repeatedly)

$$0 = [[ad, b], c] = [a[d, b], c] + [[a, b]d, c]$$
$$= (a[[d, b], c] + [a, c][d, b]) + ([a, b][d, c] + [[a, b], c]d)$$
$$= 0 - [a, c][b, d] - [a, b][c, d] + 0.$$

$\square$

Corollary 3.44. *Every algebra satisfying the Grassman identity satisfies the identity $[x_1, x_2][x_1, x_3]$.*

Proof. Just specialize $x_3 \to x_1$ to get the identity $[x_1, x_2][x_1, x_4]$. $\square$

Theorem 3.45. *Every Spechtian polynomial f can be reduced modulo the Grassman identity to a sum of polynomials of the form*

$$[x_{i_1}, x_{i_2}] \ldots [x_{i_{2k-1}}, x_{i_{2k}}],$$

where $i_1 < i_2 < \cdots < i_{2k}$.

Proof. By definition, f can be writtten as a sum of products of higher commutators. But every higher commutator of level at least 3 vanishes as a consequence of the Grassman identity, so f is a sum of products of commutators. The corollary to Latyshev's lemma enables us to assume that the letters in these commutators must be distinct, so f may be assumed to be the sum of multilinear polynomials. Thus, we may assume f is multilinear.

Suppose $i_1 < i_2$ are the two lowest indices appearing in indeterminates of f. By Latyshev's lemma we can always exchange the commutator $[x_{i_1}, x_j][x_{i_2}, x_k]$ with $-[x_{i_1}, x_{i_2}][x_j, x_k]$, so $f = [x_{i_1}, x_{i_2}]h$ and the result follows by induction. $\square$

Corollary 3.46. *Each multilinear identity of a Grassman algebra G (over an infinite dimensional vector space over a field) is a consequence of the Grassman identity.*

Proof. By Specht's lemma, any multilinear identity f can be assumed to be Spechtian and thus, by Theorem 3.45, reduced to the form

$$f = [x_1, x_2][x_3, x_4] \ldots [x_{2n-1}, x_{2n}],$$

which certainly is not an identity of G (seen by specializing $x_i \mapsto e_i$) unless $n = 0$, i.e., f reduces to 0.   □

When V is finite dimensional, then so is G, implying it satisfies extra identities. But Theorem 3.45 also helps us out here, since the extra multilinear identities can be reduced to the form $[x_1, x_2][x_3, x_4] \ldots [x_{2n-1}, x_{2n}]$.

Corollary 3.47 (Regev). *In characteristic 0, the T-ideal of G is a consequence of the Grassman identity.*

The same argument shows in characteristic 0 that any T-ideal properly containing $[[x_1, x_2], x_3]$ is generated by at most one other polynomial, $[x_1, x_2][x_3, x_4] \ldots [x_{2n-1}, x_{2n}]$ for some n.

This result also holds in all characteristics $\neq 2$, but is a bit trickier to prove, cf. Exercise 5. We shall study the Grassman algebra in characteristic p in Chapter 7; one can prove readily that it satisfies the Frobenius identity $(x+y)^p = x^p + y^p$, which must then be deducible formally from $[[x_1, x_2], x_3]$, cf. Exercise 7.7. For $p = 2$, G is commutative, and so satisfies the identity $[x_1, x_2]$, which clearly is not a consequence of $[[x_1, x_2], x_3]$. But this case is rather trivial; in Chapter 7, to make it more useful, we introduce a modification to the Grassman algebra in characteristic 2, and then the Grassman identity implies the identity $(x + y)^4 = x^4 + y^4$, cf. Exercise 7.3.

Chapter 4

Specht's Problem in the Affine Case

In this chapter and the next two chapters, we present the details of Kemer's proof of the finite basis property in characteristic 0—that the T-ideal id(W) of an arbitrary PI-algebra W is finitely based. In Chapter 7, we shall see that although Capelli identities hold for arbitrary PI-algebras of characteristic p, there are non-affine counterexamples to Specht's problem in characteristic p.

We work over an infinite field F, at first of arbitrary characteristic, although later we assume char(F) = 0. By Theorems 6.14 and 6.20, it is enough to prove the finite basis property for affine superalgebras. Since proof for the super-case differs from the ungraded case only with regard to some technical difficulties, we shall prove the finite basis property first for affine algebras of characteristic 0, and then indicate (in Chapter 6) how to modify the proof in the super-case to get the full solution of Specht's problem in characteristic 0. Thus, throughout this chapter, we assume W is affine PI, and in particular, its Jacobson radical $J(W)$ is nilpotent by the Razmyslov-Kemer-Braun theorem (Theorem 2.57).

Although Kemer's paper also handled the important case of algebras without 1, we continue to assume for convenience that our algebras contain 1. For the reader interested for algebras without 1, Remark 4.7 contains the essential observation needed for the modification of the proof.

4.1 Kemer's PI-Representability Theorem

The main step of Kemer's approach is proving:

Theorem 4.1 (Kemer). *Any affine PI-algebra of characteristic 0 is PI-representable, i.e., PI-equivalent to some finite dimensional algebra over a field.*

In fact, once one has this theorem, together with some classical results about finite dimensional algebras which we shall review below, it is not difficult to prove the finite basis property already, cf. Exercise 3, although the proof we shall give in the text is based on the techniques that are developed here. Thus, the venerable theory of finite dimensional algebras is elevated to the level of current research, and we review it now.

4.1.1 Finite dimensional algebras

Recall that we write f.d. for finite dimensional. A will always be a f.d. algebra over a field F, with $J = J(A)$ and $\bar{A} = A/J(A)$. There are two important numerical invariants of A:

(i) $t_A = \dim_F(\bar{A})$, the dimension of the semisimple part.

(ii) s_A, the nilpotency index of J.

Note that $s_A = 1$ iff A is semisimple. We shall see that these two invariants can be described in terms of polynomials, and this procedure is perhaps the single most important aspect of the proof. In the sequel, *we shall assume throughout that the base field F is algebraically closed*, in view of the following remark:

Remark 4.2. Suppose $\bar{F}$ is the algebraic closure of F. Then

(i) $\mathrm{id}(W \otimes_F \bar{F}) = \mathrm{id}(W)$, by Proposition 1.19.

(ii) Any f.d. $\bar{F}$-algebra A can be written as $A_0 \otimes_K \bar{F}$, where K is a field of finite transcendence degree over F, and A_0 is finite dimensional over K. (Indeed, take a base $b_1, \ldots, b_n$ of A over $\bar{F}$, and write

$$b_i b_j = \sum_{k=1}^{n} \alpha_{ijk} b_k,$$

and let K be the field generated by the α_{ijk}.) Thus, $\mathrm{id}(A) = \mathrm{id}(A_0)$.

Our main structure theorem about f.d. algebras is *Wedderburn's Principal Theorem* [Row88, Theorem 2.5.37], which we already used in Chapter 2:

Theorem 4.3. $\bar{A}$ *can be taken as a subalgebra of A, and $A = \bar{A} \oplus J$ as vector spaces over F.*

Thus, in checking whether a t-linear polynomial $f(x_1, \ldots, x_t; y_1, \ldots, y_m)$ is an identity A, we may assume each substitution for x_i is either in $\bar{A}$ or in J. We shall call such a substitution a *semisimple substitution* or a *radical substitution*, respectively.

$\bar{A}$ in turn can be written as a finite direct product of f.d. simple algebras R_j, $1 \le j \le q$, each of which is of the form $M_{n_j}(F)$ since F is algebraically closed. Thus,

$$A = R_1 \oplus \cdots \oplus R_q \oplus J. \qquad (4.1)$$

A word about the use of $\times$ and $\oplus$: Strictly speaking, we could write

$$R_1 \times \cdots \times R_q$$

to emphasize this as a direct product of rings, i.e., the components respect multiplication as well as addition. Having said this, we shall revert to the notation of $\oplus$ to avoid confusion with the notation in Remark 4.5.

Thus, we can (and shall) assume that any semisimple specialization is a matrix unit in a suitable simple component. Furthermore, let us write e_j for the multiplicative unit of R_j; then $e_j \bar{A} e_j = R_j$. Since $J = \oplus_{j,k=1}^n e_j J e_k$, we shall assume each radical substitution is in $e_j J e_k$ for suitable j, k, and say such a radical substitution has *type* (j, k). (Note by passing to A/J that $e_j A e_k \subseteq J$, so $e_j A e_k = e_j J e_k$.)

Remark 4.4. Under this notation, suppose that $f(x_1, \ldots, x_t; y_1, \ldots, y_m)$ is a t-alternating polynomial, nonzero under the specializations $x_i \mapsto \bar{x}_i$. If all of the $\bar{x}_1, \ldots, \bar{x}_t$ are semisimple substitutions, then $t \le \dim_F \bar{A}$.

(One might think that they must all be in the same component R_j, since $R_j R_{j'} = 0$ for all $j \ne j'$. However, this is not so, since the specializations of the y_j could transfer us from component to component.)

Viewed the other way, if $t > \dim \bar{A}$, then every nonzero substitution in f must have a radical substitution for at least one of $x_1, \ldots, x_t$.

Recall that an algebra A is *subdirectly irreducible* if the intersection of nonzero ideals is always nonzero. (In particular, a simple algebra is subdirectly irreducible.)

Remark 4.5. If $I, J \lhd A$ with $I \cap J = 0$, then $\mathrm{id}(A) = \mathrm{id}(A/I \times A/J)$. (Indeed, there is a natural injection $A \to A/I \times A/J$; on the other hand, if $f \in \mathrm{id}(A/I \times A/J)$, then checking each component we see $f(A) \subseteq I \cap J = 0$.)

Thus, by induction, if A is a subdirect product of $A_1, \ldots, A_m$, then

$$\mathrm{id}(A) = \mathrm{id}(A_1 \times \cdots \times A_m) = \bigcap_{i=1}^m \mathrm{id}(A_i). \qquad (4.2)$$

Consequently, if $\dim_F A < \infty$, we can use induction to find another algebra that is the direct product of subdirectly irreducible algebras, satisfying the same PIs as A.

When considering questions about id(A), we shall use (4.2) repeatedly to replace our f.d. algebras A by a direct product of subdirectly irreducible algebras. Consequently, the theory often reduces to considering a specific subdirectly irreducible algebra A. As usual, for notational simplicity, we assume $1 \in A$.

Remark 4.6. Recall $e \in A$ is *idempotent* if $e^2 = e$; the *trivial* idempotents are $0, 1$. Thus, the e_j defined above are idempotents.

Given a nontrivial idempotent e of A, and letting $e' = 1 - e$, we recall the *Peirce decomposition*

$$A = eAe \oplus eAe' \oplus e'Ae \oplus e'Ae'. \tag{4.3}$$

Note that $eAe, e'Ae'$ are algebras with respective multiplicative units e, e'. If $eAe' = e'Ae = 0$, then also $eAe, e'Ae' \lhd A$, which is impossible for A subdirectly irreducible.

Remark 4.7. Although our exposition deals with algebras with 1, let us digress for a moment to discuss how to generalize this for an algebra A without 1. In this case, Wedderburn's principal theorem still holds, enabling us to write $A = \bar{A} \oplus J$ where $\bar{A}$ is a semisimple subalgebra of A. Of course, the unit element e of $\bar{A}$ is not the unit of A. Let us adjoin a unit element 1 to A in the standard way, i.e., take $\hat{A} = A \oplus F$, with multiplication

$$(a_1, \alpha_1)(a_2, \alpha_2) = (a_1 a_2 + \alpha_1 a_2 + \alpha_2 a_1, \alpha_1 \alpha_2),$$

and note that $\hat{1} = (0, 1)$ is the multiplicative unit of $\hat{A}$.

Now we write $\hat{A} = (\bar{A} \oplus J) \oplus F$, and put $e_0 = \hat{1} - e$. Now we have a Peirce decomposition

$$A = (e + e_0)A(e + e_0) = eAe \oplus eAe_0 \oplus e_0 Ae \oplus e_0 Ae_0,$$

where the last three components belong to J. Thus, any substitution from one of these is a radical substitution, and now the proofs go through in the same way as in the treatment here, where we handle the $(0, 0)$ component $e_0 Ae_0$ in the same fashion as the (j, k) radical substitutions $e_i Ae_j$ for $i \neq j$; for example $e_0 Ae_0 = e_0 Je_0$. See Exercise 4, for example.

4.1.2   Sketch of Kemer's program

Theorem 4.1 will be proved in several steps, which we outline here. Let W be any affine PI-algebra, and let $\Gamma = \mathrm{id}(W)$ be its T-ideal (in an infinite number of indeterminates, so that we may make use of multilinear identities of arbitrarily large degree).

Step 1. $\Gamma \supseteq \mathrm{id}(A)$ for some f.d. algebra A. This requires Lewin's theorem 1.72. We start with an easy observation.

Proposition 4.8. *Suppose* $W = \begin{pmatrix} W_1 & * \\ 0 & W_2 \end{pmatrix}$. *Then*

$$\mathrm{id}(W) \supseteq \mathrm{id}(W_1)\,\mathrm{id}(W_2).$$

Proof. If $f \in \mathrm{id}(W_1)$ then $f(W)$ has the form

$$\begin{pmatrix} 0 & * \\ 0 & * \end{pmatrix};$$

if $g \in \mathrm{id}(W_2)$ then $g(W)$ has the form

$$\begin{pmatrix} * & * \\ 0 & 0 \end{pmatrix}.$$

Multiplying these two matrices together gives 0.   □

Corollary 4.9. *For any affine PI-algebra W over a field F, there is a f.d. F-algebra A such that $\mathrm{id}(A) \subseteq \mathrm{id}(W)$. Furthermore, we can take A to be an algebra of block upper triangular $n \times n$ matrices over F.*

Proof. We may replace W by its relatively free algebra. Now write $W = F\{X\}/I$, and let $J(W) = J/I$. Then $J(W)^k = 0$ for some k or, in other words, $J^k \subseteq I$. Furthermore, $F\{X\}/J$ is relatively free PI and semiprime, so is an algebra of generic matrices and thus representable. Thus, $F\{X\}/J^k$ is representable by Corollary 1.73, and so is PI-equivalent to some f.d. algebra A; clearly $\mathrm{id}(A) \subseteq \mathrm{id}(W)$ since there is a surjection $F\{X\}/J^k \to W$.

The last assertion follows from the (highly nontrivial) Theorem 1.72(ii) of Lewin.   □

In fact, one gets more from Lewin's Theorem 1.72:

Proposition 4.10. *If I_1, I_2 are T-ideals of $F\{X\}$ and we put*

$$A = \begin{pmatrix} F\{X\}/I_1 & M \\ 0 & F\{X\}/I_2 \end{pmatrix} \tag{4.4}$$

(in the notation of Theorem 1.72(ii)), then $\mathrm{id}(A) = I_1 I_2$.

Proof. We already showed $\mathrm{id}(A) \supseteq I_1 I_2$. But conversely, Theorem 1.72 shows $F\{X\}/I_1 I_2$ is embedded in A, so $\mathrm{id}(A) \subseteq I_1 I_2$.   □

This shows the product of T-ideals is a T-ideal.

Remark 4.11. The upper triangular matrix algebra over F (of size $nk \times nk$, with $n \times n$ blocks along and above the diagonal) is a good model for intuition, arising as the critical case in the theory of exponents of Giambruno and Zaicev [GiZa03c], although we shall not use it. See Exercise 1 for an application to $\mathrm{id}(W)$.

Theorem 1.72(ii) does *not* say that a relatively free affine PI-algebra is representable—this is one of the key results in Kemer's program and is considerably deeper, to be proved below.

Remark 4.12. This step is the only one which is not (yet) generalized to the nonassociative cases of alternative and Jordan algebras, as to be discussed in Chapter 11, because Lewin's theorem is not known to hold more generally, but nevertheless Specht's problem has a positive solution in the affine case.

Step 2. We want to describe the two invariants t_A and s_A in terms of alternating polynomials. We define the *Kemer index* of Γ, denoted $\mathrm{index}(\Gamma)$, to be a pair $(\beta_\Gamma, \gamma_\Gamma)$, ordered lexicographically, defined such that for $\Gamma = \mathrm{id}(A)$ for A f.d. subdirectly irreducible algebras satisfying certain technical conditions, we have

$$(\beta_\Gamma, \gamma_\Gamma) = (t_A, s_A).$$

Also it turns out that $\Gamma_1 \subseteq \Gamma_2$ implies $\mathrm{index}(\Gamma_1) \geq \mathrm{index}(\Gamma_2)$. Unfortunately, this correspondence is weaker than we might like (in the sense that one can have $\Gamma_1 \subset \Gamma_2$ with $\mathrm{index}(\Gamma_1) = \mathrm{index}(\Gamma_2)$; furthermore, for n large, the algebra of upper triangular $n \times n$ matrices has low Kemer index, contrary to intuition.) Nevertheless, if $\mathrm{index}(\Gamma_1) > \mathrm{index}(\Gamma_2)$, then there is a certain sort of polynomial in $\Gamma_2 \setminus \Gamma_1$, which we call a *Kemer polynomial*, and which has very useful combinatorial properties. Unfortunately, we need to work with the T-ideal generated by the Kemer polynomial, whose elements lose these nice properties, so we need the "Phoenix property," that any element of this T-ideal generates another Kemer polynomial.

Step 3. In view of Remark 4.5, we may replace A by a direct product of f.d. subdirectly irreducible PI-algebras $A_1 \times \cdots \times A_m$ satisfying the technical conditions mentioned in Step 2. Since $\Gamma \supseteq \bigcap \mathrm{id}(A_i)$, we have $\mathrm{index}(\Gamma) \leq \max \mathrm{index}(A_i)$. Taking those A_i of maximal Kemer index, we prove by induction on the Kemer index that, replacing the A_i by other f.d. algebras if necessary, their Kemer polynomials (of large enough degree) are precisely the Kemer polynomials of Γ of that degree.

Step 4. Let Γ' be the T-ideal generated by Γ and the Kemer polynomials described in Step 3. Then index$(\Gamma') <$ index(Γ), so by induction, there is a f.d. algebra A' with id$(A') = \Gamma'$. On the other hand, using the trace ring construction for representable algebras desribed in Chapter 2, one can define a representable algebra $\hat{A}'$ with $\Gamma \subseteq \hat{A}'$, but whose Kemer polynomials are not identities of $\hat{A}'$. Clearly $\hat{A}'$ is PI-equivalent to a f.d. algebra. Hence,

$$\Gamma = \Gamma' \cap \operatorname{id}(\hat{A}) = \operatorname{id}(A') \cap \operatorname{id}(\hat{A}) = \operatorname{id}(A \times A'),$$

as desired.

In brief, our strategy is to start first with

$$\bigcap_{i=1}^{m} \operatorname{id}(A_i) \subseteq \Gamma,$$

using Proposition 4.9, and then to modify the A_i to get

$$\bigcap_{i=1}^{m} \operatorname{id}(A_i) = \Gamma.$$

4.2 The Kemer Index

Since Step 1 has been taken care of via Lewin's theorem, let us work out Steps 2, 3, and 4. We work with the T-ideal $\Gamma = \operatorname{id}(W)$. We start with some important invariants described in terms of polynomials. Recall that we are looking for a way of describing $t_A = \dim_F(\bar{A})$ and s_A, the nilpotency index of $J(A)$, in terms of polynomials. In view of Theorem J, our description should involve non-identities that alternate simultaneously in many sets of indeterminates. The following definition generalizes Definition 1.20.

Definition 4.13. Given sets $X_1, \dots, X_\mu$ of indeterminates, say

$$X_i = \{x_{i,1}, \dots, x_{i,t_i}\},$$

suppose the polynomial $f(X_1, \dots, X_\mu; Y)$ is linear in all of these indeterminates as well perhaps as in extra indeterminates denoted Y. We say f is $(t_1, \dots, t_\mu)$-*alternating* if f is t_j-alternating in the indeterminates of X_j, for each $1 \le j \le \mu$.

If $t_1 = \cdots = t_\mu = t$, we say f is μ-*fold t-alternating*, and call the X_j the *folds* of f. The indeterminates $\cup X_j$ are called the *designated indeterminates*. More generally, if f is $(t_1, \dots, t_{\mu+\mu'})$-alternating where $t_1 = \cdots = t_\mu = t$ and $t_{\mu+1} = \cdots = t_{\mu+\mu'} = t'$, we say f has μ folds of t indeterminates, μ' folds of t' indeterminates, and $\mu t + \mu' t'$ designated indeterminates.

As in the alternator of Definition 1.27, there is an effective way of producing μ-fold t-alternating polynomials:

Definition 4.14. Given sets $X_1, \ldots, X_\mu$ of indeterminates, and a polynomial $f(X_1, \ldots, X_\mu; Y)$ linear in all of these indeterminates as well perhaps as extra indeterminates designated Y, define the $(t_1, \ldots, t_\mu)$-*alternator*

$$\tilde{f}_{X_1, \ldots, X_\mu}$$

of f to be

$$\tilde{f} = \sum_{\pi_1 \in S_{t_1}, \ldots, \pi_\mu \in S_{t_\mu}} \mathrm{sgn}(\pi_1) \ldots \mathrm{sgn}(\pi_\mu) \, \pi_1 \ldots \pi_\mu f,$$

where π_i acts on the indeterminates of X_i, i.e., $x_{i,\pi_i(j)}$ is substituted for $x_{ij} \in X_i$, for all i, j.

Note that the x_i do not need to appear in succession in f. For example, we could have $X_1 = \{x_{11}, x_{12}, x_{13}\}$ and $X_2 = \{x_{21}, x_{22}\}$, and

$$f = f(x_{11}, y_1, x_{21}, x_{12}, y_2, x_{13}, y_3, y_4).$$

We leave it to the reader (or his computer) to write down the 2-fold alternator of this polynomial.

Definition 4.15. For any affine PI-algebra W, $\beta(W)$ is the largest t such that for any μ, there is a μ-fold t-alternating polynomial

$$f(X_1, \ldots, X_\mu; Y) \notin \mathrm{id}(W).$$

Our motivating example is:

Example 4.16. If $A = M_n(F)$, then every (n^2+1)-alternating polynomial is in $\mathrm{id}(A)$, whereas $c_{n^2}(x_1, \ldots, x_{n^2}; y_1, \ldots, y_{n^2})^\mu$ takes on the value $e_{11}^\mu = e_{11}$, for any μ. Thus, $\beta(M_n(F)) = n^2$ in this case.

Let us generalize this result.

Remark 4.17. For any affine algebra W, letting n be the PI-class of W, we know any $(n^2 + 1)$-alternating polynomial f is an identity of $W/J(W)$. Taking s such that $J(W)^s = 0$, we see that each fold of f has values only in $J(W)$, and thus any s-fold $(n^2 + 1)$-alternating polynomial f is in $\mathrm{id}(W)$. It follows that $\beta(W) \leq n^2$.

Although, as we just saw, $\beta(W)$ exists for affine algebras W, Equation (1.13) shows that any product of Capelli identities is not an identity of the Grassman algebra G. So $\beta(G)$ does not exist. This is the main reason why in this chapter we can only prove Kemer's PI-Representability Theorem for affine algebras, and need more theory to get the full theorem in Chapter 6. Having defined $\beta(W)$, we are led naturally to a refinement:

Definition 4.18. For $\mu \in \mathbb{N}$, $\gamma(W, \mu)$ is the largest s such that some μ-fold, $\beta(W)$-alternating and $(s-1)$-fold, $(\beta(W)+1)$-alternating polynomial $f(X_1, \ldots, X_{s-1}; Y)$ is not in $\mathrm{id}(W)$. Noting that $\gamma(W, \mu)$ decreases as μ increases, we take $\gamma(W)$ to be $\gamma(W, \mu)$ for $\mu \to \infty$.

In the literature, $\gamma(W)$ is defined to be 1 less, so as to be 0 for semisimple PI-algebras. We chose this definition so that Kemer's Second Lemma below would often give equality with the nilpotence index of the Jacobson radical. For each algebra A used in Kemer's lemmas, $\gamma(A) = \gamma(A, 0)$, so Kemer's original definition is shorter.

Definition 4.19. The *Kemer index* of an affine PI-algebra W, denoted $\mathrm{index}(W)$, is the ordered pair $(\beta(W), \gamma(W))$. For a T-ideal $\Gamma = \mathrm{id}(W)$, we also write $\mathrm{index}(\Gamma)$ for the Kemer index of W, and call it the *Kemer index* of Γ.

Example 4.20. $\mathrm{index}(M_n(F)) = (n^2, 1)$. Thus, the same holds for any algebra of PI-class n.

Latent in the definition is another PI-invariant that can be bypassed in the proof of Kemer's Theorem, but which plays a role in later applications:

Definition 4.21. $\omega(W)$ is the smallest μ such that $\gamma(W, \mu) = \gamma(W)$.

The Kemer index is measured in terms of a certain class of polynomials.

Definition 4.22. Suppose $\beta(W) = t$ and $\gamma(W) = s$. A *μ-Kemer polynomial* for W is a polynomial $f = f_W \notin \mathrm{id}(W)$ in $s + \mu - 1$ sets of designated indeterminates X_j (and could also involve other indeterminates), which is $(s-1)$-fold $(t+1)$-alternating with another μ folds that are t-alternating. In other words,

$$|X_1| = \cdots = |X_{s-1}| = t+1; \qquad |X_s| = |X_{s+1}| = \cdots = |X_{s+\mu-1}| = t,$$

and f is alternating in each set of X_j, but yet is not an identity of W.

   If $\Gamma = \mathrm{id}(W)$, a μ-Kemer polynomial for W is also called a *μ-Kemer polynomial* for Γ.

We shall need to take μ "large enough" for various proofs to work; usually we want $\mu \geq \omega(W)$ in particular.

Remark 4.23. Technical problems arise when the names of the indeterminates do not match up; for example,

$$f = [x_1, x_2][x_3, x_4] + [x_1, x_4][x_2, x_3]$$

does not alternate in any pair of indeterminates, even though each of the summands is a Kemer polynomial for the algebra A of upper triangular 3×3 matrices.

Lemma 4.24. If index$(W) = \kappa = (\beta_W, \gamma_W)$ and I is the ideal generated by all evaluations in W of μ-Kemer polynomials of W, then index$(W/I) < \kappa$.

Proof. Clearly, index$(W/I) \leq \kappa$. But if index$(W/I) = \kappa$, then there would be a μ-Kemer polynomial, which by hypothesis is in id(W/I), contradiction. $\square$

Kemer indices are ordered via the lexicographic order on pairs.

Proposition 4.25. *Suppose T-ideals $\Gamma, \Gamma_1, \ldots, \Gamma_m$ satisfy* index$(\Gamma_i) <$ index(Γ) *for each i. Then there is $\mu \in \mathbb{N}$ such that every μ-Kemer polynomial for Γ is in $\bigcap_{i=1}^m \Gamma_i$.*

Proof. This observation follows at once from the definitions. Write index $\Gamma = (t, s)$ and index$(\Gamma_i) = (t_i, s_i)$. If $t_i < t$, then by definition there is some number μ_i such that any μ_i-fold t-alternating polynomial is in Γ_i. If $t_i = t$, then by definition any s_i-fold $(t+1)$-alternating polynomial is in Γ_i. Hence, taking

$$\mu = \max\{\mu_1, \ldots, \mu_m, \omega(W)\},$$

we see by definition that each μ-Kemer polynomial is in each Γ_i, and thus the T-ideal they generate also is in Γ_i. $\square$

Corollary 4.26. *If $\Gamma_1 \supseteq \Gamma_2$, then* index$(\Gamma_1) \geq$ index(Γ_2).

Proof. Otherwise, index$(\Gamma_2) <$ index(Γ_1), contrary to the proposition, since this would yield a Kemer polynomial for Γ_1 which is in Γ_2. $\square$

Corollary 4.27. index$(W_1 \times \cdots \times W_m) = \max_{i=1}^m$ index(W_i), *for any PI-algebras W_i on which the Kemer index is defined.*

Nevertheless, we must be careful.

Example 4.28. The algebra of upper triangular $n \times n$ matrices has Kemer index $(1, n)$, but does not satisfy the standard identity s_{2n-1}, which holds in $M_{n-1}(F)$.

4.2.1   Computing Kemer polynomials

Since Kemer polynomials will be the backbone of our theory, let us digress to see how they can be constructed.

Definition 4.29. A polynomial with specified alternating folds, cf. Definition 4.13 is *monotonic* if the undesignated indeterminates occur in the same order in each monomial of f.

For example, the Capelli polynomial c_n, viewed as a 1-fold t-alternating polynomial, is monotonic since the y_j always occur in the order $y_1, \ldots, y_n$.

Remark 4.30. If we call the undesignated indeterminates $y_1, \ldots, y_m$, then for $\sigma \in S_m$, we define f_σ to be the sum of those monomials in which the y_j appear in the order

$$y_{\sigma(1)}, \ldots, y_{\sigma(m)}.$$

Then clearly $f = \sum_{\sigma \in S_m} f_\sigma$, so some $f_\sigma \notin \operatorname{id}(A)$, and any such f_σ is also a Kemer polynomial. For future use (after we prove Kemer's Theorem) we call the f_σ the *monotonic components* of the Kemer polynomial f.

In particular, in Remark 4.55(vi) we could take I to be the evaluations of a monotonic component of a Kemer polynomial.

We want to apply the same idea to Kemer polynomials (or more precisely, their monotonic components), and so we must consider polynomials that alternate in several folds, cf. Definition 4.14. Let us introduce an appropriate notation, which will enable us to investigate the pumping procedure more closely. First we focus on a single fold, so suppose $f(x; y)$ alternates in $x_{i_1}, \ldots, x_{i_t}$.

Definition 4.31. Recall (Definition 1.27) the alternator $f_{\mathcal{A}(i_1, \ldots, i_t; X)}$ of a polynomial f in a set of indeterminates X is $\sum \operatorname{sgn}(\pi) f_\pi$, summed over all permutations π of $i_1, \ldots, i_t$. Up to sign, this is independent of the order of $i_1, \ldots, i_t$. If X is understood, then we delete it from the notation, writing merely $f_{\mathcal{A}(i_1, \ldots, i_t)}$. Then the $(t_1, \ldots, t_\mu)$ alternator of Definition 4.14 is

$$f_{\mathcal{A}(i_{1,1}, \ldots, i_{1, t_1}) \mathcal{A}(i_{2,1}, \ldots, i_{2, t_2}) \ldots \mathcal{A}(i_{1,u}, \ldots, i_{1, t_\mu})}.$$

To cut down on notation, we write this as

$$f_{\mathcal{A}(I_1) \mathcal{A}(I_2) \ldots \mathcal{A}(I_\mu)},$$

where $I_j = (i_{j,1}, \ldots, i_{j, t_j})$ are the sets of subscripts of the x_i to be alternated, and it is assumed throughout that the I_j are disjoint. Thus, the order in which we apply the $\mathcal{A}(I_j)$ is irrelevant.

Suppose $\kappa = (t, s)$. We define the *μ-alternator (for Kemer index κ)* to be the application of $s - 1$ folds of $(t+1)$-alternators and μ folds of t-alternators, or in the notation of the previous paragraph,

$$f_{\mathcal{A}(I_1)\ldots\mathcal{A}(I_s)\mathcal{A}(I_{s+1})\ldots\mathcal{A}(I_{s+\mu})},$$

where $I_1, \ldots, I_s$ have $t + 1$ indices each, and the other I_j have t indices each.

Remark 4.32.

(i) If f is $(t_1, \ldots, t_\mu)$-alternating, and we take f_0 to be the sum of monomials of f in which the designated x_i appear in a given order, then f is the $(t_1, \ldots, t_\mu)$-alternator of f_0.

(ii) If f is monotonic in the non-designated indeterminates, then f is the $(t_1, \ldots, t_\mu)$-alternator of some monomial. (Indeed, this follows from (i), since $y_1, \ldots, y_m$ also occur in a fixed order.)

(iii) Any μ-Kemer polynomial corresponding to Kemer index $\kappa = (t, s)$ is the sum of μ-alternators (appropriate to Kemer index κ) of suitable monomials. (Indeed, the polynomial is a sum of its monotonic Kemer components, so apply (ii).)

Finally, Remark G gives us a way to switch the sizes of folds.

Remark 4.33. Suppose $f(x_{1,1}, \ldots, x_{1,t}, x_{2,1}, \ldots, x_{2,t+1}; y_1, \ldots, y_m)$ is t-alternating in $X_1 = \{x_{1,1}, \ldots, x_{1,t}\}$ and $(t + 1)$-alternating in the $X_2 = \{x_{2,1}, \ldots, x_{2,t+1}\}$. Write $x_{2,t+1} = x'_{2,t+1}z$. Let

$$h_i = (-1)^i f(x_{1,1}, \ldots, x_{1,i-1}, x'_{2,t+1}, x_{1,i+1}, \ldots, x_t, x_{1,i}z; y).$$

By Remark G, $f + \sum_{i=1}^t h_i$ is $(t+1), t$ alternating (in the sets $X_1 \cup \{x'_{2,t+1}\}$ and $X_2 \setminus \{x_{2,t+1}\}$), whereas each h_i is $t, t+1$ alternating in $X_1 \cup \{x'_{2,t+1}\} \setminus \{x_{1,i}\}$ and $X_2 \cup \{x_{1,i}z\} \setminus \{x_{2,t+1}\}$.

In the proofs of Kemer's Theorem that already appear in the literature, the *relative* Kemer index (and relative Kemer polynomials) of an algebra is defined with respect to a T-ideal, cf. Exercise 6. However, we shall only deal with absolute Kemer indices, hopefully to avoid confusion. Although we are using the Kemer index here to prove Kemer's Theorem, the Kemer index also should be an important tool in PI-theory; we shall use it in Chapters 8 and 9.

4.3   Subdirectly Irreducible Algebras

Our first objective with the Kemer index is to find a f.d. algebra having the same Kemer index as an arbitrary affine PI-algebra W. Towards this end, we want to describe the Kemer index of a f.d. A in terms of properties instrinsic to A. We fix the notation $t_A = \dim_F \bar{A}$, and s_A equals the nilpotence index of $J = J(A)$. We write

$$\bar{A} = R_1 \oplus \cdots \oplus R_q,$$

each R_k having the form $M_{n_k}(F)$. Thus,

$$A = R_1 \oplus \cdots \oplus R_q \oplus J.$$

Kemer's lemmas will describe t_A (and later s_A) in terms of the Kemer index, under certain reasonable technical hypotheses which we introduce now. (Surprisingly, q, the number of simple components of $\bar{A}$, plays only a secondary role.)

4.3.1   Full algebras

Definition 4.34. Notation as above, we say a f.d. algebra A is *full* with respect to a multilinear polynomial $f(x_1, \ldots, x_m)$ if $f(\bar{x}_1, \ldots, \bar{x}_m) \neq 0$ for suitable substitutions $\bar{x}_i$ such that $R_k \cap \{\bar{x}_1, \ldots, \bar{x}_m\} \neq \emptyset$ for each k. (In other words, the substitutions "pass through" each component R_k.)

The f.d. algebra A is *full* if A is full with respect to some polynomial.

Remark 4.35. If A is full, we have a permutation π of $\{1, \ldots, q\}$ such that

$$R_{\pi(1)} J R_{\pi(2)} J \ldots J R_{\pi(q)} \neq 0. \tag{4.5}$$

Converesely, iff (4.5) holds, then A is full with respect to the polynomial $x_1 \ldots x_{2q-1}$.

We would like to prove every subdirectly irreducible algebra is full, but are confronted with examples such as:

Example 4.36. Let A be the subalgebra of $M_4(F)$ spanned by the matrix units

$$e_{11}, e_{22}, e_{33}, e_{44}, e_{12}, e_{24}, e_{13}, e_{34}, e_{14}.$$

Obviously any ideal contains some e_{ij}, seen via the Peirce decomposition, and thus contains e_{14}, so A is subdirectly irreducible.

The last five matrix units span $J = J(A)$, so $J^3 = 0$. But $\bar{A} = F^{(4)}$, i.e., $q = 4$, so A is not full.

Our next step is to determine how to replace A by a direct product of full subdirectly irreducible algebras. Intuitively, any identity on A must "pass through" e_{12} or e_{13}, so A is PI-equivalent to the direct product of two subalgebras of $M_4(F)$: that spanned by $e_{11}, e_{22}, e_{33}, e_{44}, e_{12}, e_{24}, e_{14}$ and that spanned by $e_{11}, e_{22}, e_{33}, e_{44}, e_{13}, e_{34}, e_{14}$. To handle the situation in general, we introduce some graph theory.

Remark 4.37. There is a well-known directed graph-theoretic interpretation of matrix units, in which one draws an edge from e_{ij} to $e_{k\ell}$ iff $j = k$; thus, the paths correspond precisely to the nonzero products of matrix units. Viewed in this way, $\bar{A}$ is a disjoint union of q components, which we shall call *stages*, corresponding to the matrix algebras $R_1, \ldots, R_q$, and this takes care of all semisimple substitutions. The situation becomes more interesting when we bring in radical substitutions. A type (j, k) radical substitution connects the j and k stages, and these are the substitutions that can connect separate stages.

Viewed in this light, any nonzero evaluation of a monomial h comes from substitutions in a connected component of the graph, and in fact in a loopless path. Indeed, suppose the evaluation of h comes from $R_{i_1} J R_{i_2} J \ldots$ Whenever a subscript duplicates, say $i_\ell = i_{\ell'}$, then we can replace $J R_{i_{\ell'}} J$ by J, thereby erasing the duplication. The most efficient way to do this is through a generic sort of construction.

Proposition 4.38. *Any f.d. subdirectly irreducible algebra that is* not *full is PI-equivalent to a direct product of subdirectly irreducible algebras, each having fewer simple components.*

Proof. Decompose A as in (4.1). Let e_i denote the unit element of R_i, thereby yielding the Peirce decomposition

$$A = \bigoplus_{i,j=1}^{n} e_i A e_j.$$

(Recall $e_i A e_j \subset J$ for $i \neq j$. In the case of algebras without 1, we also need e_0 as defined in Remark 4.7, and take the limits in the direct sum from 0 to n.) If we have $e_{i_1} J e_{i_2} \ldots e_{i_{q-1}} J e_{i_q} \neq 0$ for $i_1, \ldots, i_q$ distinct, then A is full, so we assume this is *not* the case, i.e.,

$$e_{i_1} A e_{i_2} \ldots e_{i_{q-1}} A e_{i_q} = e_{i_1} J e_{i_2} \ldots e_{i_{q-1}} J e_{i_q} = 0$$

whenever $i_1, \ldots, i_q$ are distinct.

Consider the commutative algebra $H = F[\lambda_1, \ldots, \lambda_q]/I$, where I is generated by $\lambda_i^2 - \lambda_i$ and $\lambda_1 \cdots \lambda_q$. In other words, writing $\tilde{e}_i$ for the image of λ_i, we have $\tilde{e}_i^2 = \tilde{e}_i$ and $\tilde{e}_1 \cdots \tilde{e}_q = 0$. Let $\tilde{A}$ be the subalgebra of $A \otimes_F H$ generated by all

$$e_i A e_j \otimes \tilde{e}_i \tilde{e}_j, \quad \forall i, j \leq q.$$

We claim $A \sim_{\mathrm{PI}} \tilde{A}$. Clearly

$$\mathrm{id}(A) \subseteq \mathrm{id}(A \otimes_F H) \subseteq \mathrm{id}(\tilde{A}),$$

so it suffices to prove that any nonidentity f of A is also not an identity of $\tilde{A}$; we may assume $f(x_1, \ldots, x_n)$ is multilinear.

Note. $\tilde{A}$ is graded by the number of distinct $\tilde{e}_i$ appearing in the tensor product of an element. In evaluating f on A, it suffices to consider specializations $x_k \mapsto v_k$ where $v_k \in e_{i_k} A e_{i_{k+1}}$. In order for $v_1 \ldots v_n \neq 0$, $\{i_1, \ldots, i_{n+1}\}$ must contain at most $q-1$ distinct elements, so $\tilde{e}_{i_1} \ldots \tilde{e}_{i_{n+1}} \neq 0$. Then

$$f(v_1 \otimes \tilde{e}_{i_1}, \ldots, v_n \otimes \tilde{e}_{i_n}) = f(v_1, \ldots, v_n) \otimes \tilde{e}_{i_1} \tilde{e}_{i_2} \ldots \tilde{e}_{i_n} \neq 0,$$

so $f \notin \mathrm{id}(\tilde{A})$, and we conclude $\tilde{A} \sim_{\mathrm{PI}} A$, as claimed.

Now letting $I_j = \langle e_j \otimes \tilde{e}_j \rangle \triangleleft \tilde{A}$ we see

$$\bigcap_{j=1}^n I_j = (1 \otimes \tilde{e}_1) \ldots (1 \otimes \tilde{e}_n) \left(\bigcap_{j=1}^n I_j \right) = (1 \otimes \tilde{e}_1 \cdots \tilde{e}_n)(\bigcap I_j) = 0,$$

so $\tilde{A}$ is subdirectly decomposable to the $\tilde{A}/I_j$. Furthermore the $\tilde{A}/I_j$ all have $\leq q - 1$ simple components (modulo the radical), since we eliminated one of the idempotents corresponding to a simple component. Hence we may conclude the proof by induction. $\qquad\square$

Corollary 4.39. *Every f.d. algebra is PI-equivalent to a direct product of full, subdirectly irreducible f.d. algebras.*

Proof. Apply the proposition and apply induction on the number of simple components. (A subdirectly irreducible algebra with one simple component clearly is full.) $\qquad\square$

Remark 4.40. For computational purposes, let us make Definition 4.34 (and the following remark) more specific. Assume A is full with respect to f. Rearranging the x_k if necessary we assume $\bar{x}_k \in R_k$, $1 \leq k \leq q$. Put

$$a = f(\bar{x}_1, \ldots, \bar{x}_q, \bar{x}_{q+1}, \ldots, \bar{x}_m) \neq 0.$$

Clearly we may assume each $\bar{x}_k$ is a matrix unit, which we write as $e_{ij}^{(k)} \in R_k$. But

$$e_{ij}^{(k)} = e_{i1}^{(k)} e_{11}^{(k)} e_{1j}^{(k)},$$

so defining

$$g(x_1, \ldots, x_m; y_1, \ldots, y_k; z_1, \ldots, z_k) = g(y_1 x_1 z_1, \ldots, y_q x_q z_q, x_{q+1}, \ldots x_m)$$

and taking $\bar{y}_k = e_{i1}^{(k)}$, $\bar{z}_k = e_{1j}^{(k)}$, $1 \le k \le q$, we have

$$g(e_{11}^{(1)}, \ldots, e_{11}^{(q)}, \bar{x}_{q+1}, \ldots, \bar{x}_m; \bar{y}_1, \ldots, \bar{y}_q; \bar{z}_1, \ldots \bar{z}_q)$$
$$= f(\bar{x}_1, \ldots, \bar{x}_q, \bar{x}_{q+1}, \ldots, \bar{x}_m) = a.$$

Thus, if we replace f by g (which is in the T-ideal generated by f), we may asume each $\bar{x}_k = e_{11}^{(k)}$, $1 \le k \le q$, in Definition 4.34.

This observation is useful in enabling us to keep the notation more manageable in the proof of Kemer's lemmas.

4.4   Kemer's Lemmas

Our goal is to produce non-identities that are alternating in several folds, our main tool being the t-alternator. Fortunately, there is one case when the t-alternator clearly is not an identity, namely when we have a t-compatible set of matrix units as defined in Remark 1.42. Since we shall rely heavily on this idea, let us generalize the notion.

Definition 4.41. A set of elements $u_1, \ldots, u_t; v_1, \ldots, v_m$ is *compatible with* with a polynomial

$$h(x_1, \ldots, x_t; y_1, \ldots, y_m)$$

if $h(u_1, \ldots, u_t; v_1, \ldots, v_m) \ne 0$ but

$$h(u_{\pi(1)}, \ldots, u_{\pi(t)}; v_1, \ldots, v_m) = 0, \quad \forall \pi \ne (1).$$

More generally, a polynomial $h(X_1, \ldots, X_\mu; Y)$ in μ sets of variables $X_j = \{x_{1,j}, \ldots, x_{t,j}\}$, $1 \le j \le \mu$ and in $y_1, \ldots, y_m$ is *μ-fold t-compatible* in A if there is a set of semisimple and radical substitutions $\{u_{i,j} : 1 \le i \le t, 1 \le j \le \mu\}$, $\{v_1, \ldots, v_m\}$, such that

$$h(u_{1,1}, \ldots u_{t,1}, \ldots, u_{1,\mu}, \ldots, u_{t,\mu}, v_1, \ldots, v_m) \ne 0,$$

but the specialization

$$h(u_{\pi_1(1),1}, \ldots, u_{\pi_1(t),1}, \ldots, u_{\pi_\mu(1),\mu}, \ldots, u_{\pi_\mu(t),\mu}; v_1, \ldots, v_m)$$

is 0 whenever some $\pi_j \ne (1)$. We call $\{u_{ij}\}$ and $\{v_1, \ldots, v_m\}$ a *μ-fold t-compatible substitution*; the $\{u_{ij}\}$ occupy *alternating positions*.

The point is that (in the above notation) if h has a μ-fold t-compatible substitution, then

$$\sum_{\pi_1 \in S_t} \sum_{\pi_2 \in S_t} \cdots \sum_{\pi_\mu \in S_t} h(x_{\pi_1(1),1}, \ldots, x_{\pi_1(t),1}, \ldots, x_{\pi_\mu(1),t}, \ldots, x_{\pi_\mu(t),t}; y_1, \ldots, y_m)$$

is a μ-fold t-alternating polynomial that is *not* an identity of A, since it has precisely one nonzero summand. (Note that this argument is characteristic free.)

Example 4.42. The special case we encountered in Remark 1.42 was for $h = x_1 y_1 \ldots x_t y_t$.

Given a f.d. full subdirectly irreducible algebra A, let

$$u_{j,1}, \ldots, u_{j,t_j}; v_{j,1}, \ldots, v_{j,t_j} \in R_j$$

be as in Remark 1.42, i.e., compatible with $x_1 y_1 \ldots x_{t_j} y_{t_j}$, where $t_j = n_j^2$, and whose product is e_{11} in R_j. Taking type $(j, j+1)$ radical substitutions r_j, we claim

$$u_{1,1}, \ldots, u_{1,t_1}, r_1, u_{2,1}, \ldots, u_{2,t_2}, r_2, \ldots, r_{q-1}, u_{q,1}, \ldots, u_{q,t_q}; \tag{4.6}$$
$$v_{1,1}, \ldots, v_{1,t_1}, \ldots, v_{q,1}, \ldots, v_{q,t_q}$$

is a 1-fold $(\dim_F \bar{A} + q - 1)$-compatible substitution (when the r_i are chosen to make it nonzero). Thus, A has a 1-fold $(\dim_F \bar{A} + q - 1)$-alternating polynomial that is not an identity. (Here the r_k are counted among the alternating positions.)

This is seen most readily from the graph-theoretic point of view of Remark 4.37. Namely, all substitutions in the k stage must be clustered between r_{k-1} and r_k for each k, so the only nonzero substitutions must have the r_k appearing in their original positions. But then the compatibility can be checked at each stage, which was done in the easy Remark 1.42.

Note that in order to achieve a nonzero specialization, we had to take radical substitutions among the alternating positions. This is clearly the case whenever we need more than $\dim_F \bar{A}$ alternating positions.

Remark 4.43. For any f.d. algebra A, any s_A-fold $(t_A + 1)$-alternating polynomial $f(X_1, \ldots, X_{s_A}; Y)$ is an identity of A. Indeed, by Example 4.41, any nonzero specialization of f must involve at least one radical substitution in each X_j, so has at least s_A radical substitutions in all. But then every monomial must have elements of J appearing in at least s_A positions, so belongs to $J^{s_A} = 0$, contradiction. (Compare with Example 4.42.) Thus, $\beta(A) \leq t_A$.

Proposition 4.44 (Kemer's First Lemma). *If F is algebraically closed and A is f.d., full subdirectly irreducible, then $\beta(A) = t_A$.*

Proof. Let $t = t_A$. $\beta(A) \le t$, by Remark 4.43. It remains to show that for any s we can find an s-fold t-alternating polynomial that is not an identity of A. We do this by finding an s-fold t-compatible substitution for a suitable monomial $h(X_1, \ldots, X_s; Y; Z)$, where each $|X_j| = t$, $|Y| = st$, and $|Z| = t - 1$. (We introduced a new set of variables Z that are not involved in the alternation so could be included in the set Y, but they play a special role that we want to specify for further use.)

The idea of the proof is to build folds around idempotent substitutions. Using Remark 4.40, we may choose our polynomial f so that we can multiply the various $\bar{x}_k$ by the idempotent $e_{11}^{(k)}$ and still get the same value. Our objective will be to write this $e_{11}^{(k)}$ itself as an evaluation of a Capelli polynomial, thus permitting us to add folds to f and still have a nonidentity.

For example, recall from Remark 1.42 that

$$c_4(e_{11}, e_{12}, e_{21}, e_{22}; e_{11}, e_{22}, e_{12}, e_{21}) = e_{11},$$

implying $c_4 \notin \mathrm{id}(M_2(\mathbb{Q}))$. But then we see the polynomial

$$c_4(x_1 c_4(x_5, x_6, x_7, x_8; y_5, y_6, y_7, y_8), x_2, x_3, x_4; y_1, y_2, y_3, y_4) \notin \mathrm{id}(M_2(\mathbb{Q}),$$

and we could continue this ad infinitum. The general case is more complicated since we may need to interweave substitutions in several different simple components $R_1, \ldots, R_q$ of $\bar{A}$, but the idea always remains the same.

Take the substitution of Example 4.42. By definition,

$$u_{1,1} v_{1,1} \cdots u_{1,t_1} v_{1,t_1} = e_{11}^{(1)} \in R_1;$$

since $e_{11}^{(1)}$ is idempotent, we could take s copies of these and label them with j, for $1 \le j \le s$, i.e.,

$$u_{1,1,j} v_{1,1,j} \cdots u_{1,t_1,j} v_{1,t_1,j} = e_{11}^{(1)}$$

for each j, so

$$u_{1,1,1} v_{1,1,1} \cdots u_{1,t_1,1} v_{1,t_1,1} \cdots u_{1,1,s} v_{1,1,s} \cdots u_{1,t_1,s} v_{1,t_1,s} = e_{11}^{(1)}.$$

From the graph-theoretic point view of Remark 4.37, all substitutions into the first stage R_1 still must occur together to achieve a nonzero value. This happens for each k; all $u_{k,i,j}$ appear between r_k and r_{k+1}. By the same argument as Example 4.42, fixing j, any non-identity permutation of the $u_{k,i,j}$ would yield value 0.

To write the monomial h explicitly, recall $\bar{A} = \oplus_{k=1}^q M_{n_k}(F)$ and let $t_k = n_k^2$; write

$$X_j = \{x_{k,i,j} : 1 \le k \le q, \quad 1 \le i \le t_k\}.$$

Thus,

$$|X_j| = \sum_{k=1}^q t_k = \sum_{k=1}^q n_k^2 = \dim_F \bar{A} = t.$$

Write

$$Y = \{y_{k,i,j} : 1 \le k \le q, \ 1 \le i \le t_k, \ 1 \le j \le s\}; \quad Z = \{z_k : 1 \le k \le q-1\},$$

and

$$h_k = x_{k,1,1} y_{k,1,1} \cdots x_{k,t_k,1} y_{k,t_k,1} \cdots x_{k,1,s} y_{k,1,s} \cdots x_{k,t_k,s} y_{k,t_k,s}, \quad 1 \le k \le q;$$
$$h = h_1 z_1 h_2 z_2 \ldots h_{q-1} z_{q-1} h_q.$$
$$(4.7)$$

Specializing $x_{k,i,j} \mapsto u_{k,i,j}$, $y_{k,i,j} \mapsto v_{k,i,j}$, $z_k \mapsto r_k$ yields the desired s-fold t-compatible substitution for the sets of variables $X_1, \ldots, X_s; Y \cup Z$. $\quad\square$

Remark 4.45. The above proof shows that we can tack on an extra fold of t-alternating variables to any non-identity of A, whenever $\beta(A) \ge t$.

Note that we had to retreat from $\dim_F \bar{A} + q - 1$ (in Example 4.42) to $\dim_F \bar{A}$, in order to avoid radical substitutions for the X_j. The radical substitutions come into play as we now describe intrinsically the second invariant s_A. For this we have to refine our previous analysis—fullness requires $q \le s$ in the above notation, but does not give us enough information when $s > q$. Towards this end, we need a few more definitions.

Definition 4.46. A f.d. algebra A is *PI-minimal* if the ordered pair (t_A, s_A) is minimal (lexicographically) among all f.d. algebras PI-equivalent to A. In other words, if $A' \sim_{\text{PI}} A$, and A' is f.d. with ordered pair (t', s'), then either $t' > t_A$, or $t' = t_A$ with $s' \ge s_A$.

For example, $M_n(F)$ is PI-minimal among algebras of PI-class n.

Example 4.47. Suppose $q = 1$ and $J(A)$ has nilpotence index 2. Then $A \approx M_n(H)$, with $H/J = F$. But since $H = J + F$, the fact $J^2 = 0$ implies H is commutative! Hence A is PI-equivalent to $M_n(F)$, so cannot be PI-minimal; the easiest nontrivial case is disposed of via a technicality!

Definition 4.48. A *PI-basic* algebra is a PI-minimal, full, subdirectly irreducible f.d. algebra.

The PI-basic algebras are the building blocks of our theory, as indicated by our next observation.

Proposition 4.49. *Any f.d. algebra A is PI-equivalent to the finite direct product of PI-basic algebras.*

Proof. Induction on the ordered pair (t_A, s_A). Writing A as a subdirect product of f.d. subdirectly irreducible algebras A_i, we have $A \sim_{\text{PI}} \prod A_i$, and we may assume each A_i is PI-minimal. If each A_i is full, we are done. If some A_i is not full, we use Proposition 4.38 to reduce the number of simple components (without increasing the dimensions) and thus reduce t_{A_i}, so we conclude by induction on t_A. $\qquad\qquad\square$

There is a very useful construction of relatively free algebras of PI-minimal algebras.

Example 4.50. The "generic" algebra of a PI-minimal f.d. algebra A. Take the free product $A' = \bar{A} * F\{x_1, \ldots, x_\nu\}$, where ν is some large number, at least $\dim J(A) = \dim A - \dim \bar{A}$, and for any $u > 1$, define

$$\hat{A}_u = A'/(I_1 + I_2^u),$$

where I_1 is the ideal generated by all evaluations in A' of polynomials in $\text{id}(A)$, and I_2 is the ideal generated by $x_1, \ldots, x_\nu$. In other words, we have the "freest algebra" containing $\bar{A}$ and satisfying the identities from $\text{id}(A)$, such that the radical J has nilpotence index u. Of course, $J(\hat{A}_u)$ is the image of $\langle x_1, \ldots, x_\nu \rangle$, since modding this out yields the semisimple algebra $\bar{A}$. Furthermore, letting $\bar{x}_i$ denote the image of x_i, and taking a base $\{b_1, \ldots, b_k\}$ of $\bar{A}$, we see that $\hat{A}_u$ is spanned by all terms of the form

$$b_{i_1} \bar{x}_{j_1} b_{i_2} \bar{x}_{j_2} \ldots \bar{x}_{j_{u-1}} b_{i_u}, \tag{4.8}$$

and thus is f.d. over F, of dimension $\leq k^u (\nu)^{u-1}$.

By definition, $\text{id}(A) \subseteq \text{id}(\hat{A}_u)$. On the other hand, one can clearly find surjections $\hat{A}_{s_A} \to A$ by means of the Wedderburn decomposition of A (sending the "generic nilpotent" elements to a base of $J(A)$). Thus, $\hat{A}_{s_A} \sim_{\text{PI}} A$.

We turn to the second invariant, $\gamma(A)$.

Remark 4.51. In Remark 4.43, we saw that $\gamma(A) \leq s_A$, for any f.d. algebra A.

To determine when $\gamma(A) = s_A$, we need to refine our argument of Kemer's First Lemma. One of the key steps is a technical-sounding result, which enables us both to calculate $\gamma(A)$ and also get a good hold on Kemer polynomials. We start by saying a multilinear polynomial $f(x_1, \ldots, x_m) \notin \mathrm{id}(A)$ has *Property K* if f vanishes under any specialization $\bar{x}_1, \ldots, \bar{x}_m$ with *fewer* than $s_A - 1$ radical substitutions.

This is anti-intuitional, since we know f must vanish under any specialization $\bar{x}_1, \ldots, \bar{x}_m$ with $\geq s_A$ radical substitutions, because $J^s = 0$. Since $f \notin \mathrm{id}(A)$, one concludes that f must not vanish with respect to some specialization with precisely $s_A - 1$ radical substitutions.

Lemma 4.52. *Any PI-minimal f.d. algebra A has a polynomial satisfying Property K.*

Proof. By induction on $s = s_A$. Suppose Property K always fails. As we just saw, this means for any polynomial f, if f vanishes on $s-1$ radical substitutions, then $f \in \mathrm{id}(A)$. We prove the lemma by building a f.d. algebra $\hat{A}$ of nilpotence index $s - 1$ such that $\mathrm{id}(A) = \mathrm{id}(\hat{A})$.

Indeed, take $\hat{A}_{s-1}$ as defined in Example 4.50. In other words, we have the "freest algebra" containing $\bar{A}$ and satisfying the identities from $\mathrm{id}(A)$, such that the radical J has nilpotence index $s - 1$. By definition, $\mathrm{id}(A) \subseteq \mathrm{id}(\hat{A})$. Conversely, if $f \in \mathrm{id}(\hat{A})$, then f vanishes whenever there are $s - 1$ radical substitutions, which implies by hypothesis $f \in \mathrm{id}(A)$. Hence, $\hat{A} \sim_{\mathrm{PI}} A$, contrary to the PI-minimality of A. $\square$

Lemma 4.53. *Suppose $\mathrm{char}(F) = 0$, and a given multilinear polynomial $f(x_1, \ldots, x_m) \notin \mathrm{id}(A)$ satisfies the following two properties with respect to substitutions $\bar{x}_1, \ldots, \bar{x}_m$ which are either semisimple or radical substitutions, notation as in Section 4.1.1:*

(i) f satisfies Property K;

(ii) $f(\bar{x}_1, \ldots, \bar{x}_m) = 0$ whenever there is some component R_k (in the notation of §4.1.1) for which $R_k \bar{x}_i = 0 = \bar{x}_i R_k$ for all i.

In this case, the T-ideal generated by f contains a μ-Kemer polynomial for A, for arbitrarily large μ.

Before proving the lemma, let us note that Condition (ii) is a variation of the notion of "full." It implies that A must be full with respect to f.

Proof. Let $s = s_A$. We know $\gamma(A) \leq s$, so to prove the first assertion it suffices to find an $(s - 1)$-fold $(t+1)$-alternating polynomial

$$g(X_1, \ldots, X_s; Y) \notin \mathrm{id}(A).$$

As in the proof of Kemer's First Lemma, we want to achieve this by finding an $(s-1)$-fold $(t+1)$-compatible substitution for a suitable polynomial. (In fact, we want to add another μ folds that are t-compatible, in order to obtain a μ-Kemer polynomial.)

We rewrite f as $f(z_1, \ldots, z_{s-1}, y_s, \ldots, y_m)$, in order to pinpoint the variables $z_1, \ldots, z_{s-1}$ that are to have the radical substitutions. Say

$$f(r_1, \ldots, r_{s-1}; b_s, \ldots, b_m) \neq 0 \qquad (4.9)$$

where $r_k \in e_{\ell_k} J e_{\ell'_k}$ for $1 \leq k \leq s - 1$.

For technical reasons to become clearer, we want $\ell_k \neq \ell_{k'}$ for each k. Thus, in case $\ell_k = \ell_{k'}$, we pick $\ell \neq \ell_k$ and rewrite $r_k \approx (r_k e_{\ell_k, \ell}) e_{\ell, \ell_k}$.

As in the proof of Kemer's First Lemma, we shall specialize the z_k to more complicated expressions involving Capelli polynomials, in order to create new folds. However, to do this, we need to multilinearize f, cf. Proposition 1.18. (This is the only place in this proof that we require characteristic 0, and one can get around it, with difficulty.)

Recall that $\bar{A} = R_1 \oplus \cdots \oplus R_q$, where $R_k = M_{n_k}(F)$.

Case 1. $q > 1$. As in the proof of Kemer's First Lemma, since A is full and subdirectly irreducible, the graph touches each stage (cf. Remark 4.37). Thus, $q \leq s - 1$. Reordering the indices, we may assume

$$r_k \in e_k A e_{k+1}, \quad 1 \leq k < q. \qquad (4.10)$$

Note that we may have $q < s - 1$, in which case there are "extra" r_k. This is what makes the proof more complicated than for Kemer's First Lemma.

For $k < q - 1$, we then have $r_k = e_{\ell\ell}^{(k)} r_k$, for a suitable matrix unit $e_{\ell\ell}^{(k)}$ in R_k; we may assume $\ell = 1$.

Likewise, for $k = q - 1$, we may write $r_{q-1} = r_{q-1} e_{11}^{(q)}$.

Thus, we have pasted appearances of $e_{11}^{(k)}$ in each monomial of f, for each $1 \leq k \leq q$. Let us write $e_{11}^{(k)}$ as a value of a suitable Capelli polynomial $c_{n_k^2}$, say

$$e_{11}^{(k)} = c_{n_k^2}(u_{k,1}, \ldots, u_{k,n_k^2}; v_{k,1}, \ldots, v_{k,n_k^2}),$$

under a suitable compatible substitution $u_{k,i}, v_{k_i}, 1 \leq i \leq n_k^2$ of matrix units in R_k, cf. Remark 1.42. These are all semisimple substitutions, by definition.

Replacing z_k ($1 \leq k \leq q - 1$) throughout by the monomials

$$y_{k,0}(x_{k,1,1} y_{k,1} \cdots x_{k,n_k^2,1} y_{k,n_k^2}) \cdots (x_{k,1,s+1} y_{k,1} \cdots x_{k,n_k^2,s+\mu-1} y_{k,n_k^2}) z_k,$$

and then once again replacing z_{q-1} by

$$z_{q-1}(x_{q,1,1} y_{q,1} \cdots x_{q,n_q^2,1} y_{q,n_q^2}) \cdots (x_{q,1,s+1} y_{q,1} \cdots x_{q,n_q^2,s+\mu-1} y_{q,n_q^2}) y_{q,0},$$

yields a new polynomial f', clearly in the T-ideal generated by f, for which the substitutions

$$x_{k,i,j} \mapsto u_{k,i}, \quad y_{k,i} \mapsto v_{k,i}, \quad 1 \leq i \leq n_k^2, \quad y_{k,0}, y_{k,n_k^2+1} \mapsto e_{11}^{(k)}$$

yield the original (nonzero) value of f, but any other substitution that sends $x_{k,i,j} \mapsto u_{k',i'}$ for different k' or i' (and still sending $y_{k,i} \mapsto v_{k,i}$) is 0. In other words, f' is $(s + \mu - 1)$-fold t-compatible.

Thus, for each $1 \leq k \leq q - 1$, replacing each z_k in f by

$$y_{k_0} c_{n_k^2}(x_{k,1,1}, \ldots, x_{k,n_k^2,1}; y_{k,1}, \ldots, y_{k,n_k^2}) \cdots$$
$$\cdots c_{n_k^2}(x_{k,1,s+\mu-1}, \ldots, x_{k,n_k^2,s+\mu-1}; y_{k,1}, \ldots, y_{k,n_k^2}) z_k$$

and then replacing z_{q-1} by

$$z_{q-1} c_{n_q^2}(x_{q,1}, \ldots, x_{q,n_q^2,1}; y_{q,1}, \ldots, y_{q,n_q^2}) \cdots$$
$$\cdots c_{n_q^2}(x_{q,1,s+u-1}, \ldots, x_{q,n_q^2,s+(u-1)}; y_{q_1,1}, \ldots, y_{q,n_q^2}) y_{k_0}$$

yields a new polynomial f'', which is $(s + \mu - 1)$-fold t-alternating. (Note that we tacked onto z_{q-1} on both sides.)

To complete the proof, we need to increase $|X_j|$ by 1, for $1 \leq j \leq s - 1$. Towards this end, we reassign z_j to X_j, now putting

$$X_j = \{z_j, x_{k,i,j} : 1 \leq k \leq q, \quad 1 \leq i \leq n_k^2\}$$

for $1 \leq j \leq s - 1$.

To prove compatibility, we need to show that in our evaluation of f', any permutation of the r_j with the $u_{k,i,j}$ produces 0. Since we have already seen this with permutations fixing r_j, it remains to consider the situation where we switch r_j with some $u_{k,i,j}$. But in this case, we get 0 unless $r_j \in e_k A e_k$ (because it is now bordered on either side by semisimple substitutions from R_k). In particular, $j \geq q$ in view of (4.10). The effect in our original evaluation of f in (4.9) has been to incorporate r_j into the substitution for z_k (which now is thus in J^2), and turn the substitution of z_j into $u_{k,i,j}$, a semisimple substitution. This switch has lowered the number of radical substitutions to $< s - 1$, so by hypothesis on f, the corresponding specialization of f must be 0. But the new specialization of f' (under switching r_j with $u_{k,i,j}$) has the same value as this new specialization of f, i.e., it also is 0, as desired.

Case 2. $q = 1$. This is conceptually easier, but less convenient, than the previous case. Now we do not have any place to tag on the extra substitutions; we have no r_k for $k \leq q - 1$ because $q - 1 = 0$. On the other hand, the structure of A is much nicer: $\bar{A} = M_n(F)$ where $n^2 = t = \dim_F \bar{A}$, so we can lift the matrix units and get $A = M_n(H)$ for some ring H for which $F = H/J(H)$. (This is not strictly relevant to the proof, but is included to provide intuition and keep the notation as easy as possible.) So suppose $f(r_1, \ldots, r_{s-1}, b_1, \ldots, b_m) = ce_{ij}$ for some $c \in H$. Reordering the indices, we may assume $e_{ij} = e_{1j}$, and again we take $s + \mu - 1$ t-compatible substitutions of matrix units for c_k, for which

$$c_k(u_{1,j}, \ldots, u_{t,j}; v_1, \ldots, v_t) = e_{11}.$$

Take

$$f' = x_{1,1}y_s \ldots x_{t,1}y_t \ldots x_{1,s+\mu-1}y_s \ldots x_{t,s+\mu-1}y_t f.$$

Clearly, this is $(s + \mu - 1)$-fold t-compatible, and as before, we claim that we may enlarge the first $s - 1$ sets of variables, i.e., put

$$X_j = \{x_{1,j}, \ldots, x_{1,j}, z_j\}$$

for $1 \le j \le s - 1$. Again, we need to show that interchanging r_j and some $u_{i,j}$ gives 0. But this is for the same reason: The radical substitution for z_j has been changed to a semisimple substitution, so the specialization of f now has fewer than $s - 1$ radical substitutions, so is 0, and this is also the specialization of f', as desired. $\square$

We are ready for a key result.

Proposition 4.54 (Kemer's Second Lemma). *Suppose* char$(F) = 0$. *If A is a PI-basic algebra, then $\gamma(A) = s_A$. In this case, A has a μ-Kemer polynomial for any μ. In fact, for any f satisfying property K, the T-ideal generated by f contains a μ-Kemer polynomial for A.*

Proof. Let $t = \beta(A) = t_A$ by Kemer's First Lemma. However here, instead of obtaining a suitable monomial, we utilize Lemma 4.52 to find a polynomial that satisfies Property K, and thus are able to conclude by Lemma 4.53. $\square$

Remark 4.55. Let us summarize Kemer's First and Second Lemmas, together with some related properties of Kemer polynomials. Suppose F is algebraically closed field of characteristic 0, and A is a PI-basic f.d. PI-algebra (with $\bar{A} = A/J(A)$).

(i) $\beta(A) = t = \dim_F \bar{A}$, and $\gamma(A) = s$, the nilpotence index of $J(A)$.

(ii) For any μ there is a μ-Kemer polynomial that is $(s-1)$-fold $(t+1)$-alternating and has μ extra t-alternating folds (but is not in id(A) by definition). We also recall that by definition of $\gamma(A)$, any s-fold $(t+1)$-alternating polynomial is in id(A).

(iii) Any nonzero specialization of a μ-Kemer polynomial f has precisely t semisimple substitutions in each of the first $s - 1$ folds. (Indeed, if there were $t + 1$ semisimple substitutions in some fold, then that fold would yield 0. Hence each fold has at least one radical substitution, and if any fold had two radical substitutions, the evaluations would be in $J^s = 0$. We conclude that each of the first $s - 1$ folds has precisely one radical substitution, and thus has t semisimple substitutions, and we have a total of $s - 1$ radical substitutions.

(iv) Suppose f is any non-identity of A which is in the T-ideal generated by a Kemer polynomial. Then (iii) implies f has Property K, and thus, by Lemma 4.53, its T-ideal contains a μ-Kemer polynomial, for any given μ. In other words, starting with a Kemer polynomial of A, any consequence $f \notin \mathrm{id}(A)$ itself has a consequence that is a μ-Kemer polynomial. This important observation, which one could call the *phoenix property*, enables us to recover new Kemer polynomials from the consequences of Kemer polynomials, and is essential to all proofs we know of Kemer's theorem, since it bypasses the difficulty raised by Remark 4.23.

(v) Applying (iii) to Theorem J, we see that if f is any Kemer polynomial for A, and α is a characteristic coefficient of an element of $A/J(A)$, then $\alpha f(A) \subseteq f(A)$.

(vi) If S is a T-ideal generated by a set of Kemer polynomials for A, then letting $I = \{f(A) : f \in S\}$, we have $\alpha I \subseteq I$ for any characteristic coefficient α of an element of $A/J(A)$. (This is seen by applying (v) to each substitution in a Kemer polynomial.)

Example 4.56. Conversely to Example 4.20, suppose A is a PI-basic algebra of Kemer index $(t, 1)$. By Kemer's First and Second Lemmas, $J = 0$ and $t = \dim \bar{A}_i$. But the only way for a t-alternating polynomial to be a non-identity of a semisimple algebra of dimension t is for the algebra to be simple. In particular, A has PI-class n, where $n^2 = t$, so any f.d. algebra of the same Kemer index as A also must have PI-class n.

Let us formulate the situation for f.d. algebras which need not be subdirectly irreducible.

Remark 4.57. Suppose $A = A_1 \times \cdots \times A_m$, where all A_i have the same Kemer index. A polynomial f is Kemer for A iff f is not an identity (and thus is Kemer) for some A_i.

Proposition 4.58. *Any f.d. algebra A is PI-equivalent to a direct product of f.d. algebras*

$$\tilde{A} \times A_1 \times \cdots \times A_m,$$

where $A_1, \ldots, A_m$ are PI-basic, and where $\tilde{A}$ is a direct product of PI-basic algebras $\tilde{A}_1, \ldots, \tilde{A}_u$, such that

$$\mathrm{index}(\tilde{A}_1) = \mathrm{index}(\tilde{A}_2) = \cdots = \mathrm{index}(\tilde{A}_u) > \mathrm{index}(A_1) \geq \mathrm{index}(A_2)$$
$$\geq \cdots \geq \mathrm{index}(A_m).$$

Furthermore, there exists a polynomial f which is Kemer for each of $\tilde{A}_1, \ldots, \tilde{A}_u$ but is an identity for $A_1 \times \cdots \times A_t$.

Proof. For each $j \leq m$, let f_j denote a Kemer polynomial of $\tilde{A}_j$ that is an identity of $A_1 \times \cdots \times A_m$, cf. Proposition 4.25. But (matching the indeterminates in the folds) and taking

$$f = f_1(x_1, \ldots, x_n)y_1 + f_2(x_1, \ldots, x_n)y_2 + \cdots + f_u(x_1, \ldots, x_n)y_u,$$

we see f is Kemer for each of $\tilde{A}_1, \ldots, \tilde{A}_u$, and an identity of $A_1 \times \cdots \times A_m$; specializing $y_j \mapsto 1$ and all other $y_i \mapsto 0$ shows $f \notin \mathrm{id}(\tilde{A}_j)$. □

4.4.1   Manufacturing representable algebras

To utilize Proposition 4.58, we bring in Shirshov's program for a key way to generate new representable algebras.

Proposition 4.59. *Suppose $\mathcal{A}$ is the relatively free algebra of a PI-basic algebra A over an algebraically closed field of characteristic 0, and let $\mathcal{S}$ be a set of Kemer polynomials for A. Let I denote the ideal generated by all $\{f(\mathcal{A}) : f \in \mathcal{S}\}$. Then $\mathcal{A}/I$ is representable.*

Proof. Consider the relatively free algebra $\mathcal{A}$ as in Example 3.26. Since $\mathcal{A} \subseteq A[\Lambda]$ for $\Lambda = \{\lambda_1, \ldots, \lambda_t\}$, we can extend the trace linearly to $\mathcal{A}$, to take on values in $F[\Lambda] \subset F(\Lambda)$. As in Definition 2.37, we form the trace ring: Letting $\hat{F}$ denote the commutative affine subalgebra of $F[\Lambda]$ consisting of all the traces of powers (up to $\beta(A)$) of products (of length up to the Shirshov height) of powers of the generators of $\mathcal{A}$, we define $\hat{\mathcal{A}} = \mathcal{A}\hat{F} \subseteq A[\Lambda]$.

By Proposition 2.38, $\hat{\mathcal{A}}$ is a f.g. module over $\hat{F}$, and in particular is a Noetherian algebra. In the notation of Remark 4.55(vi), I is also closed under multiplication by elements of $\hat{F}$, so in fact, I is a common ideal of $\mathcal{A}$ and $\hat{\mathcal{A}}$. Thus, $\mathcal{A}/I \subseteq \hat{\mathcal{A}}/I$, which is f.g. over the image of the affine algebra $\hat{F}$, so $\mathcal{A}/I$ is representable, by Proposition 1.92. □

Corollary 4.60. *Notation as in Proposition 4.58, let $\kappa = \operatorname{index}(\tilde{A})$, and let*

$$\mathcal{S} = \{\mu\text{-Kemer polynomials of } \tilde{A} \text{ which are identities of } A_1 \times \cdots \times A_m\}.$$

Also let U_i denote the relatively free algebra of $\tilde{A}_i$, for $1 \leq i \leq u$. Then $\operatorname{index}(U_i/\mathcal{S}(U_i)) < \kappa$.

Proof. Otherwise, some of the $\tilde{A}_j/\mathcal{S}(\tilde{A}_j)$ have Kemer index κ. For these j, there is some μ-Kemer polynomial f which is an identity of all the other factors. But obviously $f \notin \operatorname{id}(A_j)$, implying f is a Kemer polynomial of $\tilde{A}$, and thus, $f(\tilde{A}) \subseteq \mathcal{S}(\tilde{A})$, implying $f \in \operatorname{id}(\tilde{A}/\mathcal{S}(\tilde{A}))$, contrary to choice of f.   □

The point is that in some sense, the algebras $U_i/\mathcal{S}(U_i)$ can replace the $\tilde{A}_i$ and lower the Kemer index of A while preserving some of its properties. We need a more technical version of this argument, which contains our main idea. (In the following proposition, we remind the reader that $\operatorname{index}(\Gamma)$ is defined independently of A.)

4.4.2   Matching Kemer indices

Half of the proof of Kemer's PI-Representability Theorem is to find a f.d. algebra having the same Kemer index as a T-ideal Γ, such that $\Gamma \supseteq \operatorname{id}(A)$.

Proposition 4.61. *Suppose Γ is a T-ideal properly containing $\operatorname{id}(A)$ for some f.d. algebra A. Then there is a f.d. algebra A' such that $\Gamma \supseteq \operatorname{id}(A')$ but $\operatorname{index}(A') = \operatorname{index}(\Gamma)$. Furthermore, if A' is taken to be a direct product of PI-basic algebras $\prod A_i$, then for each component A'_i of highest Kemer index, and for large enough μ (to be specified in the proof), no μ-Kemer polynomial of A'_i is in Γ.*

Proof. We may assume A is a direct product $A_1 \times \cdots \times A_m$ of PI-basic f.d. algebras. Let $\kappa = \operatorname{index}(\Gamma)$ and $\kappa_0 = \operatorname{index}(A) = \max \operatorname{index}(A_i)$. Clearly, $\kappa_0 \geq \kappa$. We are done if $\Gamma = \operatorname{id}(A)$, so we may assume $\Gamma \not\subseteq \operatorname{id}(A_i)$ for some i; pick such j with $\operatorname{index}(A_j)$ maximal such. There may be several such j, say $A_{j_1}, \ldots, A_{j_u}$, all of this Kemer index, which we call κ_1. However, by Lemma 4.52, we have $f_i \notin \operatorname{id}(A_{j_i})$ which satisfies Property K, and thus can be made into a Kemer polynomial of A_{j_i}, by Kemer's Second Lemma. To ease notation, we reorder the A_j, putting $A_1, \ldots, A_u$ for these algebras of Kemer index κ_1. Picking $\mu \geq \omega(\Gamma)$, we reorder the A_i further, and assume we have $0 \leq u' \leq u$ such that Γ contains a μ-Kemer polynomial of A_i for each $i \leq u'$, and Γ does not contain a μ-Kemer polynomial for $u' < i \leq u$.

We need to reduce to $\kappa = \kappa_0 = \kappa_1$ and $u' = 0$. Our method will be a double induction, first on

$$\kappa_2 = \max\{\text{index}(A_i) : i \leq u'\}$$

and then on u'. The following observation will be very useful.

Note. If $\kappa_0 > \kappa$ then for any A_i having Kemer index κ_0, any μ-Kemer polynomial of A_i is in Γ (since $\mu \geq \omega(\Gamma)$), so in this case we have $\kappa_2 = \kappa_1 = \kappa_0$ and $u' = u$.

On the other hand, if $\kappa_1 < \kappa$ then $\kappa_0 > \kappa_1$, so by definition $\text{id}(A_i) \supseteq \Gamma$ for each A_i of Kemer index κ_0, implying (for $\mu > \max\{\omega(A_i)\}$) no polynomial of Γ can be a μ-Kemer polynomial of A. By the note, we could not have $\kappa_0 > \kappa$, so $\kappa_0 = \kappa$. In this case, $\text{id}(\tilde{A}_1) \supseteq \Gamma$, so $\text{index}(A) = \text{index}(\Gamma)$, and no μ-Kemer polynomial is an identity of $\tilde{A}$, by definition. Thus, we are done in this case.

So we may assume $\kappa_0 \geq \kappa_1 \geq \kappa$. In view of the note, $\kappa_0 = \kappa_1$, and we are done if $u' = 0$. We shall find a procedure to reduce u' in case $u' > 0$; then we are done by induction.

Recall $A_1, \ldots, A_u$ all have Kemer index κ_1. Let $A_{u+1}, \ldots, A_v$ be the algebras of Kemer index $< \kappa_1$ for which $\Gamma \not\subseteq \text{id}(A_i)$, and let $A_{v+1}, \ldots, A_m$ be the algebras for which $\Gamma \subseteq \text{id}(A_i)$. Putting

$$\tilde{A} = A_1 \times \cdots \times A_u,$$

let $\mathcal{S}$ be the set of polynomials from Γ that are μ-Kemer polynomials for $\tilde{A}$, where in view of Remark 4.45, μ is taken large enough such that

$$\mathcal{S} \subseteq \text{id}(A_{u+1} \times \cdots \times A_v).$$

Let U_i be the relatively free algebra of A_i for $1 \leq i \leq u$, and let I_i be the ideal of U_i generated by $\{f(U_i) : f \in \mathcal{S}\}$; let $\bar{U}_i = U_i/I_i$, and $\bar{U} = \bar{U}_1 \times \cdots \times \bar{U}_u$. Thus,

$$\text{id}(\bar{U}) = \mathcal{S} + \text{id}(\tilde{A}).$$

We claim

$$\left(\bigcap_{i>u} \text{id}(A_i)\right) \cap \text{id}(\bar{U}) \subseteq \Gamma. \tag{4.11}$$

Indeed, suppose

$$g \in \left(\bigcap_{i>u} \text{id}(A_i)\right) \cap \text{id}(\bar{U}).$$

Writing $g = f + h$ for $f \in \mathcal{S}$ and $h \in \mathrm{id}(\tilde{A})$, we see $f \in \mathrm{id}(A_i)$ for $i \geq u$ (for $u \leq i \leq v$ by choice of Kemer polynomial, and for $i > v$ since $\mathcal{S} \subseteq \Gamma$). Hence, $h = g - f \in \mathrm{id}(A_i)$ for $i \geq u$. But $h \in \mathrm{id}(A_i)$ for $1 \leq i \leq u$ by hypothesis; hence,

$$h \in \mathrm{id}(A_1 \times \cdots \times A_v) \subseteq \Gamma,$$

implying $g = f + h \in \Gamma$.

On the other hand, each $\bar{U}_i$ is representable, by Proposition 4.59, so $\bar{U}_i$ is PI-equivalent to a f.d. algebra A'_i and we would like to conclude the proof by taking

$$A' = A'_1 \dots A'_u \times A_{u+1} \times \cdots \times A_v.$$

We claim A' has lower u' than A. This is clear unless $\mathrm{index}(\bar{U}_i) = \kappa_2$ and Γ contains a Kemer polynomial of $\bar{U}_i$, for $1 \leq i \leq u'$. Thus, as in the proof of Proposition 4.58, Γ contains a Kemer polynomial f common to each of these $\bar{U}_i$, $1 \leq i \leq u'$, also with $f \in \mathrm{id}(\bar{U}_i)$ for all $u' < i \leq v$. Then f is also a Kemer polynomial for A_i for $1 \leq i \leq u'$, implying $f \in \mathcal{S}$ and thus, $f \in \mathrm{id}(\bar{U}_i)$, $1 \leq i \leq u'$, contrary to our choice of f.

Having established the claim, we have a contradiction to our assumption on κ_2 unless $u' = 0$, as desired.   □

We need to enhance this result with an easy observation.

Remark 4.62. If f is a μ-Kemer polynomial of $\Gamma + \mathrm{id}(A)$, then Γ contains a μ-Kemer polynomial of A. (Indeed, write $f = g + h$ where $g \in \Gamma$ and $h \in \mathrm{id}(A)$. Applying the alternator (cf. Definition 1.27) to each of the folds of f in turn yields

$$\tilde{f} = \tilde{g} + \tilde{h},$$

where $\tilde{f}$ is a scalar multiple of f, cf. Remark 1.28), and leaves $\tilde{g} \in \Gamma$, but $\tilde{h}$ is an identity of A that alternates in each fold. Thus, $\tilde{f} - \tilde{h} = \tilde{g}$ is a Kemer polynomial of A that belongs to Γ.

We have not finished with the idea of Proposition 4.59.

Definition 4.63. Given a PI-basic algebra A, consider the generic algebra $\hat{A}_s$, constructed as in Example 4.50, where $s = s_A$. $\hat{A}_s$ being finite dimensional, we can now take the idea of Example 3.26 and construct $\mathcal{A}_1$ to be the subalgebra of $\hat{A}_{s_A}[\Lambda]$ generated by $\mathcal{A}$ (the algebra generated by generic elements) and the $\bar{x}_i$; then $J(\mathcal{A}_1)$ contains the ideal J_0 generated by the $\bar{x}_i$, and $\mathcal{A}_1/J_0$ is isomorphic to $\mathcal{A}$.

We can define a trace on $\mathcal{A}_1$ by taking the given trace on $\mathcal{A}$, and defining the trace of every word including some $\bar{x}_i$ to be 0 (all elements of $J(\mathcal{A}_1)$ are nilpotent.) Adjoining all these traces to $\mathcal{A}_1$ gives a representable algebra

$\mathcal{A}_2$. We call $\mathcal{A}_1$ the *expediting algebra* of A, and $\mathcal{A}_2$ the *trace expediting algebra*, since they expedite our proof.

Having a f.d. algebra of the same Kemer index already has interesting consequences.

Proposition 4.64. *Any affine algebra W (over a field of arbitrary characteristic) satisfies one of the following two cases:*

(i) There exists a Kemer polynomial f such that $f(W)Wf(W) = 0$ (and thus, $\langle f(W) \rangle$ is nilpotent); or

(ii) index$(W) = (t, 1)$ *for some t. In this case, if* char$(F) = 0$, W *has some PI-class n, and in particular W has an n^2-alternating central polynomial $\hat{f}$ in the T-ideal generated by f, taking on non-nilpotent values.*

Proof. Let $\kappa = (t, s) = $ index(W). First assume $s > 1$. Then any Kemer polynomial $f(x_1, \ldots, x_m)$ is $(s-1)$-fold $(t+1)$-alternating, so any polynomial in

$$f(x_1, \ldots, x_m)F\{X\}f(x_{m+1}, \ldots, x_{2m})$$

has at least $2s - 2$ folds that are $(t+1)$-alternating, and by definition is in id(W) since $2s - 2 \geq s$. Thus, $f(W)Wf(W) = 0$, and we have (i).

Since one cannot have $s = 0$, we are left with $s = 1$. In this case, some t-alternating polynomial is a Kemer polynomial; since this is a consequence of the Capelli polynomial c_t, we see c_t itself must be a Kemer polynomial.

Let A be a PI-basic algebra of the same Kemer index $(t, 1)$. By Example 4.56, A is simple of some PI-class n. Thus, W has the same central polynomials as A, and the values are non-nilpotent (or else we pass back to (i)). $\qquad \square$

Remark 4.65. This result underlines the difference between the approach of this monograph and the approach of, say, [Row80], whose foundation was central polynomials. This is our first use of central polynomials. Note that index$(A) = (t, 0)$ implies c_{t+1} is an identity of A, and localizing A by a regular value of a t-alternating central polynomial yields an Azumaya algebra of rank t, cf. [Row80, Theorem 1.8.48]. So one might well say that Proposition 4.64 is where the East (Kemer's theory) meets the West (Artin-Procesi theorem). The conclusion of (ii) holds in all characteristics, but we have not developed the theory to present the proof.

4.4.3   Kemer's PI-Representability Theorem concluded

We are finally ready to prove Theorem 4.1, which we rephrase for convenience.

Theorem 4.66 (Kemer). *Suppose* $\mathrm{char}(F) = 0$. *Any* $\Gamma \lhd_T F\{x_1, \ldots, x_\ell\}$ *is the set of identities of a suitable f.d. algebra over a field.*

Proof. Let $\kappa = \mathrm{index}(\Gamma)$. $\Gamma \supseteq \mathrm{id}(A)$ for suitable f.d. A, cf. Corollary 4.9. By Proposition 4.61, we may assume $\mathrm{index}(A) = \kappa$, and furthermore, for μ large enough, Γ and A satisfy precisely the same μ-Kemer polynomials (since no μ-Kemer polynomial of one can be an identity of the other).

Let $\mathcal{S}$ be the T-ideal generated by the μ-Kemer polynomials of Γ, and let $\Gamma' = \Gamma + \mathcal{S}$. Writing $\kappa = (t, s)$, we see by definition that

$$\mathrm{index}(\Gamma') \le (t - 1, \mu) < \kappa,$$

so by induction there is some f.d. algebra A' with $\mathrm{id}(A') = \Gamma'$.

Our objective is to find some algebra A'' for which all elements of Γ are identities, but none of $\mathcal{S} \setminus \Gamma$ are identities. Once we have such an algebra, we note that

$$\Gamma = (\Gamma + \mathcal{S}) \cap \mathrm{id}(A'') = \mathrm{id}(A' \times A''),$$

as desired. (Indeed, if $f \in \mathrm{id}(A'')$ and $f = g + h$ for $g \in \Gamma$ and $h \in \mathcal{S}$, then $h = f - g \in \mathrm{id}(A'')$, implying by hypothesis that $h \in \Gamma$.)

To find A'', we need the expediting algebras defined in Definition 4.63. Write $A = A_1 \times \cdots \times A_m$ where $A_1, \ldots, A_u$ have Kemer index κ and the other A_i have lower Kemer index. Let $\mathcal{S}_i$ be the set of μ-Kemer polynomials of A_i; then $\cup \mathcal{S}_i = \mathcal{S}$.

For each $i \le u$, let U_i denote the expediting algebra of A_i, and let $\hat{U}_i$ denote the trace expediting algebra of A_i. In constructing U_i, recall that we have a set of nilpotent variables $\bar{x}_j : 1 \le j \le \nu$; we take this set large enough to include all of the designated variables of a μ-Kemer polynomial. (Thus, $\nu \ge (t + 1)s + t\mu$.) Let $\bar{\mathcal{S}}_i$ be the set of specializations of $\mathcal{S}_i$ to U_i, for which each designated variable x_i is sent to $\bar{x}_i$.

Claim. $\Gamma(\hat{U}_i) \cap \bar{\mathcal{S}}_i = 0$. Indeed, otherwise $\Gamma \cap \mathcal{S}_i$ contains some $f \notin \mathrm{id}(A)$ (seen by evaluating on the $\bar{x}_i$); in view of the phoenix property of Remark 4.55(iv), we may assume f is a Kemer polynomial, with the folds specialized to the $\bar{x}_i$. Suppose

$$f = \sum_j \mathrm{tr}(v_j) w_j,$$

where $v_j \in U_i$ and $w_j \in \Gamma$. We may assume no $\bar{x}_i$ appears in v_j, since otherwise $\mathrm{tr}(v_j) = 0$. But then all the $\bar{x}_1, \ldots, \bar{x}_\nu$ appearing in f must appear in each w_j. Applying the alternator (for each fold) to f yields a multiple of f, but on the right side we get a sum of alternators, each of which is 0—indeed, the alternator of w_j if nonzero would be a μ-Kemer polynomial for A_i in Γ, which is impossible by Proposition 4.61.

Let $\mathcal{A}'_i = \hat{U}_i/\Gamma(\hat{U}_i)$. Thus, $\mathrm{id}(\mathcal{A}'_i) = \Gamma + \mathrm{id}(A_i)$. The claim shows that the Kemer polynomials $\mathcal{S}_i$ remain nonidentities of $\mathcal{A}'_i$. By Proposition 4.61 again, Γ does not contain a Kemer polynomial of A_i, so, by Remark 4.62, $\mathrm{id}(\mathcal{A}'_i)$ does not contain a Kemer polynomial of A_i. In view of the phoenix property, no polynomial in $\mathcal{S}_i$ is an identity of $\mathcal{A}'_i$, and consequently no polynomial in $\mathcal{S}$ is an identity of $\mathcal{A}'_1 \times \cdots \times \mathcal{A}'_u$. The canonical map $U_i \to \hat{U}_i/\Gamma(\hat{U}_i)$ has kernel $U_i \cap \Gamma(\hat{U}_i)$, which is disjoint from $\bar{\mathcal{S}}_i$. Thus, we have a map $A_i \to \hat{\mathcal{A}}'_i$ that preserves the non-identities of $\bar{\mathcal{S}}_i$.

On the other hand, by definition, Kemer polynomials are alternating, so Proposition 2.38 shows that $\hat{U}_i$ is f.g. over its center. Hence $\mathcal{A}'_i$ is f.g. over its center and thus, by Proposition 1.92 is representable, say PI-equivalent to a f.d. algebra A''_i, so we can take $A'' = A''_1 \times \cdots \times A''_u$. $\qquad\square$

Corollary 4.67. *In characteristic 0, any relatively free affine PI-algebra U is representable.*

Proof. By Theorem 4.1, U is PI-equivalent to a finite dimensional algebra A, so Example 3.26 enables us to view U as a subalgebra of $A[\Lambda] \subset A \otimes K$, where $K = F(\Lambda)$. $\qquad\square$

4.5 Specht's Problem Completed in the Affine Case

The solution to Specht's problem is an easy corollary to Kemer's PI-Representability Theorem. Kemer [Kem91, pp. 66-67] proves the ACC on T-ideals by modifying the original chain and appealing to semisimple substitutions; his argument is given in Exercise 3. However, we choose a more prosaic, structural method based on the techniques already developed.

Remark 4.68. Let us refine Proposition 4.58 even further, using the proof of Lemma 4.52. Given any f.d. algebra A of Kemer index $\kappa = (t, s)$, we replace A by a direct product of PI-basic algebras, notation as in Proposition 4.58, so $\mathrm{index}(\tilde{A}_1) = \kappa$. Let us call $\tilde{A} = \tilde{A}_1 \times \cdots \times \tilde{A}_u$ the *leading component* of A.

Write $\bar{A}_1 = \tilde{A}_1/J(\tilde{A}_1)$. Then $\dim \bar{A}_1 = t$. Since we have assumed all along that F is algebraically closed, $\bar{A}_1$ is a direct product of matrix algebras the sum of whose dimensions is t. Thus, the algebra $\bar{A}_1$ is determined completely by the way t is partitioned into a sum of squares. Noting there are only a finite number of such partitions, and thus a finite number of possibilities for $\bar{A}_1$, let us call such $\bar{A}_1$ a *foundation* of A.

Now, analogously to the proof of Lemma 4.52, we take the free product $A' = \bar{A}_1 * F\{x_1, \ldots, x_s\}$, and

$$\hat{A}_1 = A'/(I_1 + I_2^s),$$

where I_1 is the ideal generated by all evaluations in A' of polynomials in id(A), and I_2 is the ideal generated by $x_1, \ldots, x_{s-1}$. Letting $\bar{x}_i$ denote the image of x_i, and taking a base $\{b_1, \ldots, b_t\}$ of $\bar{A}$, we see that $\hat{A}_1$ is spanned by all terms of the form

$$b_{i_1} \bar{x}_{j_1} b_{i_2} \bar{x}_{j_2} \ldots \bar{x}_{j_{s-1}} b_{j_s}, \tag{4.12}$$

and thus, has dimension $\leq t^s s^{s-1}$.

By definition, id$(A) \subseteq$ id$(\hat{A}_1)$. Conversely, note that $\gamma(\tilde{A}_i) = s$ for each $1 \leq i \leq u$. If $f \in$ id$(\hat{A}_1)$ is multilinear, then f vanishes whenever there are s radical substitutions, which implies $f \in$ id$(\tilde{A}_i)$ for each i such that $\tilde{A}_i$ has the same foundation as $\hat{A}_1$, by the same argument as in Lemma 4.52. Hence, we can replace each such $\tilde{A}_i$ by $\hat{A}_1$. Thus, the leading component may be assumed to be a direct product of $\hat{A}_i$, formed in this way, each having a different foundation.

If id$(A) \subset$ id(A') and A' has some $\tilde{A}'_i$ with the same foundation as $\tilde{A}_i$, then, applying the parallel construction, $\hat{A}'_i$ is a homomorphic image of $\hat{A}_i$ and thus, either $\hat{A}'_i = \hat{A}_i$ or dim $\hat{A}'_i <$ dim $\hat{A}_i$. The number of such inequalities can be at most $t^s s^{s-1}$; we conclude that for any infinite ascending chain of T-ideals of f.d. algebras, infinitely many of them must have the *same* leading component.

Theorem 4.69. *Any T-ideal of $F\{x_1, \ldots, x_\ell\}$ is finitely based (assuming* char$(F) = 0$).

Proof. In view of Remark 1.95, it suffices to verify the ACC for T-ideals, i.e., any chain

$$\text{id}(A_1) \subseteq \text{id}(A_2) \subseteq \text{id}(A_3) \subseteq \ldots \tag{4.13}$$

of T-ideals terminates. In view of Theorem 4.66, we may assume each A_i is f.d. Since ordered pairs of natural numbers satisfy the descending chain condition, we may start far enough along our chain of T-ideals and assume that all the A_j in (4.13) have the same Kemer index κ. We do this for κ minimal possible, and induct on κ.

By Proposition 4.58, we can write each

$$A_j = \tilde{A}_j \times A_{j,1} \times \cdots \times A_{j,t_j},$$

where $\tilde{A}_j$ is the leading component and each index$(A_{j,i}) < \kappa$. Furthermore, by Remark 4.68, some particular leading component must repeat infinitely often throughout the chain. Selecting only these entries, we may assume $\tilde{A}_j = \tilde{A}$, where $\tilde{A}$ is a direct product of PI-basic algebras of Kemer index κ.

Let $\mathcal{I}$ be the T-ideal generated by all Kemer polynomials of $\tilde{A}$ that are identities of $A_{1,1} \times \cdots \times A_{1,t}$. As we argued before, $\tilde{A}/\mathcal{I}(\tilde{A})$ cannot contain any Kemer polynomials (for κ) that are identities of $A_{1,1} \times \cdots \times A_{1,t}$, so its Kemer index must decrease and the chain

$$\text{id}(A_1) + \mathcal{I} \subseteq \text{id}(A_2) + \mathcal{I} \subseteq \text{id}(A_3) + \mathcal{I} \ldots \tag{4.14}$$

must stabilize. On the other hand, for each j,

$$\mathcal{I} \cap \text{id}(\tilde{A}) = \mathcal{I} \cap \text{id}(A_1) \subseteq \mathcal{I} \cap \text{id}(A_j) \subseteq \mathcal{I} \cap \text{id}(\tilde{A}),$$

so equality must hold at each stage, i.e., $\mathcal{I} \cap \text{id}(A_1) = \mathcal{I} \cap \text{id}(A_j)$ for each j. But coupled with (4.14), this means (4.13) stabilizes, contradiction. $\square$

4.6   Appendix: Applying Kemer's Theory

The reader who has waded through the proofs of this chapter now has an unexpected dividend: The techniques introduced in these proofs provide a host of results which will be used later, after we conclude the solution of Specht's problem. Thus, those who would like to push on directly to Kemer's solution should proceed directly to Chapter 5, but meanwhile we shall pause to prepare the groundwork for some rather sophisticated applications of Kemer's theory. Our objective, as usual, is to develop as many properties of the Capelli polynomial in as general a situation as possible.

4.6.1   The trace ring revisited

We finally have the means to understand the true role of representability in the theory of affine PI-algebras of characteristic 0. On the one hand, we have Kemer's PI-Represenatibility Theorem 4.1. On the other hand, we can develop a general theory of representable affine algebras via adjoining characteristic coefficients (in characteristic 0, this means traces).

Definition 4.70. An arbitrary affine PI-algebra A is *varietally reducible* if there are affine PI-algebras A_1, A_2 with $\text{id}(A_1) \cap \text{id}(A_2) = \text{id}(A)$ but $\text{id}(A_i) \supset \text{id}(A)$ for $i = 1, 2$.

Remark 4.71. Any affine PI-algebra A is PI-equivalent to a subdirect product of varietally irreducible PI-algebras, by induction on the Kemer index. We shall see in Proposition 6.73 that if A is relatively free, these varietally irreducible algebras also may be taken to be relatively free.

Definition 4.72. A K-*admissible* affine algebra $A = F\{a_1, \ldots, a_\ell\}$ is a varietally irreducible subalgebra of $M_n(K)$ (for a suitable algebraically closed field K) such that AK is PI-minimal.

Example 4.73. Some examples of K-admissible algebras:

(i) Any prime affine PI-algebra is K-admissible, since one can take K to be the algebraic closure of the field of fractions of A, and then $AK = M_n(K)$.

(ii) Any relatively free PI-algebra $\tilde{A}$ that is varietally irreducible is K-admissible. To see this, first we pass to the algebraic closure of F and assume F is algebraically closed. By Kemer's Theorem 4.1, $\tilde{A}$ is PI-equivalent to a f.d. algebra A. Replacing A by the algebra constructed in the proof of Lemma 4.52, we may assume A is PI-minimal. But then $\tilde{A}$ is the relatively free algebra of A, so by Example 3.26, $\tilde{A}$ can be constructed as a subalgebra of $A[\Lambda]$, which can be viewed as a subalgebra of $A \otimes_F K$, where K is the algebraic closure of the field of fractions of $F(\Lambda)$. Of course,

$$\dim_K A \otimes_F K = \dim_F A < \infty.$$

If $J = J(A)$ has nilpotence index s, then $J \otimes_F K$ also has nilpotence index s in $A \otimes K$, and

$$A \otimes K / J \otimes K \approx \tilde{A} \otimes_F K,$$

which is semisimple (since F is algebraically closed), so $A \otimes K$ also is PI-minimal, as desired.

Definition 4.74. Suppose $\mathrm{char}(F) = 0$. The K-*trace ring* of a K-admissible affine algebra $A = F\{a_1, \ldots, a_\ell\}$ is defined as follows:

Viewing $A \subseteq M_n(K)$ as in Definition 4.72, we are assuming that AK is varietally irreducible and PI-minimal. Let $\bar{A} = AK/J(AK)$, written as a direct sum of matrix subalgebras $R_1 \oplus \cdots \oplus R_q$, where $R_k = M_{n_k}(K)$; then we can define tr_k to be the trace in the k component and then for any $a = (a_1, \ldots, a_q) \in \bar{A}$, define

$$\mathrm{tr}(a) = \sum n_k \, \mathrm{tr}_k(a_k) \in K.$$

Consider the relatively free algebra $\mathcal{A}$ as constructed in Example 3.26. (We may use this particular construction since A is f.d.) Since $\mathcal{A} \subseteq A[\Lambda]$ for $\Lambda = \{\lambda_1, \ldots, \lambda_t\}$, we can extend the trace linearly to $\mathcal{A}$, taking values in $K[\Lambda]$. Let $\hat{F}$ denote the commutative affine subalgebra of $K[\Lambda]$ consisting of all the traces of powers ($\leq n_k$) of products (of length up to the Shirshov height) of the generators of $\mathcal{A}$, we let $\hat{A} = \mathcal{A}\hat{F} \subseteq A[\Lambda]$.

Remark 4.75. As in Proposition 4.59, we see $\hat{A}$ is f.g. over the commutative affine algebra $\hat{F}$.

By Remark 4.55(vi), in its notation, I is also closed under multiplication by elements of $\hat{F}$, so in fact, I is a common ideal between A and $\hat{A}$. Thus, $A/I \subseteq \hat{A}/I$, a representable algebra by Proposition 1.92, so A/I is representable.

Thus, Definition 4.74 contains all the main features of the characteristic closure, but works much more generally.

4.7   Razmyslov-Zubrilin Traces Revisited

Now that we have one proof of Kemer's Theorem under our belt, let us consider a more formal approach, one which might be more amenable to generalizations to nonassociative algebra. The idea, as improbable as it may seem, is to free ourselves from the associative structure theory on which we have relied so heavily, and to translate even more of the proof to properties of polynomials. We already introduced Razmyslov-Zubrilin traces in Chapter 2, but the main deficiency there was that the formal traces do not commute with the other elements of the algebra. Our objective here is to see how to remedy that, and thereby obtain the seeds of an alternate proof of Kemer's Theorem 4.66. At the same time, we shall improve our techniques for further use.

4.7.1   Applying Razmyslov-Zubrilin traces
to Kemer polynomials

The Razmyslov-Zubrilin theory meshes well with Kemer polynomials because of the following observation.

Remark 4.76. If f is a μ-Kemer polynomial for a PI-algebra A, where $\mu > \omega(A)$, cf. Definition 4.21, then the polynomial $\tilde{f}$ of Proposition 2.44 is in $\mathrm{id}(A)$. (Indeed, if $\mathrm{index}(A) = (t, s)$, then $\tilde{f}$ is $(t + 1)$-fold $(s + 1)$-alternating and t-fold $(\mu - 1)$-alternating.) Consequently, the conclusion of Proposition 2.44 applies.

We also have a useful finiteness test for Kemer index.

Remark 4.77. Suppose $\mathcal{S}$ is a set of monotonic components of Kemer polynomials of a PI-algebra A. Let $\tilde{A}'$ be the algebra described in Remark 2.50, and let I be the ideal of A generated by all $\{f(A) : f \in \mathcal{S}\}$. Then the image $\bar{I}$ of I in $\tilde{A}$ is an ideal in $\tilde{A}'$. Since $\tilde{A}'$ is Noetherian, $\bar{I}$ is f.g. as an $\tilde{A}'$-module. Write $\bar{I} = \sum_{i=1}^{t} \tilde{A}' f_i$. In view of Corollary 2.49, if we can show that $f_1, \ldots, f_t \in \mathrm{id}(A)$, then $\mathcal{S} \subseteq \mathrm{id}(A)$, contradiction.

To summarize, for any potential Kemer index κ and any PI-algebra A, a finite set of monotonic polynomials tests whether $\mathrm{index}(A) \geq \kappa$.

4.7.2   Making Razmyslov-Zubrilin traces commute

As stated above, one unpleasant aspect of the Razmyslov-Zubrilin traces is that they need not commute with each other. It turns out this can be resolved by using μ-Kemer polynomials for $\mu \geq 2$.

Definition 4.78. An *RZ polynomial* (for Razmyslov and Zubrilin) for an algebra A is a non-identity $f(x_1, \ldots, x_{2n}, x_{2n+1}; y)$ that n-alternates both in $x_1, \ldots, x_n$ and in $x_{n+1}, \ldots, x_{2n}$ and is linear in x_{2n+1}, but such that any alternator of f in $n+1$ indeterminates from $x_1, \ldots, x_{2n+1}$ (cf. Definition 1.24) is an identity of A.

The obvious situation for this to be applied is when $c_{n+1} \in \mathrm{id}(A)$ but $c_n^2 \notin \mathrm{id}(A)$, which was the case used by Zubrilin. However, this result can be made applicable for all PI-algebras.

Proposition 4.79. *Every affine PI-algebra A has an RZ-polynomial.*

Proof. Let $n = \beta(A)$. Take a μ-Kemer polynomial k_A (cf. Definition 4.22) for any $\mu \geq 2$. Since we are interested mainly in the two n-alternating folds, we rewrite k_A as

$$k_A(x_1, \ldots, x_n; x_{n+1}, \ldots, x_{2n}; y),$$

where k_A is n-alternating in the indeterminates $x_1, \ldots, x_n$, and also is n-alternating in $x_{n+1}, \ldots, x_{2n}$, but the other folds of $(n+1)$-alternating and n-alternating indeterminates are notated now within the y. Let $f = x_{2n+1} k_A$.

Clearly, the alternator of f with respect to any $n+1$ indeterminates from $x_1, \ldots, x_{2n}$ (leaving the y_j alone) adds an extra fold of $(t+1)$-alternating indeterminates to the $s-1$ folds that already exist, and thus is in $\mathrm{id}(A)$. Hence, f is an RZ polynomial. $\square$

Thus, Proposition 2.44 is available (where in the notation of Definition 2.40 we replace x_{n+1} by x_{2n+1}), but we can do much better. We write formally $\mathrm{tr}(z)$ for δ_z^1. Note that $\mathrm{tr}(z_1 + z_2) = \mathrm{tr}(z_1) + \mathrm{tr}(z_2)$.

We need some computations in the group algebra. We fix two disjoint subsets $I, J \subset \mathbb{N}$ of n elements each, with the symmetric group S_{2n} acting naturally on $I \cup J$. For $I' \subseteq I$, we define elements $T(I')$ in the group algebra $\mathbb{Z}[S_{2n}]$, as follows:

$$T(\emptyset) = \sum_{\sigma \in S_{2n}} \text{sgn}(\sigma)\sigma; \quad T(I') = \sum_{\sigma(I') \subseteq J} \text{sgn}(\sigma)\sigma, \quad I' \neq \emptyset.$$

Note that $\sigma(I) \subseteq J$ iff $\sigma(I) = J$, since $|I| = |J| = n$; thus,

$$T(I) = \sum_{\sigma(I)=J} \text{sgn}(\sigma)\sigma.$$

Lemma 4.80. *The equation*

$$\sum_{I' \subseteq I} (-1)^{|I'|} T(I') = \sum_{\sigma(I)=I} \text{sgn}(\sigma)\sigma \qquad (4.15)$$

holds in $\mathbb{Z}[S_{2n}]$, where the sum on the left hand is taken over all subsets I' of I including I and $\emptyset$.

Proof. Let $I_\sigma = \sigma^{-1}(J) \cap I$. A subset $I' \subseteq I$ satisfies $\sigma(I') \subseteq J$ iff $I' \subseteq I_\sigma$. Also the binomial expansion shows

$$\sum_{I' \subseteq I_\sigma} (-1)^{|I'|} = (1-1)^{|I_\sigma|} = 0$$

unless $I_\sigma = \emptyset$, in which case it is 1. Hence, reversing the order of summation in the left side of (4.15) yields

$$\sum_{I' \subseteq I} \sum_{\sigma \in S_{2n}} (-1)^{|I'|}\text{sgn}(\sigma)\sigma = \sum_{\sigma \in S_{2n}} \Big(\sum_{I' \subseteq I_\sigma} (-1)^{|I'|} \Big)\text{sgn}(\sigma)\sigma = \sum_{I_\sigma=\emptyset} \text{sgn}(\sigma)\sigma,$$

which is the right side of (4.15) (since $I_\sigma = \emptyset$ iff $\sigma(I) = I$). $\qquad \square$

Now we apply this to the polynomial f with respect to the action of Definition 1.23.

Remark 4.81. Suppose $I = \{1, \ldots, n\}$; $J = \{n+1, \ldots, 2n\}$. Then $T(I')f$ alternates in the variables $\{x_{\sigma(i)} : i \in I'\}$, and also alternates in the variables $\{x_{\sigma(i)} : i \in J \cup (I \setminus I')\}$. In particular, if f is an RZ-polynomial for A, then $T(I')f$ is an identity unless $I' = I$.

Theorem 4.82. *Suppose $f(x_1, \ldots, x_{2n+1}; y)$ is an RZ-polynomial for A. Then*

$$f(x_1, \ldots, x_n, x_{n+1}, \ldots, x_{2n}, x_{2n+1}; y)$$
$$- f(x_{n+1}, \ldots, x_{2n}, x_1, \ldots, x_n, x_{2n+1}; y) \qquad (4.16)$$

is an identity of A.

Proof. Write $f_{(2n)}$ to be the sum of monomials in f in which $x_1, x_2, \ldots, x_{2n}$ appear in ascending order. Then

$$f = \sum_{\sigma(I)=I, \, \tau(J)=J} \mathrm{sgn}(\sigma)\mathrm{sgn}(\tau)\sigma\tau\left(f_{(2n)}\right).$$

By Lemma 4.80, the right side is $(-1)^{|I|+|J|}T(I)T(J)f_{(2n)}$, since all other terms drop out by Remark 4.81. This translates to

$$f(x_{n+1}, \ldots, x_{2n}, x_1, \ldots, x_n, x_{2n+1}; y),$$

thereby yielding (4.16). $\qquad\square$

Corollary 4.83. *Hypotheses as in the theorem.*

(i) *Then $\delta_z^k(f)$ does not depend on the set of variables used ($\{x_1, \ldots, x_n\}$ or $\{x_{n+1}, \ldots, x_{2n}\}$); hence, the operators $\delta_{z_1}^k$ and $\delta_{z_2}^s$ commute for any z_1, z_2.*

(ii) *Furthermore, $\mathrm{tr}(z_1 z_2) = \mathrm{tr}(z_2 z_1)$.*

Proof. The first assertion follows immediately from (4.16), and the second assertion is a consequence since

$$\mathrm{tr}(z_1 z_2) = \delta_{z_1}^1 \delta_{z_2}^1 - \tilde{\delta}^2(z_1, z_2),$$

where $\tilde{\delta}^2$ is the linearization of the form $\delta_{z_1}^2$. $\qquad\square$

Remark 4.84. Suppose f is an RZ polynomial for A. Then the δ_z^i commute, by Corollary 4.83 so we can form the *commutative* algebra $\tilde{C}$ generated by the operators δ_z^k acting on the values of f, via the $2n+1$ position, i.e., as in Definition 2.40.

Let I_f be the ideal of A generated by the evaluations of f. I_f is a module over $\tilde{C}$. But $\tilde{C}$ is commutative affine and thus Noetherian, over which $\hat{A}$ is f.g., so $\hat{A}$ is a Noetherian ring, and thus, its ideal I_f is Noetherian.

This theory gains in power from a theorems of Donkin [Do94]and Zubkov [Zu96]. One deep application of the theory is given in Exercise 7.

4.7.3 Another approach to Kemer's PI-Representability Theorem

The Razmyslov-Zubrilin theory gives us a rather elegant method of reworking the key step in the proof of Kemer's Theorem 4.66. Recall that we had added on a set of μ-Kemer polynomials to Γ to get a T-ideal Γ', which by induction is PI-equivalent to a f.d. algebra A', and then had to take the direct product with suitable f.d. algebras that satisfy the identities of Γ but whose identities do not include the original Kemer polynomials. This step could also be accomplished via the following argument:

Remark 4.85. Taking $\mu > 2$, let $\mathcal{A}$ be the relatively free algebra of the T-ideal Γ. Then, reminiscent of Remark 4.68, take the free product $A' = \mathcal{A} * F\{x_1, \ldots, x_\nu\}$, where ν is large enough to include all the variables needed in folds of μ-Kemer polynomials, and

$$\hat{A} = A'/(I_1 + I_2),$$

where I_1 is the ideal generated by all evaluations in A' of polynomials in Γ, and I_2 is the ideal generated by all $x_i A' x_i$. (This is to make the image $\bar{x}_i$ have degree at most 1 in any nonzero evaluation of a polynomial; also note that the ideal generated by the $\bar{x}_i$ is nilpotent of index at most ν, since taking the product of $\mu + 1$ words means some $\bar{x}_i$ must have degree at least 2, so is zero.)

We make the elements in the image $\bar{A}$ of $\hat{A}$ integral by the formal procedure of Remark 2.47. Thus, there is a map $\phi : \hat{A} \to R$ where R is representable, and where $\ker \phi$ is the obstruction of integrality of the elements of $\bar{A}$. On the other hand, we can define Zubrilin traces by means of μ-Kemer polynomials, and these provide integrality. Thus, $\ker \phi$ has trivial intersection with the space S generated by evaluations of the Kemer polynomials in which all variables of folds are specialized to the $\bar{x}_i$.

This is precisely the property we needed in the proof of Theorem 4.66 to return from Γ' to Γ, so we have an alternate conclusion to the proof.

Although this resembles the idea used in our earlier proof of Theorem 4.66, the main difference is that we did not need to anchor our adjunction of traces to the algebras A_i, since we now can adjoin the traces using only the abstract properties of the Razmyslov-Zubrilin traces, which in turn are developed in terms of intrinsic properties of Capelli polynomials.

Note that rather than requiring the representation $\ker \phi$ to be faithful, we only required that it be disjoint from S. This leads one to the notion of *representable subspace* of an algebra A, which is a subspace that intersects trivially the kernel of a suitable representation to a finite dimensional algebra. Representable subspaces are used in proofs in Chapter 9.

The Razmyslov-Zubrilin theory also indicates how one may prove versions of Kemer's Theorem where the associative structure theory is not available. In his papers [Zub95a], [Zub95b], [Zub97], Zubrilin actually used the natural generalization of Capelli identities to nonassociative algebras (namely, taking each possible arrangement of parentheses), and so this part of the proof of Kemer's Theorem "works" in a very general context. Ironically, the part that does not yet work for arbitrary nonassociative algebras is the first step, in which Lewin's theorem yields an algebra A for which $\Gamma \supseteq \mathrm{id}(A)$, cf. for example [VaZel89].

4.8 Pumping and d-Identities

In Chapter 2, we introduced Belov's powerful technique of changing substitutions in a polynomial so that occurrences of the largest letter are moved closer together. We present the basic technique here, called *pumping* (which we already used in Lemma 2.59), first as applied to the Capelli polynomial, and then indicate how it could be generalized.

Before pumping in full force, let us introduce an interesting variant of PI and see how we can pump with it.

Definition 4.86. A polynomial $f(x_1, \ldots, x_n; y_1, \ldots, y_m)$ is a *d-identity* of $A = F\{a_1, \ldots, a_\ell\}$ (with respect to designated indeterminates $x_1, \ldots, x_n$) if f vanishes for all specializations of $x_1, \ldots, x_n$ to words in the a_i of length $\leq d$.

Remark 4.87. Let $m = n = (\ell+1)^d$. The Capelli polynomial c_n is always a d-identity of $A = F\{a_1, \ldots, a_\ell\}$ (with respect to the n alternating variables $x_1, \ldots, x_n$), since there are fewer than n words of length $\leq d$ in $a_1, \ldots, a_\ell$.

Thus, when n is large with respect to d, the notion of d-identity can be vacuous. Nevertheless, d-identities fit in well with pumping, in view of the following observation.

Lemma 4.88. *Let A be a PI-algebra for which c_d is a d-identity. Then for each homogeneous d-linear polynomial $f(x_1, \ldots, x_n; y)$, any specializations $\bar{x}_i, \bar{y}$ of the x_i and y, and any words v_i in the $\bar{x}_i$,*

$$f(v_1, \ldots, v_n; \bar{y})$$

is a linear combination of elements of the form $f(v_1', \ldots, v_n'; \bar{y})$ where at most $d - 1$ of the v_i' have length $\geq d$.

Proof. Copy the proof of Lemma 2.59, taking $g = c_d$ and weakening the hypothesis to c_d being a d-identity, since the only place it was used in the proof was in evaluating substitutions to the u_i, which were constructed to have length $< d$. $\qquad \square$

We already can get an interesting result about Capelli identities.

Theorem 4.89. *If c_d is a d-identity of A (with respect to the d alternating variables $x_1, \ldots, x_d$), then $c_{2d-1} \in \mathrm{id}(A)$.*

Proof. Since any d-alternating polynomial is a consequence of c_d, the hypothesis implies any d-alternating polynomial is a d-identity of A with respect to those d variables. We want to show

$$c_{2d-1}(\bar{x}_1, \ldots, \bar{x}_{2d-1}, \bar{y}_1, \ldots, \bar{y}_{2d-1}) = 0$$

evaluated on any words $\bar{x}_i, \bar{y}_i$ in A. If at least d of the $\bar{x}_i$ have length $\leq d$, then we view c_{2d-1} as alternating in these particular $\bar{x}_i$, and are done by the first sentence. Thus, we may assume at most $d - 1$ of the $\bar{x}_i$ have length $\leq d$, and thus, at least d of the $\bar{x}_i$ have length $> d$; for notational convenience, we assume $\bar{x}_1, \ldots, \bar{x}_d$ have length $> d$. Assume we have some

$$a = c_{2d-1}(\bar{x}_1, \ldots, \bar{x}_{2d-1}, \bar{y}_1, \ldots, \bar{y}_{2d-1}) \neq 0,$$

with the smallest possible number of $\bar{x}_i$ of length $> d$.

Given these restrictions, we choose a with $\bar{x}_1$ of minimal length; given this $\bar{x}_1$, we choose a with $|\bar{x}_2|$ minimal, and so forth. Let us induct on the d-tuple $(|\bar{x}_1|, \ldots, |\bar{x}_d|)$. Viewing c_{2d-1} as a d-alternating polynomial in $x_{i_1}, \ldots, x_{i_d}$, we apply Lemma 2.59, to show a is a linear combination of smaller terms, each of which is 0 by induction, contradiction. □

The proof worked because we had $d - 1$ "extra" alternating indeterminates x_i which could absorb the excess length of the substitutions and then be ignored (i.e., considered as part of the indeterminates y). This process of "pumping" all length $\geq d$ into a small number of indeterminates is delicate, but produces surprising results, as we shall see soon. First a game.

Remark 4.90. Pumping game: There are n piles of counters, denoted $v_1, \ldots, v_t$, where pile v_i contains m_i counters. Player 1 can select d piles $v_{i_1}, \ldots, v_{i_t}$, and from each pile v_{i_j} he selects a certain number $k_{i_j} < m_{i_j}$ counters. Player 2 must permute these selected sub-piles (with a nonidentical permutation). For example, if $t = 5$ and $n = 3$, suppose the 5 piles respectively have 11,3,5,7,4 counters, and Player 1 selects sub-piles of 4,2,3, counters from piles 1,2, and 5. Then Player 2 could perform the cyclic permutation in the sub-piles to get 3,4,2 and the new piles have respective sizes

$$11 - 4 + 3 = 10, \quad 3 - 2 + 4 = 5, \quad 5, \quad 7, \quad 4 - 3 + 2 = 3.$$

Problem: Find a strategy such that Player 1 can eventually force $n - d$ of the piles to have at most d counters.

Solution: Each time, Player 1 takes d counters from the largest pile, $d - 1$ from the second largest pile, and so on. Player 2 must then reduce the d-tuple of the d largest piles lexicographically, so, arguing by induction, we see that Player 1 can force the desired outcome.

The proof of Theorem 4.89 could be viewed as a way of playing the game with Capelli polynomials. Here is a slight generalization.

Lemma 4.91. *Suppose c_d is a d-identity of A with respect to the d alternating variables.*

For each homogeneous n-linear polynomial $f(x_1, \ldots, x_n; y)$, any specializations $\bar{x}_i, \bar{y}$ of the x_i and y, and any words v_i in the $\bar{x}_1, \ldots, \bar{x}_n$ such that $m_i = |v_i| \geq d$, we can rewrite $f(v_1, \ldots, v_n; \bar{y})$ as a linear combination

$$\sum_{1 \neq \sigma \in S_d} \alpha_\sigma f(v_{1,\sigma}, \ldots, v_{d,\sigma}, v_{d+1}, \ldots, v_n; \bar{y}),$$

where $v_{1,\sigma}, \ldots v_{d,\sigma}$ are words formed by rearranging the letters of $v_1, \ldots, v_d$ and lexicographically the d-tuple

$$(|v_{1,\sigma}|, |v_{2,\sigma}|, \ldots, |v_{d,\sigma}|) < (m_1, m_2, \ldots, m_d).$$

Proof. Identical, word for word, to the proof in the induction procedure of Lemma 2.59. □

4.8.1 Digression: Pumping Kemer polynomials

We continue with an intricate analysis of the pumping procedure in conjunction with Kemer polynomials. In order to pump with Kemer polynomials, first we want them to look more like Capelli polynomials.

Now that we have a way of constructing Kemer polynomials for arbitrary Kemer index, let us see how the pumping procedure fits into this. Let A be any affine algebra. By Theorem 2.56, A satisfies a suitable Capelli identity c_d.

Proposition 4.92. *Suppose an algebra A has Kemer index $\kappa = (t, s)$ and satisfies c_d. Then for any $\mu \geq d$, A has some μ-Kemer polynomial which is not a $d + 1$-identity of A.*

Proof. Take any μ-Kemer polynomial f for A. By definition, f is μ-fold t-alternating and s-fold $(t+1)$-alternating. Let us write $X_j = \{x_{j,1}, x_{j,2}, \ldots\}$ for the designated indeterminates in the j-th fold; thus $|X_j| = t$ for $j \leq \mu$, and $|X_j| = t+1$ for $\mu < j \leq \mu+s-1$. So we write $f = f(X_1, \ldots, X_{\mu+s-1}; Y)$.

Take a nonzero specialization $x_{j,i} \mapsto \bar{x}_{j,i}$ and $y_i \mapsto \bar{y}_i$ (where the x_i are the designated indeterminates). We call a specialization *short* if it has length $\leq d$; the specialization is *long* if it has length $> d$. We call one of the folds X_j *acceptable* if all of the $\bar{x}_{j,i}$ of this fold are short. Our first objective is to make as many folds acceptable as possible. We select one $x_{j,i}$ from each set X_j, and to emphasize their special role, we write $v_{j,i}$ for $\bar{x}_{j,i}$, $1 \leq i \leq \mu$; $\bar{x}$ denotes the specializations of the $x_{j,i}$ for the other i. By the Sparse Pumping lemma (2.59), we may write

$$f(v_1, \ldots, v_\mu; \bar{x}; \bar{y})$$

as a linear combination of elements of the form $f(v_1', \ldots, v_\mu'; \bar{x}; \bar{y})$ where at most $d - 1$ of the v_i' have length $\geq d$, and the $\bar{y}$ remain unchanged. Thus, replacing v_i by v_i', we have made all but $d-1$ of these specializations short, without affecting the other specializations. Using this procedure wherever possible, we may assume that *every* specialization in the components $X_1, \ldots, X_\mu$ is short, except for at most $d - 1$ components; indeed, if d components contain long specializations, we choose the variables of those long specializations and apply the pumping procedure while leaving the other specializations alone.

Thus, renumbering the X_j, we may assume that every specialization in $X_1, \ldots, X_{\mu-d+1}$ is short. So far we have not considered the specializations in $X_{\mu+1}, \ldots, X_{\mu+s}$. Towards this end we assume that, given $k > \mu$, we have a Kemer polynomial that is nonzero with respect to as few as possible long specializations for the X_k, and induct on the length of such long specializations. Let us call a specialization *moderate* if it has length precisely $d + 1$. We claim that we can reduce the length of all long specializations in the X_k to moderate or short specializations. So we assume this is impossible— i.e., any nonzero specialization for f must involve a specialization of length $> d + 1$ in some X_k.

We appeal to Remark 4.33, in which we focus on some X_j and X_k, where $j \leq \mu - d + 1$ and $\mu < k \leq \mu + s - 1$. In other words, suppose X_k has a long specialization of length $> d + 1$, which we may as well assume is $\bar{x}_{k,t+1}$. Write

$$\bar{x}_{k,t+1} = \bar{x}_{k,t+1}' \bar{z},$$

where $\bar{x}_{k,t+1}'$ is moderate, i.e., of length $d + 1$. Let

$$h_i = (-1)^i f(x_{j,1}, \ldots, x_{j,i-1}, x_{k,t'+1}, x_{j,i+1}, \ldots, x_{j,t}, x_{k,1}, \ldots, x_{k,t}, x_{j,i}z; y).$$

Then $g = f + \sum_{i=1}^t h_i$ is $(t+1), t$ alternating (in the sets $X_1 \cup \{x_{2,t+1}'\}$ and $X_2 \setminus \{x_{2,t+1}\}$), whereas each h_i is $t, t+1$ alternating in $X_1 \cup \{x_{2,t+1}'\} \setminus \{x_{1,i}\}$ and $X_2 \cup \{x_{1,i}z\} \setminus \{x_{2,t+1}\}$.

In each h_i we have interchanged the specialization $\bar{x}_{k,t+1}$ with the moderate specialization $\bar{x}_{j,i}\bar{z}$, contrary to our assumption unless the specialization of h_i is 0. Since

$$f = g + \sum h_i,$$

we must have a nonzero specialization for g. But in g we have created a new fold using one specialization of length $d+1$ and t short specializations, contrary to hypothesis.

Having proved that all X_k can be made to have specializations of length $\leq d+1$, we then reapply the argument at the beginning to make all the specializations of X_j short, for $1 \leq j \leq \mu - d + 1$. $\qquad\Box$

Corollary 4.93. *Suppose A is an affine algebra, of Kemer index κ. Taking d such that A satisfies the Capelli identity c_d, let $\mu \geq d$ be arbitrary and let I be the ideal generated by all evaluations of μ-Kemer polynomials, where each designated indeterminate is specialized to a word in A of length $\leq d+1$. Then $\mathrm{index}(A/I) < \kappa$.*

This result will be used to prove Noetherian PI-algebras are finitely presented (Theorem 8.2), and to obtain a converse to Shirshov's Theorem, cf. Exercise 9.20.

Chapter 5

Representations of S_n and Their Applications

In this chapter we turn to a sublime connection, first exploited by Regev, of PI-theory with classical representation theory (via the group algebra $F[S_n]$), leading to many interesting results (including Regev's exponential bound theorem). Perhaps the key result of this chapter, discovered independently by Amitsur-Regev and Kemer, shows in characteristic 0 that any PI-algebra satisfies a sparse identity. This is the key hypothesis in translating identities to affine superidentities in Chapter 6, leading to Kemer's solution of Specht's problem.

5.1   Passage to $F[S_n]$

As usual, S_n denotes the symmetric group on $1, \ldots, n$. Since any multilinear polynomial in $x_1, \ldots, x_n$ is an F-linear combination of the monomials $x_{\sigma(1)} x_{\sigma(2)} \cdots x_{\sigma(n)}$, $\sigma \in S_n$, we are led to introduce the following notation. Assume $X = \{x_1, x_2, \ldots\}$ is infinite, and for each n, $V_n = V_n(x_1, \ldots, x_n)$ is the F-vector space of multilinear polynomials in $x_1, \ldots, x_n$, i.e.,

$$V_n(x_1, \ldots, x_n) = \mathrm{span}_F\{x_{\sigma(1)} x_{\sigma(2)} \cdots x_{\sigma(n)} \mid \sigma \in S_n\}.$$

Given a PI-algebra A, we shall study the multilinear identities $\mathrm{id}(A) \cap V_n$, $n = 1, 2, \ldots$ Note that in characteristic zero, this sequence of spaces determines $\mathrm{id}(A)$, by Proposition 1.18.

Definition 5.1. Identify a permutation $\sigma \in S_n$ with its corresponding monomial (in $x_1, x_2, \ldots, x_n$):

$$\sigma \leftrightarrow M_\sigma(x_1, \ldots, x_n) = x_{\sigma(1)} \cdots x_{\sigma(n)}.$$

Any polynomial $\sum \alpha_\sigma x_{\sigma(1)} \cdots x_{\sigma(n)}$ corresponds to $\sum \alpha_\sigma \sigma \in F[S_n]$, and conversely, $\sum \alpha_\sigma \sigma$ corresponds to the polynomial

$$\left(\sum \alpha_\sigma \sigma \right) x_1 \cdots x_n = \sum \alpha_\sigma x_{\sigma(1)} \cdots x_{\sigma(n)}.$$

Definition 5.1 identifies V_n with the group algebra $F[S_n]$. Let us study some basic properties of this identification.

Given $\sigma, \pi \in S_n$, the product $\sigma\pi$ which corresponds to the monomial

$$M_{\sigma\pi} = x_{\sigma\pi(1)} \cdots x_{\sigma\pi(n)}$$

can be viewed in two ways, corresponding to *left* and *right* actions of S_n on V_n, described respectively as follows:

(i) $\sigma M_\pi(x_1 \ldots, x_n) = M_{\sigma\pi},$ and

(ii) $M_\sigma(x_1 \ldots, x_n)\pi = M_{\sigma\pi}.$

Lemma 5.2. *Let $\sigma, \pi \in S_n$, and let $y_i = x_{\sigma(i)}$. Then*

(i) $\sigma M_\pi(x_1 \ldots, x_n) = M_\pi(x_{\sigma(1)}, \ldots, x_{\sigma(n)});$

(ii) $M_\sigma(x_1 \ldots, x_n)\pi = (y_1 \cdots y_n)\pi = y_{\pi(1)} \cdots y_{\pi(n)}.$

Thus, the effect of the right action of π on a monomial is to permute the places of the variables according to π.

Proof. $y_{\pi(j)} = x_{\sigma\pi(j)}$. Then

$$M_{\sigma\pi} = x_{\sigma\pi(1)} \cdots x_{\sigma\pi(n)} = y_{\pi(1)} \cdots y_{\pi(n)} =$$

(which already proves (ii))

$$= M_\pi(y_1, \ldots, y_n) = M_\pi(x_{\sigma(1)}, \ldots, x_{\sigma(n)}),$$

which proves (i).   □

Extending by linearity, we obtain

Corollary 5.3. *If $p = p(x_1, \ldots, x_n) \in V_n$ and $\sigma, \pi \in S_n$, then*

(i) $\sigma p(x_1, \ldots, x_n) = p(x_{\sigma(1)}, \ldots, x_{\sigma(n)});$

(ii) $p(x_1, \ldots, x_n)\pi = q(x_1, \ldots, x_n)$, *where $q(x_1, \ldots, x_n)$ is obtained from $p(x_1, \ldots, x_n)$ by place-permuting all the monomials of p according to the permutation π.*

For any finite group G and field F, there is a well-known correspondence between the $F[G]$-modules and the representations of G. The simple modules correspond to the irreducible representations, which in turn correspond to the irreducible characters; in fact, there is a 1:1 correspondence between the isomorphic classes of simple modules and the irreducible characters, which we shall use freely.

Remark 5.4. If $p \in \mathrm{id}(A)$, then $\sigma p \in \mathrm{id}(A)$ since the left action is just a change of variables.

Hence, for any PI-algebra A, the spaces

$$\mathrm{id}(A) \cap V_n \subseteq V_n$$

are in fact left ideals of $F[S_n]$, thereby affording certain S_n representations. We shall study the quotient modules

$$V_n / (\mathrm{id}(A) \cap V_n)$$

and their corresponding S_n-characters. Of course, these characters are independent of the particular choice of $x_1, \ldots, x_n$. The above obviously leads us into the theory of group representations, which we shall review in the next section.

However, $p\sigma$ need not be an identity. Let us consider some examples.

Example 5.5. Applying the transposition (12) on the left to the Grassman identity

$$f = [[x_1, x_2], x_3] = x_1 x_2 x_3 - x_2 x_1 x_3 - x_3 x_1 x_2 + x_3 x_2 x_1$$

yields $-f$; applying (23) yields $[[x_1, x_3], x_2]$, which also is a PI of the Grassman algebra G.

However, applying (12) to f on the right yields

$$x_2 x_1 x_3 - x_1 x_2 x_3 - x_1 x_3 x_2 + x_2 x_3 x_1 = -[x_2, x_1 x_3] - [x_1, x_2 x_3],$$

which clearly is *not* an identity of G (seen by specializing $x_2 \mapsto 1$).

Here is a slightly more complicated example, which is much more to the point.

Example 5.6. The Capelli polynomial c_n is

$$(s_n(x_1, \ldots, x_n) x_{n+1} \cdots x_{2n})\pi,$$

where $\pi \in S_{2n}$ is described below in Remark 5.9.

This example is so crucial that we formalize it.

Definition 5.7. (The embedding $V_n \subseteq V_m$) Let $n \leq m$. The natural embedding $S_n \subseteq S_m$ (namely, fixing all indices $> n$) induces an embedding $V_n \subseteq V_m$, given by $p(x_1, \ldots, x_n) \mapsto p(x_1, \ldots, x_n) x_{n+1} \cdots x_m$.

Definition 5.8. Given $p(x_1, \ldots, x_n) = \sum_{\sigma \in S_n} a_\sigma x_{\sigma(1)} \cdots x_{\sigma(n)} \in V_n$ we form the *Capelli-type* polynomial in V_{2n} :

$$p^*(x_1, \ldots, x_n; x_{n+1}, \ldots, x_{2n})$$
$$= \sum_{\sigma \in S_n} a_\sigma x_{\sigma(1)} x_{n+1} x_{\sigma(2)} x_{n+2} \cdots x_{\sigma(n-1)} x_{2n-1} x_{\sigma(n)} x_{2n}.$$

Thus, $c_n = s_n^*$. Note that the last indeterminate x_{2n} is superfluous, since we could specialize it to 1, and so strictly speaking we need only pass to V_{2n-1}. However, the notation is more symmetric (with n permuted variables and n nonpermuted variables), so we keep x_{2n}. The following observation is a straightforward application of Corollary 5.3.

Remark 5.9. For any polynomial $p(x_1, \ldots, x_n) \in V_n$,

$$p^* = (p(x_1, \ldots, x_n) x_{n+1} \cdots x_{2n}) \pi$$
$$= \sum_{\sigma \in S_n} a_\sigma x_{\sigma(1)} x_{n+1} x_{\sigma(2)} x_{n+2} \cdots x_{\sigma(n)} x_{2n},$$

where the permutation π is given explicitly as

$$\begin{pmatrix} 1 & 2 & 3 & 4 & \ldots & 2n-1 & 2n \\ 1 & n+1 & 2 & n+2 & \ldots & n & 2n \end{pmatrix}.$$

Note that if $p^* \in \mathrm{id}(A)$, then $p(x) \in \mathrm{id}(A)$, seen by specializing $x_j \mapsto 1$ for all $j > n$. In particular, c_n implies s_n, but s_n need not imply c_n, cf. Example 3.34.

5.2 Review of the Representation Theory of S_n

In this section we describe, without proof, the classical theory of Frobenius and Young about the representations of S_n. Although formulated here in characteristic 0, there is a characteristic p version given in Green [Gre80].

5.2.1 Structure of semisimple algebras

The structure of the group algebra $F[G]$ is particularly nice in characteristic 0; for the reader's convenience, we shall review some of the main structural results, especially those involving idempotents.

Note that an algebra is a left module over itself, in which case a submodule is the same as a left ideal, and a simple submodule is the same as a minimal left ideal. We start with Maschke's theorem and its consequences.

Theorem 5.10 (Maschke). *Let $A = F[G]$, where G is a finite group and F is a field of characteristic 0. Then A is semisimple, implying any module M over A is semisimple; in particular, any submodule N of M has a complement N' such that $N \oplus N' = M$. Furthermore,*

$$A = \bigoplus_{i=1}^{r} I_i$$

where the I_is are the minimal two-sided ideals of A. When F is algebraically closed, r equals the number of conjugacy classes of G,

$$I_i \approx M_{n_i}(F),$$

the $n_i \times n_i$ matrices over F, and the numbers n_i satisfy

$$n_1^2 + \cdots + n_r^2 = |G|.$$

Maschke's theorem thereby leads us to a closer study of direct sums of matrix rings.

Remark 5.11. Notation as above, let $J_i \subseteq I_i$ be minimal left ideals, so for $i = 1, \ldots r$, $\dim J_i = n_i$ and $\mathrm{End}_{I_i} J_i = F$. Then any f.g. module M over A can be decomposed as

$$M = \bigoplus_{i=1}^{r} M_i \quad \text{where} \quad M_i = I_i M \quad \text{and} \quad M_i \approx J_i^{\oplus m_i}$$

for suitable multiplicities m_i, $i = 1, \ldots, r$. This is called the *isotypic decomposition* of M as an A-module.

Definition 5.12. An idempotent $e = e^2$ is *primitive* if e cannot be written $e = e_1 + e_2$ for idempotents $e_1, e_2 \neq 0$ with $e_1 e_2 = e_2 e_1 = 0$.

Proposition 5.13. *Suppose A is a semisimple ring. For any left ideal $J \subseteq A$, there exists an idempotent $e \in J$ such that $J = Ae$. Moreover, such e is primitive if and only if J is a minimal left ideal.*

Proof. Take a complement J' of J in A and write $1 = e + e'$ where $e \in J$, $e' \in J'$. Then $e = e1 = e^2 + ee'$; matching components in $J \oplus J'$ shows $e = e^2$ and $ee' = 0$. Likewise, e' is idempotent. Clearly, $J \supseteq Ae$. But if $a \in J$, then $a = a1 = ae + ae'$, implying $a = ae$ (and $ae' = 0$); hence, $J = Ae$. The last assertion is now clear. $\square$

Remark 5.14. In practice, it is more convenient to work with *semi-idempotents*, i.e., elements e such that $0 \neq e^2 \in Fe$. Of course, if $e^2 = \alpha e$ for $\alpha \neq 0$, then $\frac{e}{\alpha}$ is idempotent and generates the same left ideal as e.

Remark 5.15. Even in characteristic $p > 0$, enough structure theory carries over to enable us to obtain useful results. For G finite, $F[G]$ is finite dimensional and thus as a left module over itself has a composition series

$$L_0 = F[G] \supset L_1 \supset L_2 \supset \cdots \supset L_r = 0, \tag{5.1}$$

whose simple factors L_i/L_{i+1} comprise the set of irreducible representations of G. (Indeed, any irreducible representation is a simple $F[G]$-module, thus of the form $F[G]/L$ for L a suitable maximal left ideal; conversely, for any maximal left ideal L, the chain $F[G] \supset L$ can be refined to a composition series $F[G] \supset L \supset \ldots$, and by the Jordan-Holder theorem, the factors $F[G]/L$ and L_i/L_{i+1} are isomorphic for some i.)

5.2.2 Young's Theory

Partitions. In the case of S_n, Theorem 5.10 holds where F is any field of characteristic zero, for example, $F = \mathbb{Q}$. The conjugacy classes of S_n, and therefore the minimal two-sided ideals of $F[S_n]$, are indexed by the partitions of n: $\lambda = (\lambda_1, \ldots \lambda_k)$ is a partition of n, denoted $\lambda \vdash n$, if $\lambda_1, \ldots, \lambda_k$ are integers satisfying

(i) $\lambda_1 \geq \lambda_2 \geq \cdots \geq \lambda_k \geq 0$, and

(ii) $\lambda_1 + \cdots + \lambda_k = n$.

There is a theory, due to Frobenius and to A. Young, that each partition $\lambda \vdash n$ allows the explicit construction of a corresponding minimal two-sided ideal $I_\lambda \subseteq F[S_n]$, as well as minimal left ideals in the I_λ. Here is a brief account of that theory.

Diagrams. The *Young diagram* of $\lambda = (\lambda_1, \ldots \lambda_k)$ is a left justified array of empty boxes such that the i-th row contains λ_i boxes. For example,

$$x\ x\ x$$
$$x$$

is the Young diagram of $\lambda = (3, 1)$, where each x denotes a box. We identify a partition with its Young diagram. For the partitions $\mu = (\mu_1, \mu_2, \ldots)$ and $\lambda = (\lambda_1, \lambda_2, \ldots)$, we say that $\mu \geq \lambda$ if the Young diagram of μ contains that of λ, i.e., if $\mu_i \geq \lambda_i$ for all i.

Tableaux. Given $\lambda \vdash n$, fill its diagram with $1, \ldots, n$ to obtain a *Young tableau* T_λ of *shape* λ. Such a Young tableau is called a *standard tableau* if its entries increase in each row from left to right and in each column from top to bottom. For example,

$$
\begin{array}{cc}
1 & 2 \\
3 &
\end{array}
\qquad \text{and} \qquad
\begin{array}{cc}
1 & 3 \\
2 &
\end{array}
$$

are the only standard tableaux of shape $\lambda = (2, 1)$.

The number of standard tableaux of shape λ is denoted by f^λ. For example, $f^{(2,1)} = 2$. The Young-Frobenius formula and the hook formula, given below, are convenient formulas for calculating f^λ.

Left ideals. The tableau T_λ defines two subgroups of S_n, where $\lambda \vdash n$:

   (i) The row permutations R_{T_λ}, i.e., those permutations that leave each row invariant.

   (ii) The column permutations C_{T_λ} i.e., those permutations that leave each column invariant.

Define

$$
R^+_{T_\lambda} = \sum_{p \in R_{T_\lambda}} p, \qquad C^-_{T_\lambda} = \sum_{q \in C_{T_\lambda}} \mathrm{sgn}(q) q.
$$

Then T_λ defines the semi-idempotent

$$
e_{T_\lambda} := \sum_{q \in C_{T_\lambda}} \sum_{p \in R_{T_\lambda}} \mathrm{sgn}(q) q p = C^-_{T_\lambda} R^+_{T_\lambda}.
$$

The left ideal $F[S_n] e_{T_\lambda}$ is minimal in $F[S_n]$. For the tableaux T_λ of shape λ, these left ideals are all isomorphic as left $F[S_n]$ modules. In particular, they give rise to equivalent representations, and thus afford the same character, which we denote χ^λ. However, if $\lambda \neq \mu \vdash n$, with corresponding tableaux T_λ and T_μ, then

$$
F[S_n] e_{T_\lambda} \not\approx F[S_n] e_{T_\mu}.
$$

Details can be found in [Jac80, pp. 266ff].

Remark 5.16. One can always reverse directions by working with π^{-1} instead of π, and thereby replace e_{T_λ} by the semi-idempotent $e'_{T_\lambda} = R^+_{T_\lambda} C^-_{T_\lambda}$, which often appears in the PI literature because of Example 5.47.

Note. In the literature (see for example [Sag01]), there is a construction of the $F[S_n]$-module $S^\lambda \approx F[S_n] e_{T_\lambda}$, called the *Specht module* (not directly connected to Specht's problem). We use S^λ as a representative of the isomorphism class of these modules.

Two-sided ideals. Given $\lambda \vdash n$, the sum of the minimal left ideals corresponding to the T_λ is a minimal two-sided ideal, denoted

$$\sum_{T_\lambda} F[S_n]e_{T_\lambda} := I_\lambda. \tag{5.2}$$

It can be shown that

$$I_\lambda = \bigoplus_{T_\lambda \text{ is standard}} F[S_n]e_{T_\lambda},$$

and as an F-algebra, I_λ is isomorphic to $M_{f^\lambda}(F)$, the $f^\lambda \times f^\lambda$ matrices over F. This leads to the following extended S_n-version of Theorem 5.10.

Remark 5.17.

(i) There is a 1:1 correspondence between the ideals I_λ and the irreducible S_n-characters χ^λ.

(ii) The unit element E_λ in I_λ is a central semi-idempotent in $F[S_n]$. By general representation theory, it is given by the formula

$$E_\lambda = \sum_{\sigma \in S_n} \chi^\lambda(\sigma)\sigma.$$

(Hence, $E_\lambda^2 = n!E_\lambda$.)

(iii) For example, let $\lambda = (1^n)$. Then $\chi = \chi^{(1^n)}$ is the sign character given by $\chi(\sigma) = \text{sgn}(\sigma)$, and $E_{(1^n)}$ is the n-th standard polynomial s_n.

Theorem 5.18. *We have the following decompositions, notation as above:*

(i)

$$F[S_n] = \bigoplus_{\lambda \vdash n} I_\lambda = \bigoplus_{\lambda \vdash n} \left(\bigoplus_{T_\lambda \text{ is standard}} F[S_n]e_{T_\lambda} \right).$$

(ii) Up to isomorphism,

$$I_\lambda \approx (S^\lambda)^{\oplus f^\lambda},$$

and hence,

$$F[S_n] \approx \bigoplus_{\lambda \vdash n} (S^\lambda)^{\oplus f^\lambda}.$$

Remark 5.19. Since F can be any field of characteristic 0, it follows that the decomposition of an S_n-module (resp. character) into simple submodules (resp. irreducible characters) is the same for any field extension of the rationals.

Example 5.20. We can view the standard identity

$$s_n = \sum_{\sigma \in S_n} \text{sgn}(\sigma)\sigma$$

as the (central) unit in the two-sided ideal I_λ, where $\lambda = (1^n)$.

The Branching Theorem. Let $\lambda = (\lambda_1, \lambda_2, \ldots)$ be a partition of n, and let λ^+ denote those tuples of the form $(\lambda_1, \ldots, \lambda_{i-1}, \lambda_i + 1, \lambda_{i+1}, \ldots)$; in other words, λ^+ are the partitions of $n + 1$ that contain λ. For example, if $\lambda = (3, 3, 1, 1)$, then

$$(3, 3, 1, 1)^+ = \{(4, 3, 1, 1), \ (3, 3, 2, 1), \ (3, 3, 1, 1, 1)\}.$$

Similarly, if $\mu = (\mu_1, \mu_2, \ldots)$ is a partition of $n + 1$, we let μ^- denote the tuples $(\mu_1, \ldots, \mu_i - 1, \ldots)$ which are partitions (of n). For example,

$$(3, 3, 1, 1, 1)^- = \{3, 2, 1, 1, 1), (3, 3, 1, 1)\}.$$

We have the following important theorem. Recall that if χ^λ is a character corresponding to an $F[S_n]$-module M, then viewing $F[S_n] \hookrightarrow F[S_{n+1}]$ naturally, we have the $F[S_{n+1}]$-module $F[S_{n+1}] \otimes_{F[S_n]} M$, whose character is called the *induced character*, denoted $\chi^\lambda \uparrow^{S_{n+1}}$. In the other direction, if χ^μ is a character corresponding to an $F[S_{n+1}]$-module M, then we can view M a fortiori as an $F[S_n]$-module, and call the corresponding character the *restricted character* denoted as $\chi^\mu \downarrow_{S_n}$.

Theorem 5.21 (Branching).

(i) *(Branching-up) If $\lambda \vdash n$, then*

$$\chi^\lambda \uparrow^{S_{n+1}} = \sum_{\mu \in \lambda^+} \chi^\mu.$$

(ii) *(Branching-down) If $\mu \vdash n + 1$, then*

$$\chi^\mu \downarrow_{S_n} = \sum_{\lambda \in \mu^-} \chi^\lambda.$$

Note by Frobenius reciprocity that both parts of Theorem 5.21 are equivalent. The following are obvious consequences of the Branching Theorem.

Corollary 5.22.

(i) Let $\lambda \vdash n$ with I_λ the corresponding minimal two-sided ideal in $F[S_n]$, and let $m \geq n$. Then

$$F[S_m]I_\lambda F[S_m] = \bigoplus_{\mu \vdash m; \mu \geq \lambda} I_\mu.$$

(ii) Let $\lambda \vdash n$, $\mu \vdash m$ with $m \geq n$, and assume $\mu \geq \lambda$. Then $f^\mu \geq f^\lambda$; furthermore, if $E_\lambda \in \mathrm{id}(A)$, then $E_\mu \in \mathrm{id}(A)$.

5.2.3 The RSK correspondence

The Robinson-Schensted-Knuth (RSK) correspondence is an algorithm that corresponds a permutation $\sigma \in S_n$ with a pair of standard tableaux (P_λ, Q_λ), both of the same shape $\lambda \vdash n$. See Knuth [Kn73] or Stanley [St99] for a description of the RSK correspondence.

This algorithm and its generalizations have become important tools in combinatorics. For example, it instantly yields a bijective proof of the formula

$$n! = \sum_{\lambda \vdash n} (f^\lambda)^2.$$

This formula can be obtained by calculating the dimensions on both sides of Theorem 5.18. However, in the development of the S_n-representation theory, this formula is usually proved earlier, as a first step towards proving Theorem 5.18.

Among other properties, the RSK correspondence $\sigma \to (P_\lambda, Q_\lambda)$ has the following property. Recall that the *length* $\ell(\lambda) = \ell(\lambda_1, \lambda_2, \ldots) = d$ if $\lambda_d > 0$ and $\lambda_{d+1} = 0$.

Theorem 5.23. *Let $\sigma \in S_n$ and*

$$\sigma \mapsto (P_\lambda, Q_\lambda)$$

in the RSK correspondence, so $\lambda \vdash n$, and let $d = \ell(\lambda)$. Then d is the length of a maximal chain $1 \leq i_1 < \cdots < i_d \leq n$ such that $\sigma(i_1) > \cdots > \sigma(i_d)$. In other words, $\ell(\lambda)$ is the length of a maximal descending subsequence in σ.

To understand the condition, recall Shirshov's notion of d-decomposability and Remark 2.7(ii).

Remark 5.24. A multilinear monomial $x_{\sigma(1)} \ldots x_{\sigma(n)}$ is d-decomposable iff the sequence $\{\sigma(1), \ldots, \sigma(n)\}$ contains a descending subsequence of length d, i.e., iff there are $i_1 < \cdots < i_d$ such that $\sigma(i_1) > \cdots > \sigma(i_d)$.

This led Latyshev to the following definition.

Definition 5.25. A permutation $\sigma \in S_n$ is d-*good* if its corresponding monomial $x_{\sigma(1)} \ldots x_{\sigma(n)}$ is d-decomposable. We denote the number of d-good permutations in S_n as $g_d(n)$.

(For $d = 3$, $g_3(n)$ can be shown to be the so-called "Catalan number" of n.) Theorem 5.23 clearly gives a bijective proof of the following fact:

Lemma 5.26. *Notation as above,*

$$g_d(n) = \sum_{\lambda \vdash n,\, \ell(\lambda) \leq d-1} (f^\lambda)^2.$$

5.2.4 Dimension formulas

Here are some formulas for calculating f^λ. First, for commuting variables $\xi_1, \ldots, \xi_k$ consider the discriminant

$$D_k(\xi) = D_k(\xi_1, \ldots, \xi_k) = \prod_{1 \leq i < j \leq k} (\xi_i - \xi_j).$$

Theorem 5.27 (The Young-Frobenius formula). *Given*

$$\lambda = (\lambda_1, \ldots, \lambda_k) \vdash n,$$

let $t_i = \lambda_i + k - i$, $1 \leq i \leq k$. *Then*

$$f^\lambda = \frac{n!}{t_1! \cdots t_k!} D_k(t_1, \ldots, t_k).$$

A very convenient formula for calculating f^λ is the hook formula, due to J. S. Frame, G. de B. Robinson, and R. M. Thrall.

Definition 5.28. Let $\lambda = (\lambda_1, \lambda_2, \ldots)$ be a partition. Define its *conjugate partition* $\lambda' = (\lambda_1', \lambda_2', \ldots)$, where λ_j' is the length of the j column of the Young diagram of λ. For $(i, j) \in \lambda$, the (i, j)-*hook number* h_{ij} in λ is the length of the "hook" from (i, j) to the right border and to the bottom of the Young diagram of λ, i.e.,

$$h_{i,j} = h_{i,j}(\lambda) = \lambda_i - j + \lambda_j' - i + 1.$$

Theorem 5.29 (The S_n hook formula). *If $\lambda \vdash n$, then*

$$f^\lambda = \frac{n!}{\prod_{(i,j)\in\lambda} h_{i,j}(\lambda)}.$$

We need one more formula concerning f^λ, for whose preparation we introduce some more structure.

5.3 S_n-Actions on $T^n(V)$

Each of our two actions on multilinear polynomials extends naturally to the space of homogeneous polynomials—the left action yields more generally the polynomial representations of the general linear group $GL(V)$ (to be described in Chapter 10), and the right action has some remarkable applications, especially for superalgebras. It is this second action on which we focus here.

Definition 5.30. Let V denote the k-dimensional vector space with base $\{x_1, \ldots, x_k\}$. Define

$$T^n(V) := V^{\otimes n} = V \otimes \cdots \otimes V \quad (n \text{ times})$$

and

$$T(V) = \bigoplus_{n=0}^{\infty} T^n(V),$$

the *tensor algebra* of V. We define multiplication

$$T^m(V) \times T^n(V) \to T^{m+n}(V)$$

by $(f, g) \mapsto f \otimes g$, which extends naturally to a multiplication on $T(V)$ that makes it an algebra. Then clearly

$$F\{x_1, \ldots, x_k\} \approx T(V),$$

with the set of homogeneous polynomials in $x_1, \ldots, x_k$ of total degree n being isomorphic to $T^n(V)$.

We denote $\operatorname{End}_F(T^n(V))$ by $E_{n,k}$.

There is a natural action of S_n on $T^n(V)$ by permuting coordinates—or places—in "monomials":

Definition 5.31. Let $\sigma \in S_n$ and define $\hat{\sigma} \in E_{n,k}$ via

$$\hat{\sigma}(v_1 \otimes \cdots \otimes v_n) = v_{\sigma(1)} \otimes \cdots \otimes v_{\sigma(n)},$$

extended by linearity to all of $T^n(V)$. It is routine to verify that $\hat{\sigma}$ is well defined. Denote by $\varphi_{n,k} : S_n \to E_{n,k}$ the map given by $\varphi_{n,k}(\sigma) = \hat{\sigma}$, and extend $\varphi_{n,k}$ by linearity to $\varphi_{n,k} : F[S_n] \to E_{n,k}$. Define

$$A(n,k) = \varphi_{n,k}(F[S_n]) \subseteq E_{n,k}.$$

Let $a \in F[S_n]$ and let $w \in T^n(V)$; by abuse of notation we write $a \cdot w$ for $\hat{a}(w)$.

Although this action extends the right action defined in Definition 5.2(ii), we write it now on the left, to conform with the literature.

Remark 5.32.

(i) $\widehat{\sigma\pi} = \hat{\pi}\hat{\sigma}$ for $\sigma, \pi \in S_n$; namely, $\varphi_{n,k}$ is an anti-homomorphism. Thus, if $a, b \in F[S_n]$ and $w \in T^n(V)$, then $(ab) \cdot w = b \cdot (a \cdot w)$. Therefore, $T^n(V)$ is a left module over the opposite algebra $(F[S_n])^{op}$.

(ii) If $f = f(x_1, \ldots, x_n) \in V_n = F[S_n]$ and $y_1 \otimes \cdots \otimes y_n \in T^n(V)$, then

$$f \cdot (y_1 \otimes \cdots \otimes y_n) = f(y_1, \ldots, y_n).$$

(iii) In the literature, the action of S_n on $W_{n,k}$ is usually defined via

$$\sigma'(v_1 \otimes \cdots \otimes v_n) = v_{\sigma^{-1}(1)} \otimes \cdots \otimes v_{\sigma^{-1}(n)},$$

with corresponding $\varphi'_{n,k} : \sigma \mapsto \sigma'$ which is a homomorphism, since $\sigma'\pi' = (\sigma\pi)'$. However, since $\varphi_{n,k}(F[S_n]) = \varphi'_{n,k}(F[S_n]) = A(n,k)$, the two approaches lead to the same results. The reason we chose the action $\hat{\sigma}$ is because of the formula (ii) above.

Recall that $F[S_n] = \oplus_{\lambda \vdash n} I_\lambda$ with each I_λ isomorphic to a matrix algebra. Recall also that via the transpose map, a matrix algebra is anti-isomorphic with itself. It follows that each I_λ is mapped by $\varphi_{n,k}$ either to zero or to $\varphi_{n,k}(I_\lambda) \approx I_\lambda$. The following theorem, due to Schur and Weyl, describes $A(n,k) = \varphi_{n,k}(F[S_n])$, and in particular, it describes those λs satisfying $\varphi_{n,k}(I_\lambda) \approx I_\lambda$. We first introduce another notion.

Definition 5.33. $H(k, 0; n)$ denotes those partitions of n contained in the infinite k-strip, namely with at most k rows:

$$H(k, 0; n) = \{\lambda \vdash n \mid \lambda_{k+1} = 0\}.$$

Theorem 5.34. *Notation as above,*

  (i) *If λ is a partition of n, then $\varphi_{n,k}(I_\lambda) \approx I_\lambda$ if $\lambda \in H(k,0;n)$, and $\varphi_{n,k}(I_\lambda) = 0$ otherwise.*

  (ii)
  $$A(n,k) = \varphi_{n,k}(F[S_n]) = \bigoplus_{\lambda \in H(k,0;n)} A_\lambda \approx \bigoplus_{\lambda \in H(k,0;n)} I_\lambda$$

  where for each $\lambda \in H(k,0;n)$, $A_\lambda = \varphi_{n,k}(I_\lambda) \approx I_\lambda$, a simple algebra.

  (iii) *By Remark 5.11,*
  $$T^n(V) = \bigoplus_{\lambda \in H(k,0;n)} W_\lambda,$$

  where $W_\lambda = I_\lambda T^n(V) = A_\lambda T^n(V)$, and this is the isotypic decomposition of $T^n(V)$ as an $A(n,k)$-module, hence, as an $F[S_n]$-module.

Corollary 5.35. *Notation as above,*

  (i)
  $$\dim A(n,k) = \sum_{\lambda \in H(k,0;n)} (f^\lambda)^2.$$

  (ii) *In particular, since $A(n,k) \subseteq E_{n,k} = \mathrm{End}_F(T^n(V))$, and $\dim E_{n,k} = k^{2n}$,*
  $$\sum_{\lambda \in H(k,0;n)} (f^\lambda)^2 \leq k^{2n}. \tag{5.3}$$

Note that the inequality (5.3) is a combinatoric fact about Young diagrams, and does not refer to $E_{n,k}$ or $A(n,k)$.

5.4 Codimensions and Regev's Exponential Bound Theorem

Having brought in some basic tools from representation theory, we already can apply them to obtain some deep results in PI-theory. Throughout, we shall take an algebra A over an arbitrary field F. Define $\Gamma_n = \mathrm{id}(A) \cap V_n$, cf. Section 5.1. Suppose for a moment that $\mathrm{char}(F) = 0$. The sequence $\{\Gamma_n : n \in \mathbb{N}\}$ determines $\mathrm{id}(A)$. Being a left ideal of the semisimple ring $F[S_n]$, Γ_n has a direct sum complement (isomorphic as an $F[S_n]$-module to V_n/Γ_n), which is a direct sum of minimal left ideals, each of which can be described as $F[S_n]e_{T_\lambda}$, for a suitable Young tableau T_λ. This already constitutes a major step, providing a finite algorithm for determining the multilinear identities of degree n. The V_n/Γ_n have other remarkable properties to be described below. We shall now bring in a technique that applies in all characteristics, although we focus on characteristic 0.

Definition 5.36. We call

$$c_n(A) := \dim_F (V_n/\Gamma_n)$$

the n-th *codimension* of the algebra A. The character of the representation corresponding to the module V_n/Γ_n is called the n-th *cocharacter*.

Thus, the algebra A defines a sequence of codimensions $c_n(A)$, for $n = 1, 2, \ldots$ (Later, in Chapter 10, we shall study the cocharacters.)

Codimensions have played a key role in PI-theory since Regev used them in his dissertation to prove his tensor product theorem (Theorem 5.42). Our notation is a bit unfortunate, since we already have used c_n to denote the Capelli polynomial. However, the contexts are completely different, since here $c_n(A)$ is a number, so there should not be ambiguity.

Observation 5.37. A satisfies a PI of degree $\leq n$ iff $\Gamma_n \neq 0$, which holds iff $c_n(A) < n!$.

The surprising basic fact Regev discovered about codimensions of a PI-algebra, which lies at the foundation of quantitative PI-theory, is their exponential bound, which presently we shall prove.

Theorem 5.38. *If A satisfies a PI of degree d, then*

$$c_n(A) \leq (d-1)^{2n}.$$

An important step towards proving this theorem is the next lemma. Recall from Remark 5.24 and Definition 5.25 that $g_d(n)$ is the number of d-indecomposable monomials in S_n.

Lemma 5.39 (Latyshev). *Let A be an algebra satisfying a polynomial identity of degree d. Then*

$$c_n(A) \leq g_d(n).$$

(This holds in any characteristic.)

Proof. By Proposition 2.8, V_n is spanned modulo $\mathrm{id}(A) \cap V_n$ by the d-indecomposable monomials, so

$$c_n(A) = [V_n : \mathrm{id}(A) \cap V_n] \leq g_d(n).$$

$\square$

Proof of Theorem 5.38 when $\text{char}(F) = 0$. By Lemma 5.39,

$$c_n(A) \le g_d(n),$$

and by Lemma 5.26,

$$g_d(n) = \sum_{\lambda \vdash n,\, \ell(\lambda) \le d-1} (f^\lambda)^2.$$

Note that $\{\lambda \vdash n : \ell(\lambda) \le d - 1\} = \{\lambda \in H(d - 1, 0; n)\}$. Thus, by Corollary 5.35(ii), $g_d(n) \le (d-1)^{2n}$. $\qquad\qquad\qquad\qquad\qquad\qquad\square$

This proof is an immediate application of standard results about combinatorics. (In [Row80], the use of Corollary 5.35(ii) was bypassed by means of Amitsur's version of Dilworth's theorem, which works in all characteristics, and even over arbitrary commutative rings. Thus, Theorem 5.38 is true over any field.)

As a first application, we prove Regev's tensor product theorem (Theorem 5.42). The key is

Remark 5.40. Let $a_1, \ldots, a_n \in A$ be ml-generic elements (cf. Definition 3.18). Denote

$$V_n(a) = V_n(a_1, \ldots, a_n) = \{f(a_1, \ldots, a_n) \mid f \in V_n(x_1, \ldots, x_n)\}.$$

Then $V_n(a)$ is a subspace of A. Also,

$$V_n(a) \approx \frac{V_n(x)}{\text{id}(A) \cap V_n(x)}$$

via the canonical map $f(x_1, \ldots, x_n) \mapsto f(a_1, \ldots, a_n)$, the kernel of which is $\text{id}(A) \cap V_n$, and therefore

$$c_n(A) = \dim V_n(a).$$

In particular, if $W \subseteq A$ is a subspace such that $V_n(a) \subseteq W$, then $c_n(A) \le \dim W$.

We can now prove a key result.

Proposition 5.41. $c_n(A \otimes B) \le c_n(A) \cdot c_n(B)$.

Proof. Let $U_A = F\{\bar{x}_1, \bar{x}_2, \ldots\}$, $U_B = F\{\bar{x}'_1, \bar{x}'_2, \ldots\}$ be the respective relatively free algebras, and write $\bar{x}$ for $(\bar{x}_1, \ldots, \bar{x}_n)$ and $\bar{x}'$ for $(\bar{x}'_1, \ldots, \bar{x}'_n)$.

Then $c_n(A) = \dim V_n(\bar{x})$ and $c_n(B) = \dim V_n(\bar{x}')$, as well as $\bar{x} \otimes \bar{x}'$ for $(\bar{x}_1 \otimes \bar{x}'_1, \ldots, \bar{x}_n \otimes \bar{x}'_n)$. It easily follows that

$$V_n(\bar{x} \otimes \bar{x}') \subseteq V_n(\bar{x}) \otimes V_n(\bar{x}').$$

Since $\bar{x}_1 \otimes \bar{x}'_1$, $\bar{x}_2 \otimes \bar{x}'_2$, ... are ml-generic,

$$c_n(A \otimes B) = c_n(U_A \otimes U_B) = \dim V_n(\bar{x} \otimes \bar{x}')$$
$$\dim V_n(\bar{x}) \dim V_n(\bar{x}'). \tag{5.4}$$

$\square$

As an immediate application, we have Regev's landmark theorem.

Theorem 5.42. *If A and B are two PI-algebras over a field F, then $A \otimes B$ is also PI, where a bound for the degree of the PI is given in Remark 5.43.*

Proof. By Theorem 5.38, the codimension sequences $c_n(A)$, $c_n(B)$ are bounded exponentially, and therefore by Proposition 5.41, so are the codimensions $c_n(A \otimes B)$. Thus, $c_n(A \otimes B) < n!$ for large enough n, and we conclude with Observation 5.37. $\square$

Remark 5.43. Combining the bound of Theorem 5.38 with the inequality of Proposition 5.41 and the fact that $n! > (n/e)^n$ yields a bound

$$\text{PI-}\deg(A \otimes B) \le e(\text{PI-}\deg(A) - 1)^2(\text{PI-}\deg(B) - 1)^2.$$

In certain cases, a much better bound can be given for the degree. For example, if $A_i \sim_{\text{PI}} M_{n_i}(F)$ for $i = 1, 2$, then $\text{PI-}\deg(A_i) = 2n_i$ by the Amitsur-Levitzki theorem, whereas $\text{PI-}\deg(A_1 \otimes A_2) \le 2n_1 n_2$ since $A_1 \otimes A_2$ satisfies the identities of $n_1 n_2 \times n_1 n_2$ matrices, cf. Exercise 7. On the other hand, one also must contend with Exercise 6.2.

Remark 5.44. Theorem 5.38 implies $\sqrt[n]{c_n(A)} \le (d-1)^2$, leading us to define the *PI-exponent* of a PI-algebra A as $\varlimsup_{n \to \infty} \sqrt[n]{c_n(A)}$. In the affine case, Kemer's PI-Represenatibility Theorem 4.1 enables us to assume that A is finite dimensional (and, in fact, PI-basic), and Giambruno and Zaicev succeeded in proving the exponent is an integer; using the techniques of Chapter 6, they extended this result to arbitrary PI-algebras and obtained other asymptotic results about $c_n(A)$, cf. [GiZa03a]. Unfortunately, since the proof of Theorem 4.1 is nonconstructive, it may be difficult to study the exponent of a given PI-algebra. From a different point of view, in Chapter 10, we discuss methods for obtaining finer information about specific T-ideals.

5.4.1   The Kemer-Regev-Amitsur trick

Applying the same methods with care provides a powerful computational technique discovered independently by Kemer [Kem80] and Amitsur-Regev [AmRe82] following Regev [Reg78], which was used to great effect by Kemer. Let A be any PI-algebra, and $\Gamma_n = \mathrm{id}(A) \cap V_n$. The basic idea is contained in:

Lemma 5.45. *If $f^\lambda > c_n(A)$, then $I_\lambda \subseteq \Gamma_n$.*

Proof. Let $J = F[S_n]e_{T_\lambda} \subseteq I_\lambda$ be a minimal left ideal of $F[S_n]$. Then $\dim J = f^\lambda$. If $J \not\subseteq \Gamma_n$, then $J \cap \Gamma_n = 0$, which implies that $f^\lambda \leq c_n(A)$, a contradiction. Hence, $J \subseteq \Gamma_n$, and the result follows since I_λ is the sum of such Js. $\qquad\square$

Next, we confront the codimension c_n with the following computation. Here $e = 2.718281828\ldots$ is the base of the natural logarithms.

Lemma 5.46. *Let u, v be integers, $n = uv$ and let $\lambda = (v^u) \vdash n$ be the $u \times v$ rectangle. Then*

$$f^\lambda > \left(\frac{uv}{u+v}\right)^n \cdot \left(\frac{2}{e}\right)^n. \qquad (5.5)$$

In particular, if $\frac{uv}{u+v} > \frac{e}{2}\alpha$, then $f^\lambda > \alpha^n$.

Proof. We estimate

$$f^\lambda = \frac{n!}{\prod_{x \in \lambda} h_x}.$$

First, it is well known that the numerator $n! > (n/e)^n$. Concerning the denominator, it is easy to see that $\sum_{x \in \lambda} h_x = uv(u+v)/2 = n(u+v)/2$ (see, for example, [MacD95, I, Ex 2]). Since the arithmetic mean is greater than the geometric mean, we deduce that

$$\left(\prod_{x \in \lambda} h_x\right)^{1/n} \leq \frac{1}{n}\sum_{x \in \lambda} h_x = \frac{u+v}{2},$$

and therefore $(\prod_{x \in \lambda} h_x)^{-1} \geq (2/(u+v))^n$. Thus,

$$f^\lambda > \left(\frac{n}{e}\right)^n \cdot \left(\frac{2}{u+v}\right)^n = \left(\frac{uv}{u+v}\right)^n \left(\frac{2}{e}\right)^n,$$

and the formula (5.5) follows.

$\qquad\square$

Example 5.47. Let us write out explicitly the polynomial corresponding to e'_{T_λ}, where we take T_λ to be

$$\begin{pmatrix} 1 & u+1 & 2u+1 & \cdots & (v-1)u+1 \\ 2 & u+2 & 2u+2 & \cdots & (v-1)u+2 \\ \cdots & & & & \\ u & 2u & 3u & \cdots & vu \end{pmatrix}.$$

$$C^-_{T_\lambda} x_1 \ldots x_{uv} = \sum_{\pi \in C_{T_\lambda}} \mathrm{sgn}(\pi) x_{\pi(1)} \cdots x_{\pi(u)} \cdots x_{\pi((v-1)u+1)} \cdots x_{\pi(vu)}$$

$$= s_u(x_1, \ldots, x_u) \cdots s_u(x_{(v-1)u+1}, \ldots, x_{vu}).$$

Thus, $e'_{T_\lambda} = R^+_{T_\lambda} C^-_{T_\lambda}$ is the multilinearization of s^v_u. The element e_{T_λ} is more difficult to describe but also plays an important role in the sequel.

Now we bring in the exponential bound of $c_n(A)$.

Proposition 5.48. *Suppose A has PI-degree d. Take $u \geq (d-1)^4 e$,*

$$v \geq \frac{(d-1)^4 eu}{2u - (d-1)^4 e},$$

$n = uv$, and any m between n and $2n$. Let λ_0 denote the rectangular shape (v^u), i.e., u rows of v columns each. For any $\lambda \vdash m$ containing λ_0 and any Young tableau T_λ, the semi-idempotents $e_{T_\lambda}, e'_{T_\lambda} \in \Gamma_m$.

Proof. Noting $\frac{uv}{u+v} \geq \frac{1}{2}(d-1)^4 e$ and using Lemma 5.46, we have

$$f^\lambda \geq f^{\lambda_0} > \left(\frac{uv}{u+v}\right)^n \left(\frac{2}{e}\right)^n \geq \left(\frac{1}{2}(d-1)^4 e \frac{2}{e}\right)^n$$

$$= (d-1)^{4n} \geq c_{2n}(A) \geq c_m(A),$$

implying $I_\lambda \subseteq \Gamma_m$ by Lemma 5.45; in particular $e_{T_\lambda}, e'_{T_\lambda} \in \Gamma_m$. $\qquad\square$

Remark 5.49 (Regev). In view of Example 5.47, this provides a second proof of Theorem 3.38, that every PI-algebra satisfies a power of the standard identity, but now we have the explicit identity s^v_u (estimated in terms of the PI-degree of A).

So far we have a way of passing from a certain identity of degree n to degree m where $m \leq 2n$. Our next theorem will remove the restriction on m. This is an amazing result. It gives us an ideal I (generated by $e_{T_{\lambda_0}}$) which, viewed inside *any* $F[S_m]$, will generate a two-sided ideal of multilinear identities of A. Clearly, we can then replace I by any element f in such an ideal, so f will remain an identity after left or right actions by permutations. The key to this improvement is the $*$ construction from Definition 5.8.

Theorem 5.50. *Let A be a PI-algebra and let $I \subseteq F[S_n]$ be a two-sided ideal in $F[S_n] = V_n$.*

(i) *Let $f = f(x) \in I$ and assume for $n \leq m \leq 2n - 1$ the elements of the two-sided ideal generated by f in $F[S_m]$ are identities of A. Then $f^* \in \mathrm{id}(A)$.*

(ii) *Assume $I \subseteq \mathrm{id}(A)$ and that for each $f \in I$, $f^* \in \mathrm{id}(A)$.*

Then for all $m \geq n$, all the elements of the two-sided ideal generated by I in $F[S_m]$ are identities of A: $F[S_m]IF[S_m] \subseteq \mathrm{id}(A)$.

Proof. (i) follows easily from Remark 5.9.

(ii) Since $\mathrm{id}(A) \cap V_m$ is a left ideal, it suffices to prove $f\pi \in \mathrm{id}(A)$ for all $f \in I$ and $\pi \in S_m$. Write

$$(x_1 \ldots x_n \ldots x_m)\pi = p_0 x_{\eta(1)} p_1 x_{\eta(2)} p_2 \cdots p_{n-1} x_{\eta(n)} p_n,$$

where $\eta \in S_n$ and p_j are either 1 or are monomials in $x_{n+1}, \ldots, x_m$. Then

$$(x_1 \ldots x_m)\eta^{-1}\pi = p_0 x_1 p_1 x_2 p_2 \cdots p_{n-1} x_n p_n. \qquad (5.6)$$

Since $I \lhd F[S_r]$, we see $f\eta \in I$; writing $f\eta = \sum_{\sigma \in S_n} a_\sigma \sigma$, we have

$$f\pi = (f\eta)(\eta^{-1}\pi) = \sum_{\sigma \in S_n} a_\sigma p_0 x_{\sigma(1)} p_1 x_{\sigma(2)} \cdots p_{n-1} x_{\sigma(n)} p_n$$

$$= p_0 (f\eta)^* (x_1, \ldots, x_n; p_1, \ldots, p_n).$$

By assumption $(f\eta)^* \in \mathrm{id}(A)$. Substituting $x_{n+j} \mapsto p_j$ and multiplying by p_0 and p_n, we conclude that $f\eta \in \mathrm{id}(A)$. $\qquad \square$

Finally, we bring in the branching theorem, Corollary 5.22, to get a powerful result.

Theorem 5.51. *Suppose A has PI-degree d, and u, v are as in Proposition 5.48. Then:*

(i) *$F[S_m]e_{T_{\lambda_0}} F[S_m] \subseteq \Gamma_m$, for any tableau T_{λ_0} of $\lambda_0 = (v^u)$.*

(ii) *Γ_m contains every two-sided ideal of S_m corresponding to a shape containing the rectangle (v^u).*

Proof. In view of Corollary 5.22, it suffices to prove (ii). By Theorem 5.50, we need to show that if

$$f \in I = F[S_n]e_{T_{\lambda_0}} F[S_n],$$

then $f^* \in \mathrm{id}(A)$. This is true by Proposition 5.48, taking $m = 2n$. $\qquad \square$

Corollary 5.52. *Hypotheses as in the theorem,* $e_{T_{\lambda_0}} \pi^* \in \mathrm{id}(A)$, *for any* $\pi \in S_{uv}$.

Proof. We just displayed $e_{T_{\lambda_0}} \pi^*$ in the right ideal generated by $e_{T_{\lambda_0}}$, so we apply Theorem 5.51(ii). $\square$

5.4.2 Hooks

In Theorem 5.51 we proved that if A has PI-degree d, then any multilinear polynomial whose Young tableau has a shape containing the rectangle v^u is an identity of A. Let us restate the contrapositive, starting with a definition generalizing Definition 5.33.

Definition 5.53. The (k, ℓ)-*hook*

$$H(k, \ell; n) = \{\lambda \vdash n \mid \lambda_{k+1} \leq \ell\}$$

is the collection of all shapes such that the $k+1$ row (and thus, all subseqent rows) has length $\leq \ell$.

Equivalently, the $\ell + 1$ column (and thus, all subsequent columns) has length $\leq k$.

Proposition 5.54. *Suppose A has PI-degree d. Then there are k and ℓ such that $I_\lambda \subseteq \mathrm{id}(A)$, for any shape λ not lying in the (k, ℓ)-hook.*

Proof. Take $k = u - 1$ and $\ell = v - 1$ for u, v as in Proposition 5.48. If the shape does not lie in the (k, ℓ) hook, then it must contain an entry in the $(k+1, \ell+1)$-position, and thus have the $u \times v$ rectangle (which we denoted v^u). But then we are done by Theorem 5.51. $\square$

5.4.3 Sparse identities

We have not finished exploiting the Kemer-Regev-Amitsur trick. In this section we show that every PI-algebra in characteristic 0 satisfies a sparse identity. In Chapter 2, we defined sparse identities as tools for providing reduction procedures. Here we shall show first that sparse identities can be viewed in terms of group actions.

In Corollary 5.3, we described two group actions of $F[S_n]$, where the left $F[S_n]$ action is preserved under identities. We are interested in the right action, which does not necessarily preserve identities, cf. Example 5.5, and characterize the sparse identities as those identities that *are* preserved under this action.

Remark 5.55. $g \in F[S_n]$ is a sparse identity of A (cf. Definition 2.58) iff it satisfies the following condition:

If $\pi \in S_m$ for $m \geq n$, then the polynomial $g\pi(x) \in \mathrm{id}(A)$.

In other words, in view of Theorem 5.50, $g \in F[S_n]$ is a sparse identity of A iff $g^* \in \mathrm{id}(A)$.

Remark 5.56. In characteristic 0, a multilinear identity of degree n is sparse iff it generates a two-sided ideal in $F[S_m]$ for all $m \geq n$.

Example 5.57. (i) Clearly, $s_n\pi = \pm s_n$ for any $\pi \in S_n$. Nevertheless, s_n is not sparse for $n > 2$, in view of Remark 5.6, since c_n is not a consequence of s_n.

(ii) On the other hand, for any $\pi \in S_m$, $c_n\pi$ is a consequence of c_n, implying any Capelli identity is sparse.

Thus, sparse identities generalize Capelli identities. Their advantage over the Capelli identities is that they exist for *all* PI-algebras, whereas we recall that arbitrary PI-algebras (such as the Grassman algebra) need not satisfy a Capelli identity.

Theorem 5.58. *Let A be a PI-algebra (not necessarily affine) of characteristic 0. There exists an integer n and nonzero element $g \in F[S_n]$ that is a sparse identity of A, and which can be calculated explicitly.*

Proof. Take $n = uv$ as in Proposition 5.48, and $g = e_{T_{\lambda_0}}\pi$. We claim g is a sparse identity. Indeed, take any monomial

$$M = p_0 x_{\eta(1)} p_1 x_{\eta(2)} p_2 \cdots p_{n-1} x_{\eta(n)} p_n = \eta M_0(x_1, \ldots, x_m),$$

for some $\eta \in S_n$, where $M_0 = p_0 x_1 p_1 x_2 p_2 \cdots p_{n-1} x_n p_n$, the p_j being either 1 or monomials in $x_{n+1}, \ldots, x_m$. As in the proof of Theorem 5.50(ii), $e_\lambda M_0 = p_0 e_\lambda^*(x; p_1, \ldots, p_{n-1})p_n$, which implies that

$$gM = e_{T_{\lambda_0}}\pi M = e_{T_{\lambda_0}}\pi\eta M_0 = p_0 e_{T_{\lambda_0}}^*\pi\eta(x; p_1, \ldots, p_{n-1})p_n,$$

and the proof follows, since, by Corollary 5.52, $e_{T_{\lambda_0}}\pi\eta M_0 \in \mathrm{id}(A)$. $\square$

Besides their important combinatoric properties already exploited in Chapter 2 and their structural properties described above, sparse identities provide us with the following cheap version of Specht's conjecture. Formanek-Lawrence [ForL76] proved for any field F of characteristic 0 that $F[S_\infty]$, the group algebra of the infinite symmetric group, is weakly Noetherian, i.e., satisifies the ACC on two-sided ideals. Since, in characteristic 0, sparse identities generate two-sided ideals in the group algebra, the Formanek-Lawrence result gives a quick proof of:

Proposition 5.59. *Any T-ideal in characteristic 0 generated by sparse identities is finitely based (by sparse identities).*

We expect that this weaker result holds in arbitrary characteristic (cf. the discussion at the end of this chapter), despite the counterexamples given in Chapter 7 to Specht's problem in nonzero characteristic.

5.5 Multilinearization

In the proof of the key Theorem 5.34, one moves back and forth between multilinear polynomials (corresponding to V_n) and homogeneous polynomials (corresponding to $T^n(V)$). Our objective here is to correlate the two theories, explaining why they coincide in characteristic 0. To do this, we must first translate the multilinearization process (Definition 1.11) to the S_n-actions we have been studying. First, we recall the action of Definition 5.31, also cf. Remark 5.32, which is a "place-permutation."

Remark 5.60. If $f = f(x_1, \ldots, x_n) \in V_n$, then

$$f \cdot (y_1^{d_1} \ldots y_k^{d_k}) = f(y_1, \ldots, y_1, \ldots, y_k, \ldots, y_k),$$

where each y_j is repeated d_j times. It follows that if $f \in \mathrm{id}(A)$, then $f \cdot y_1^{d_1} \ldots y_k^{d_k} \in \mathrm{id}(A)$. Working backwards, if $n = d_1 + \cdots + d_\ell$ and $p = p(x_1, \ldots, x_\ell) \in T^n(V)$ is homogeneous of degree $(d_1, \ldots, d_\ell)$, then there exists an element $a \in F[S_n]$ such that $p(x_1, \ldots, x_\ell) = a \cdot (x_1^{d_1} \cdots x_\ell^{d_\ell})$.

Given i, j write $S_{i,\ldots,j}$ for the subgroup of S_n fixing all k other than for $i \leq k \leq j$. Thus, $S_{1,\ldots,n} = S_n$.

Definition 5.61. If $n = d_1 + \cdots + d_\ell$, the subgroup $Y_{\underline{d}} = Y_{d_1,\ldots,d_\ell} \subseteq S_n$, given by

$$Y_{\underline{d}} = S_{1,\ldots,d_1} \times S_{d_1+1,\ldots,d_1+d_2} \times \cdots \times S_{n-d_\ell+1,\ldots,n}$$

is called a *Young subgroup*.

Note that

$$\left(\sum_{\sigma \in Y_{\underline{d}}} \sigma \right) \cdot (x_1 \cdots x_n) = \left(\sum_{\sigma \in Y_{d_1,\ldots,d_\ell}} \sigma \right) \cdot (x_1 \cdots x_n) =$$

$$\left(\sum_{\sigma \in S_{1,\ldots,d_1}} x_{\sigma(1)} \cdots x_{\sigma(d_1)} \right) \left(\sum_{\sigma \in S_{d_1+1,\ldots,d_1+d_2}} x_{\sigma(d_1+1)} \cdots x_{\sigma(d_1+d_2)} \right) \cdots$$

Let A be a PI-algebra over an infinite field. By Proposition 1.17, every homogeneous part of an identity is an identity, so the study of the identities of A is reduced to the study of its homogeneous identities. Given a homogeneous polynomial p, we now study its multilinearization $L(p)$ in terms of our S_n-action on $T^n(V)$, showing that if p is an identity of A, then $L(p)$ is also an identity of A. Furthermore, if the characteristic of the base field F is zero, we reprove that the multilinear identities of A determine all other identities. In Chapter 1 we defined the multilinear process inductively, in terms of a sequence of substitutions. Here we define it in terms of a single operator L. We write y_i for the indeterminates used in the polynomial before the application of L, and x_i for the indeterminates after L is applied.

Definition 5.62. Let $\underline{d} = (d_1, \ldots, d_k)$ with $d_1 + \cdots + d_k = n$, and let p be a homogeneous polynomial of degree $\underline{d}$, so that $p(y) = a \cdot (y_1^{d_1} \cdots y_k^{d_k})$ where $a \in F[S_n]$. Define the *multilinearization operator* L as follows:

$$L(y_1^{d_1} \cdots y_k^{d_k}) = \left(\sum_{\sigma \in Y_{d_1, \ldots, d_k}} \sigma \right) \in F[S_n],$$

and, in general,

$$L(p) = L(a \cdot (y_1^{d_1} \cdots y_k^{d_k})) = L(y_1^{d_1} \cdots y_k^{d_k})a$$

$$= \left(\left(\sum_{\sigma \in Y_{d_1, \ldots, d_k}} \sigma \right) \cdot (x_1 \cdots x_n) \right) a \in F[S_n].$$

Lemma 5.63. *Let $\underline{d} = (d_1, \ldots, d_k)$ with $d_1 + \cdots + d_k = n$, and let p be a homogeneous polynomial of degree $\underline{d}$. Then*

(i) $L(p) \cdot (y_1^{d_1} \cdots y_k^{d_k}) = d_1! \cdots d_k! p(y_1, \ldots, y_k).$

(ii) *Under the specializations*

$$y_1 \mapsto \bar{y}_1 = x_1 + \cdots + x_{d_1}, \quad y_2 \mapsto \bar{y}_2 = x_{d_1+1} + \cdots + x_{d_1+d_2}, \quad \ldots,$$

we have $p(\bar{y}_1, \ldots, \bar{y}_k) = L(p) + h$, where h is a linear combination of non-multilinear monomials.

Proof. (i) Write $\bar{y}$ for $y_1^{d_1} \ldots y_k^{d_k}$, and b for $\sum_{\sigma \in Y_{d_1, \ldots, d_k}} \sigma$. Writing $p = a \cdot bar y$, we have $L(p) = ba$. Hence,

$$L(p) \cdot \bar{y} = (ba) \cdot \bar{y} = a \cdot (b \cdot y),$$

and the proof follows from the obvious fact that

$$\left(\sum_{\sigma \in S_d} \sigma\right) \cdot y^d = d! y^d.$$

(ii) First, $(x_1 + \cdots + x_d)^d = \sum_{\sigma \in S_d} x_{\sigma(1)} \cdots x_{\sigma(d)} + g(x_1, \ldots, x_d)$, where $g(x_1, \ldots, x_d)$ is a linear combination of non-multilinear monomials. By a similar argument,

$$\bar{y}_1^{d_1} \cdots \bar{y}_k^{d_k} = \ell(x_1, \ldots, x_n) + H(x_1, \ldots, x_n)$$

where $\ell(x_1, \ldots, x_n) = \left(\sum_{\sigma \in Y_{\underline{d}}} \sigma\right) \cdot (x_1 \cdots x_n)$ and $H(x_1, \ldots, x_d)$ is a linear combination of non-multilinear monomials. Thus,

$$\bar{y}_1^{d_1} \cdots \bar{y}_k^{d_k} = L(y_1^{d_1} \cdots y_k^{d_k}) + H(x_1, \ldots, x_n).$$

Let $p(y_1, \ldots, y_k) = a \cdot (y_1^{d_1} \cdots y_k^{d_k})$ where $a \in V_n = F[S_n]$, then $p(\bar{y}_1, \ldots, \bar{y}_k) = a \cdot (\bar{y}_1^{d_1} \cdots \bar{y}_k^{d_k}) = L(y_1^{d_1} \cdots y_k^{d_k})a + a \cdot H(x_1, \ldots, x_n)$. By definition, $L(y_1^{d_1} \cdots y_k^{d_k})a = L(p)$ and also, $a \cdot H(x_1, \ldots, x_n) = h(x_1, \ldots, x_n)$ is a linear combination of non-multilinear monomials. □

Let us apply this result to identities.

Proposition 5.64. *Let A be a PI-algebra over F, and let p be a homogeneous polynomial of degree $\underline{d} = (d_1, \ldots, d_k)$, with multilinearization $L(p)$.*

 (i) *Assume the field F is infinite, of arbitrary characteristic. If $p \in \mathrm{id}(A)$, then $L(p) \in \mathrm{id}(A)$.*

 (ii) *Assume in addition that $\mathrm{char}(F) = 0$ (or $\mathrm{char}(F) > \max\{d_1, \ldots, d_k\}$). If $L(p) \in \mathrm{id}(A)$, then $p \in \mathrm{id}(A)$.*

Proof. (i) Recall that $p(\bar{y}_1, \ldots, \bar{y}_k) = L(p)(x_1, \ldots, x_n) + H(x_1, \ldots, x_n)$ as above, with $\bar{y}_1 = x_1 + \cdots + x_{d_1}$, etc. Since $p(\bar{y}_1, \ldots, \bar{y}_k)$ is an identity of A, the right-hand side is also an identity of A; hence, $L(p)(x_1, \ldots, x_n)$, its homogeneous part of degree 1 in each i, is an identity of A.
    (ii) This part clearly follows from Lemma 5.63(i): If $L(p) \in \mathrm{id}(A)$, then by Remark 5.60

$$L(p) \cdot (y_1^{d_1} \cdots y_k^{d_k}) = d_1! \cdots d_k! p(y_1, \ldots, y_k) \in \mathrm{id}(A),$$

implying $p(y_1, \ldots, y_k) \in \mathrm{id}(A)$ since $d_1! \cdots d_k!$ is invertible in F.

□

Proposition 5.65. *If* char$(F) = 0$, *then*

$$(\mathrm{id}(A) \cap V_n) \cdot T^n(V) = \mathrm{id}(A) \cap T^n(V).$$

Thus, by homogeneity,

$$\mathrm{id}(A) \cap T(V) = \bigoplus_n ((\mathrm{id}(A) \cap V_n) \cdot T^n(V)).$$

Proof. ($\subseteq$) If $f \in \mathrm{id}(A) \cap V_n$ and $w \in T^n(V)$ is a monomial, then by Remark 5.32(ii), $fw \in \mathrm{id}(A) \cap T^n(V)$, which implies the inclusion $\subseteq$.

($\supseteq$) Suppose $q(y_1, \ldots, y_k) \in \mathrm{id}(A) \cap T^n(V)$. By Proposition 1.17, we may assume that q is homogeneous, say of degree $\underline{d} = (d_1, \ldots, d_k)$. By Proposition 5.64(i), its multilinearization $L(q) \in \mathrm{id}(A) \cap V_n$, and by Lemma 5.63(i), $L(q) \cdot (y_1^{d_1} \cdots y_k^{d_k}) = d_1! \cdots d_k! q(y_1, \ldots, y_k)$. Since char$(F) = 0$, it follows that $q \in (\mathrm{id}(A) \cap V_n) \cdot T^n(V)$, which completes the proof. □

5.6　Appendix: Representation Theory in Characteristic $p > 0$

Much of the theory developed in this chapter works in characteristic $p > 0$, with some modifications. We can obtain analogous results by using modular representation theory in an analogous fashion to the Young theory. The Branching Theorem 5.21 and Hook Theorem 5.29 remain applicable. Although Remark 5.17 is no longer true, there is a workable analog in characteristic p:

Define a tableau to be *p-regular* if it does not have p rows of the same length. Then each p-regular tableau provides a simple module. Although for $n \geq p$, the group algebra $F[S_n]$ is not semisimple, we noted in Remark 5.15 that it still has a composition series whose composition factors are simple modules thereby corresponding to p-regular tableaux. There also is a characteristic free proof of Lemma 5.39, and thus of Theorem 5.42, given in [Row80, p. 241] (and in fact, the linear algebraic computations in Regev's original proof are characteristic free).

In order to obtain a p-version of Theorem 5.51, we need to find some p-regular shape whose hook number grows faster than the codimension. This role is satisfied by the "staircase tableau" T_n of n rows, with the j-th row having j boxes (a total of $\binom{n}{2}$ boxes); this tableau is p-regular for each p, so its module M is simple, and exponential bounds for the dimension can be calculated easily using the hook formula.

In fact, one can get the inequalities

$$\sqrt{\frac{\pi}{(2n)^{1/2}e}} 2^{-n} n^{(n-\sqrt{n})/2} e^{-\sqrt{18n}} < \dim M < \sqrt{\frac{\pi}{(2n)^{1/2}e}} 2^{-n} n^{(n-\sqrt{n})/2}.$$

In characteristic p, we use Remark 5.56 as a definition: A multilinear identity of degree n is called *strongly sparse* sparse if it generates a two-sided ideal in $F[S_m]$ for all $m \geq n$.

In analogy to Theorem 5.51, we see that if A satisfies a PI of degree d and $m > 10e^3 d^8$, then every polynomial corresponding to a tableau properly containing T_d is an identity of A, and arguing exactly as in Theorem 5.58, we conclude that A satisfies a strongly sparse identity. However, Kemer[Kem95] surpassed this result, as noted in Chapter 7.

Zalesskii [Za96] generalized the Formanek-Lawrence theorem quoted before Proposition 5.59 to characteristic $p > 0$, so the same argument shows that any T-ideal generated by strongly sparse identities is finitely based. Note that this theorem works in the non-affine case in all characteristics (whereas Kemer's theorem does not, cf. Chapter 7), and also is much much easier to prove; Belov's proof of Kemer's conjecture for affine algebras in characteristic p remains to be written in English.

Chapter 6

Superidentities and Kemer's Main Theorem

As mentioned in Chapter 1, "super" means "$\mathbb{Z}_2$-graded." In this chapter we develop a theory of superidentities of superalgebras, and prove Kemer's correspondence, that any PI-algebra in characteristic 0 corresponds to a suitable affine superalgebra. This assertion is false in nonzero characteristic. (Indeed, if true, it would verify Specht's conjecture, but we shall present counterexamples in Chapter 7.)

Following Kemer, we then backtrack and modify the results of Chapter 4 to prove Specht's conjecture for all algebras of characteristic 0. We shall need the following key observation about superalgebras.

Remark 6.1. We can view F in A_0. If A, R are F-superalgebras, then $A \otimes_F R$ is also a superalgebra, via

$$(A \otimes R)_0 = A_0 \otimes R_0 \oplus A_1 \otimes R_1; \qquad (A \otimes R)_1 = A_0 \otimes R_1 \oplus A_1 \otimes R_0.$$

In particular, viewing any algebra $A = A_0$ as a superalgebra under the trivial grade, $A_0 \otimes R$ becomes a superalgebra under the grade

$$(A_0 \otimes R)_i = A_0 \otimes R_i.$$

On the other hand, if K is any field extension of F, then we view $A \otimes_F K$ as a superalgebra by grading K trivially.

6.1 Superidentities

In Chapter 11, we develop a general theory of identities that also encompasses the graded situation. Meanwhile, let us do this in an ad hoc fashion, motivated by the need of having the identity keep track both of the even and odd parts. Consider polynomials in two sets of variables, $Y = \{y_1, y_2, \ldots\}$ and $Z = \{z_1, z_2, \ldots\}$. We say that the ys are even and the zs are odd. Then $F\{Y, Z\} = F\{y_1, y_2 \ldots; z_1, z_2 \ldots\}$ is the free superalgebra, in the sense of homomorphisms to superalgebras sending y_i to even elements and z_i to odd elements. The elements of $F\{Y, Z\}$ are called *superpolynomials*.

Definition 6.2. A *superideal* of a superalgebra is an ideal I each of whose elements is a sum of homogeneous elements of I, i.e., $I = I_0 \oplus I_1$. In this case, we write $I \triangleleft_2 A$.

Definition 6.3.

(i) The superpolynomial $p(y_1, \ldots, y_k; z_1, \ldots, z_\ell)$ is a *superidentity* of the superalgebra A if for any $\bar{y}_i \in A_0$ and $\bar{z}_j \in A_1$,

$$p(\bar{y}_1, \ldots, \bar{y}_k; \bar{z}_1, \ldots, \bar{z}_\ell) = 0.$$

We denote by $\mathrm{id}_2(A) \subseteq F\{Y, Z\}$ the set of superidentities of A; we say superalgebras A and A' are *PI$_2$-equivalent*, written $A \sim_{\mathrm{PI}_2} A'$, if $\mathrm{id}_2(A) = \mathrm{id}_2(A')$.

(ii) Similarly, the multilinear superidentities are the multilinear superpolynomials that are superidentities.

(iii) Define a *T_2-ideal* of $F\{Y, Z\}$ to be a superideal that is invariant under all graded homomorphisms $F\{Y, Z\} \to F\{Y, Z\}$ (sending even elements to even elements and odd elements to odd elements). Then $\mathrm{id}_2(A)$ is a T_2-ideal of $F\{Y, Z\}$, for any superalgebra A.

(iv) As in the ungraded case, given any T_2-ideal $\mathcal{I}$, we have the *relatively free* superalgebra

$$\mathcal{U} = F\{\bar{Y}, \bar{Z}\} = F\{Y, Z\}/\mathcal{I},$$

which is generated by even-generic elements $\bar{y}_i$ and odd-generic elements $\bar{z}_j$. Clearly, $\mathrm{id}_2(\mathcal{U}) = \mathcal{I}$.

(v) Often we want to view some multilinear polynomial $f = f(x_1, \ldots, x_n)$ as a superpolynomial; to do this, we specify $I \subseteq \{1, \ldots, n\}$. Then for $i \notin I$, we rewrite y_i for x_i as an even variable, and for $i \in I$, we rewrite z_i for x_i as an odd variable. Denote this by $f_I(y; z)$.

(vi) At times we shall want to permit multilinear superidentities to include ungraded variables, i.e.,

$$f(x_1, \ldots, x_j; y_1, \ldots, y_k; z_1, \ldots, z_\ell)$$

vanishes for all substitutions of x_i, all even substitutions of y_i, and all odd substitutions of z_i. We call this a *mixed superidentity*. Since any substitution for x_i is the sum of its even and odd parts, we see that f is an identity of A iff every superpolynomial obtained by substituting each x_i to an even or odd variable is in $\mathrm{id}_2(A)$. (To express all these superidentities we need $j + k$ odd variables and $j + \ell$ even variables.) Abusing language slightly, we get the corresponding relatively free mixed superalgebra $F\{X, Y, Z\}/\mathrm{id}_2(A)$.

(vii) One gets a theory of mixed superidentities for affine superalgebras (i.e., superalgebras that are generated by a finite number of homogeneous elements), by taking finite sets of indeterminates X,Y,Z.

Most of the elementary facts about ordinary identities extend easily to (mixed) superidentities. This in particular includes the process of multilinearization described in (1.2) of Chapter 1 (and in more detail in Chapter 5). The operator Δ is defined exactly as before, with the proviso that odd elements linearize to odd elements, and even elements linearize to even elements.

Remark 6.4. Let $A = A_0 \oplus A_1$ be a superalgebra and let $w = w(y; z) = w(y_1, \dots, y_k; z_1, \dots, z_\ell)$ be any word in the ys and zs. Specialize $y_i \mapsto \bar{y}_i \in A_0$ and $z_i \mapsto \bar{z}_i \in A_1$, and denote $\bar{w} = w(\bar{y}; \bar{z})$. Let d denote the total z-degree of w (for example, if $w = z_1 y_2 z_2 y_1 z_1$, then $d = 3$). If d is even (odd), then $\bar{w} \in A_0$ ($\bar{w} \in A_1$).

6.1.1   The role of odd elements

One of the main roles is played, perhaps surprisingly, by the Grassman algebra G on a countably infinite dimensional F-vector space with base $e_1, e_2, \dots$, cf. Example 1.36.

Remark 6.5. The Grassman identity $[[x, y], z]$ can be translated to the superidentities

$$[y_1, y_2], \qquad [y_1, z_1], \qquad z_1 z_2 + z_2 z_1, \qquad (6.1)$$

for even y_i and odd z_i. Moreover, Remark 1.38 translates into the superidentities

$$z_1 x z_2 + z_2 x z_1$$

where z_i are odd variables.

Taking $z_1 = z_2$, we get the superidentity $z_1 x z_1$ whenever $\mathrm{char}(F) \neq 2$.

Kemer's contribution at this stage was to incorporate this structure of G into the PI-theory, by passing from A_0 to the superalgebra $A_0 \otimes G$, cf. Remark 6.1. We study the superidentities in a procedure formalized by Berele, [Ber85b]. Recall from Remark 1.36 the base

$$B = \{e_{i_1} \cdots e_{i_m} \mid i_1 < \cdots < i_m, \quad m = 1, 2, \ldots\} \qquad (6.2)$$

of G. Generalizing Remark 6.5, we introduce the sign function $\varepsilon(\sigma, I)$.

Remark 6.6. Given an n-tuple $(b) = (b_1, \ldots, b_n)$ of such base elements such that $b_1 \ldots b_n \neq 0$, denote $I = \text{Odd}(b) = \{i \mid b_i \text{ is odd}\}$. Then for any $\sigma \in S_n$, the equation

$$b_{\sigma(1)} \cdots b_{\sigma(n)} = \varepsilon(\sigma, I) b_1 \cdots b_n \qquad (6.3)$$

uniquely defines the function $\varepsilon(\sigma, I)$, independent of the particular choice of the b_is; note that $\varepsilon(\sigma, I) = \pm 1$.

Remark 6.7. If $\text{Odd}(b) = \text{Odd}(b_1, \ldots, b_n) = I$, then

$$\text{Odd}(b_{\sigma(1)}, \ldots, b_{\sigma(n)}) = \sigma^{-1}(I),$$

where $\sigma^{-1}(I) = \{\sigma^{-1}(i) \mid i \in I\}$. Indeed, $j \in \text{Odd}(b_{\sigma(1)}, \ldots, b_{\sigma(n)})$ if and only if $b_{\sigma(j)}$ is odd. Write $j = \sigma^{-1}(i)$. Then $\sigma^{-1}(i) \in \text{Odd}(b_{\sigma(1)}, \ldots, b_{\sigma(n)})$ if and only if b_i is odd, namely, if and only if $i \in I$; that is, $\sigma^{-1}(i) \in \sigma^{-1}(I)$.

We can show now that $\varepsilon(-, I) : S_n \to \{\pm 1\}$ is almost a homomorphism: It is one precisely when either I is empty or $I = \{1, \ldots, n\}$.

Lemma 6.8. Let $\sigma, \eta \in S_n$ and $I \subseteq \{1, \ldots, n\}$, then

$$\varepsilon(\sigma\eta, I) = \varepsilon(\eta, \sigma^{-1}(I))\varepsilon(\sigma, I).$$

Proof. Choose $(b) = (b_1, \ldots, b_n)$, where $b_1, \ldots, b_n \in B$ are base elements such that $b_1 \cdots b_n \neq 0$ and $\text{Odd}(b) = I$. Denote $a_j = b_{\sigma(j)}$. By Remark 6.7, $\text{Odd}(a_1, \ldots, a_n) = \sigma^{-1}(I)$ and also $b_{\sigma\eta(i)} = a_{\eta(i)}$. Thus,

$$\varepsilon(\sigma\eta, I)b_1 \cdots b_n = b_{\sigma\eta(1)} \cdots b_{\sigma\eta(n)} = a_{\eta(1)} \cdots a_{\eta(n)}$$
$$= \varepsilon(\eta, \sigma^{-1}(I))a_1 \cdots a_n = \varepsilon(\eta, \sigma^{-1}(I))b_{\sigma(1)} \cdots b_{\sigma(n)}$$
$$= \varepsilon(\eta, \sigma^{-1}(I))\varepsilon(\sigma, I)b_1 \cdots b_n.$$

Matching coefficients yields the desired result. □

6.1.2   The Grassman involution on polynomials

Assume throughout that $\text{char}(F) = 0$.

Definition 6.9. Given a multilinear polynomial

$$p = p(x_1, \ldots, x_n) = \sum_{\sigma \in S_n} \alpha_\sigma x_{\sigma(1)} \cdots x_{\sigma(n)}$$

and $I \subseteq \{1, \ldots, n\}$, define

$$p_I^* = \sum_{\sigma \in S_n} \alpha_\sigma \varepsilon(\sigma, I) x_{\sigma(1)} \cdots x_{\sigma(n)}.$$

We call $(*)$ the *Grassman involution* (although it also depends on our choice of I).

Also, given $p(x_1, \ldots, x_n)$ and $I \subseteq \{1, \ldots, n\}$, define $\tilde{p}_I^*$ to be the superpolynomial obtained from p_I^* by substituting z_i for x_i whenever $i \in I$, and substituting y_i for x_i whenever $i \notin I$.

Theorem 6.10. *Let A_0 be a PI-algebra, and let $p = p(x_1, \ldots, x_n) = \sum_{\sigma \in S_n} \alpha_\sigma x_{\sigma(1)} \cdots x_{\sigma(n)}$ be a multilinear polynomial. Then the following assertions are equivalent:*

(i) $p \in \text{id}(A_0)$.

(ii) $\tilde{p}_I^*$ *is a superidentity of* $A = A_0 \otimes G$ *for some subset* $I \subseteq \{1, \ldots, n\}$.

(iii) $\tilde{p}_I^*$ *is a superidentity of* $A = A_0 \otimes G$ *for every subset* $I \subseteq \{1, \ldots, n\}$.

Proof. It clearly is enough to show $(i) \Leftrightarrow (ii)$, since then we apply this for each I to get (iii). Let B denote the canonical base of G given in (6.2). Both directions of the proof easily follow from the following equality:

$$\tilde{p}_I^*(a_1 \otimes b_1, \ldots, a_n \otimes b_n) = p(a_1, \ldots, a_n) \otimes b_1 \cdots b_n \qquad (6.4)$$

whenever $a_1, \ldots, a_n \in A_0$ and $b_1, \ldots, b_n \in B$ with $\text{Odd}(b_1, \ldots, b_n) = I$. This holds since

$$\tilde{p}_I^*(a_1 \otimes b_1, \ldots, a_n \otimes b_n) = \sum_\sigma \varepsilon(\sigma, I) \alpha_\sigma (a_{\sigma(1)} \otimes b_{\sigma(1)}) \cdots (a_{\sigma(n)} \otimes b_{\sigma(n)})$$

$$= \sum_\sigma \varepsilon(\sigma, I) \alpha_\sigma (a_{\sigma(1)} \cdots a_{\sigma(n)}) \otimes (b_{\sigma(1)} \cdots b_{\sigma(n)})$$

$$= \left(\sum_\sigma \alpha_\sigma a_{\sigma(1)} \cdots a_{\sigma(n)} \right) \otimes b_1 \cdots b_n.$$

Hence, if $p \in \text{id}(A_0)$, then each $\tilde{p}_I^* \in \text{id}_2(A_0 \otimes G)$.

Conversely, assume $\tilde{p}_I^*$ is a superidentity of $A_0 \otimes G$. Pick $b_1, \ldots, b_n \in B$ such that $\text{Odd}(b_1, \ldots, b_n) = I$ and $b_1 \cdots b_n \neq 0$; then (6.4) shows that $p(a_1, \ldots, a_n) = 0$ for any elements $a_1, \ldots, a_n \in A_0$.   □

Corollary 6.11. *Suppose* $\text{char}(F) = 0$. *There is an inclusion-preserving map* $\Phi : \{T\text{-ideals}\} \to \{T_2\text{-ideals}\}$ *given by* $\text{id}(A_0) \to \text{id}_2(A_0 \otimes G)$.

Proof. Every T-ideal is determined by its multilinear polynomials.   □

Better yet, Φ could be viewed as a functor from varieties to supervarieties; this can be done in any characteristic, cf. Exercise 1.

6.1.3   The Grassman envelope

Kemer's next observation was to find an inverse functor, thereby elevating the Grassman algebra to a position of prominence in the PI-theory.

Definition 6.12. The *Grassman envelope* of a superalgebra A is the subalgebra

$$G(A) = (A \otimes G)_0 = A_0 \otimes G_0 \oplus A_1 \otimes G_1.$$

Theorem 6.13. *Let* A *be a PI-superalgebra, and let* $p = p(x_1, \ldots, x_n) = \sum_{\sigma \in S_n} \alpha_\sigma x_{\sigma(1)} \cdots x_{\sigma(n)}$ *be a multilinear polynomial. Then* $p \in \text{id}(G(A))$ *if and only if* $\tilde{p}_I^* \in \text{id}_2(A)$ *for every subset* $I \subseteq \{1, \ldots, n\}$.

Proof. As in the proof of Theorem 6.10. To check whether $p \in \text{id}(G(A))$ we check substitutions $a \otimes b$, where $a \in A$ and $b \in G$ are both homogeneous of the same parity (i.e., $a_i \otimes b_i$ is in $A_u \otimes G_u$ for $u \in \{0,1\}$). But then the following analog of Equation (6.4) is applicable, where $I = \{i : a_i \text{ is odd}\}$:

$$p(a_1 \otimes b_1, \ldots, a_n \otimes b_n) = \sum_\sigma \alpha_\sigma (a_{\sigma(1)} \otimes b_{\sigma(1)}) \cdots (a_{\sigma(n)} \otimes b_{\sigma(n)}) \qquad (6.5)$$

$$= \sum_\sigma \alpha_\sigma (a_{\sigma(1)} \cdots a_{\sigma(n)}) \otimes (b_{\sigma(1)} \cdots b_{\sigma(n)})$$

$$= \left(\sum_\sigma \varepsilon(\sigma, I) \alpha_\sigma a_{\sigma(1)} \cdots a_{\sigma(n)} \right) \otimes b_1 \cdots b_n, \qquad (6.6)$$

i.e.,

$$p(a_1 \otimes b_1, \ldots, a_n \otimes b_n) = \tilde{p}_I^*(a_1, \ldots, a_n) \otimes b_1 \ldots b_n, \qquad (6.7)$$

which now shows that if each $\tilde{p}_I^* \in \text{id}_2(A)$, then $p \in \text{id}(G(A))$. Conversely, assume $p \in \text{id}(G(A))$ and let $I \subseteq \{1, \ldots, n\}$. Pick $b_1, \ldots, b_n \in B$ such that $\text{Odd}(b_1, \ldots, b_n) = I$ and $b_1 \cdots b_n \neq 0$; then Equation (6.7) shows that $p_I^*(a_1, \ldots, a_n) = 0$ for any homogeneous $a_1, \ldots, a_n \in A$, where a_i is odd iff $i \in I$.   □

Theorem 6.14. *There is a correspondence* Ψ *from* {*varieties of superalgebras*} *to* {*varieties of algebras*} *given by* $A \mapsto G(A)$, *which is the inverse of* Φ *of Corollary 6.11, in the sense that*

$$A_0 \sim_{\mathrm{PI}} G(A_0 \otimes G), \qquad A_0 \quad any \; algebra \tag{6.8}$$

and

$$A \sim_{\mathrm{PI}_2} G(A) \otimes G, \qquad A \quad any \; superalgebra. \tag{6.9}$$

Proof. We have proved everything except that they are inverses, i.e., the two displayed equations. But

$$G(A_0 \otimes G) \approx ((A_0 \otimes G_0) \otimes G_0) \oplus ((A_0 \otimes G_1) \otimes G_1) \approx A_0 \otimes (G_0 \otimes G_0 \oplus G_1 \otimes G_1)).$$

Now $G_0 \otimes G_0 \oplus G_1 \otimes G_1$ is a commutative algebra, yielding (6.8) in view of Proposition 1.16.

To see (6.9), it suffices by Theorem 6.13 to show

$$G(A) \sim_{\mathrm{PI}} G(G(A) \otimes G) \approx G(A) \otimes (G_0 \otimes G_0 + G_1 \otimes G_1),$$

which again is true by Proposition 1.16. $\square$

6.1.4 The $\bullet$-action of S_n on polynomials

In order to understand the map Φ of Corollarly 6.11 and its inverse Ψ, we introduce a new action of the symmetric group, on the multilinear polynomials $T^n(V)$, cf. Section 5.3, but now taking into account the indices I of odd (i.e., strictly anticommuting) Grassman substitutions, by means of the sign function $\varepsilon(\sigma, I)$.

Definition 6.15.

(i) Let $V = V_0 \oplus V_1$ be vector spaces with $\dim V_0 = k$ and $\dim V_1 = \ell$. For $i \in \{0, 1\}$, we say that $x \in V$ is *homogeneous of degree* i if $x \in V_i$, and then denote $\delta(x) = i$. An element $w \in T^n(V)$ is *homogeneous* if $w = x_1 \otimes \cdots \otimes x_n = x_1 \cdots x_n$ where each x_j is homogeneous; in that case, denote $\delta(w) = \delta(x_1) + \cdots + \delta(x_n) \pmod 2$.

(ii) Given $(x) = (x_1, \ldots, x_n)$ where $x_1, \ldots, x_n \in V_0 \cup V_1$, denote

$$\mathrm{Odd}(x) = \{i \mid \delta(x_i) = 1\}.$$

(iii) Define the $\bullet$-action of S_n on $T^n(V)$ as follows. Let $x_1 \cdots x_n \in T^n(V)$ be homogeneous elements with $I = \mathrm{Odd}(x)$, and let $\sigma \in S_n$. Then

$$\sigma \bullet (x_1 \cdots x_n) = \varepsilon(\sigma, I)(x_{\sigma(1)} \cdots x_{\sigma(n)}).$$

Extend that action to all of $T^n(V)$ by linearity. In fact, we can choose bases $y_1, \ldots, y_k \in V_0$, $z_1, \ldots z_\ell \in V_1$, and assume above that $x_1, \ldots, x_n \in \{y_1, \ldots, y_k, z_1, \ldots, z_\ell\}$; then extend by linearity.

Remark 6.16. Notice that here $\varepsilon(\sigma, I) = \pm 1$ is calculated as follows. σ induces a permutation τ on the elements of degree 1, and $\varepsilon(\sigma, I) = \operatorname{sgn}(\tau)$.

For example, let $w = x_1 x_2 x_3$ (homogeneous) and let σ be the 3-cycle $\sigma = (1, 2, 3)$, so that $\sigma w = x_2 x_3 x_1$. If $\delta(x_1) = \delta(x_2) = 1$ and $\delta(x_3) = 0$, then $\tau = (1, 2)$; and hence $\varepsilon = -1$ and $\sigma \bullet w = -\sigma w = -x_2 x_3 x_1$. However, if $\delta(x_2) = \delta(x_3) = 1$ and $\delta(x_1) = 0$, then $\tau = 1$ hence $\varepsilon = +1$.

Proposition 6.17. *Let $\sigma, \eta \in S_n$ and let $w \in V = V_0 \oplus V_1$. Then*

$$(\sigma\eta) \bullet w = \eta \bullet (\sigma \bullet w).$$

Hence, the $\bullet$-action makes $T^n(V)$ into a left module over the opposite algebra $(F[S_n])^{op}$.

Proof. By linearity we may assume that $w = x_1 \cdots x_n$ with $x_1, \ldots, x_n$ homogeneous. Let $I = \operatorname{Odd}(x_1, \ldots, x_n)$. As in Remark 6.7,

$$\operatorname{Odd}(x_{\sigma(1)}, \ldots, x_{\sigma(n)}) = \sigma^{-1}(I).$$

Now calculate:

$$(\sigma\eta) \bullet (x_1 \cdots x_n) = \varepsilon(\sigma\eta, I) x_{\sigma\eta(1)} \cdots x_{\sigma\eta(n)},$$

while

$$\eta \bullet (\sigma \bullet (x_1 \cdots x_n)) = \varepsilon(\sigma, I)(\eta \bullet (x_{\sigma(1)} \cdots x_{\sigma(n)}))$$
$$\varepsilon(\sigma, I)\varepsilon(\eta, \sigma^{-1}(I)) x_{\sigma\eta(1)} \cdots x_{\sigma\eta(n)}.$$

The proof now follows from Lemma 6.7. $\qquad\qquad\square$

Corollary 6.18. *The map Φ of Corollary 6.11 is a homomorphism; i.e., if J_1, J_2 are T-ideals, then $\Phi(J_1 J_2) = \Phi(J_1)\Phi(J_2)$.*

This $\bullet$-action has a very important property:

Remark 6.19. Suppose σ is a transposition (i, j). If $i, j \in I$ then $\sigma \bullet f = -\sigma f$; otherwise, $\sigma \bullet f = +\sigma f$.

6.2 Kemer's PI-Superrepresentability Theorem

We are ready to tackle Kemer's PI-Superrepresentability Theorem. The groundwork is laid in the next theorem. Our proof takes elements both from Kemer's original proof and from Berele [Ber85b].

Theorem 6.20 (Kemer's correspondence). *Let A be a PI-algebra over a field of characteristic 0. Then $\mathrm{id}_2(A \otimes G) = \mathrm{id}_2(\tilde{A})$, for a suitable affine superalgebra $\tilde{A}$ to be described in the proof.*

Proof. By Proposition 5.54, there are integers k, ℓ such that the Young frame of every multilinear nonidentity of A lies in the (k, ℓ) hook. We already saw $\mathcal{I} = \mathrm{id}_2(A \otimes G)$ is the T_2-ideal determined by the multilinear superpolynomials $\{\tilde{f}_I^* : f \in \mathrm{id}(A)\}$. Let

$$\mathcal{A} = F\{X, Y, Z\}/\mathcal{I}$$

be the relatively free (mixed) superalgebra determined by these superpolynomials, which by definition is T_2-equivalent to $A \otimes G$, and let $\tilde{A}$ be the affine superalgebra generated by $\bar{y}_1, \ldots, \bar{y}_{k\ell+k}, \bar{z}_1, \ldots, \bar{z}_{k\ell+\ell}$.

In other words,

$$\tilde{A} = F\{y_1, \ldots, y_{k\ell+k}, z_1, \ldots, z_{k\ell+\ell}\}/\mathcal{I} \cap F\{y_1, \ldots, y_{k\ell+k}, z_1, \ldots, z_{k\ell+\ell}\}.$$

We claim $\tilde{A} \sim_{\mathrm{PI}_2} \mathcal{A}$.

Since any superidentity of $\mathcal{A}$ obviously is a superidentity of the sub-superalgebra $\tilde{A}$, it suffices to prove that $\mathrm{id}_2(\tilde{A}) \subseteq \mathrm{id}_2(\mathcal{A})$, or equivalently, $\mathrm{id}(G(\tilde{A})) \subseteq \mathrm{id}(A)$. So given multilinear $f(x_1, \ldots, x_n) \in \mathrm{id}(G(\tilde{A}))$, we need to prove f is an identity of A. But now we can bring in the Young theory. Suppose f corresponds to a Young tableau T_λ. As noted above, we are done unless T_λ lies in the (k, ℓ) hook. We subdivide $T_\lambda = T_0 \cup T_1 \cup T_2$, where:

1. T_0 is the horizontal strip of width at most k lying past the ℓ column.

2. T_1 is the vertical strip of width at most ℓ lying past the k row.

3. T_2 is the part of T lying in ℓ^k, the $k \times \ell$ rectangle.

By Proposition 6.17, we may replace f by any element in the left ideal it generates in $F[S_n]$, and thus may take f to be the semi-idempotent

$$e_{T_\lambda} = \sum_{q \in C_{T_\lambda}} \sum_{p \in R_{T_\lambda}} \mathrm{sgn}(q)qp = C_{T_\lambda}^- R_{T_\lambda}^+.$$

In particular, if i, j lie in the same row in T_0, then $(i,j)f = f$; likewise if i, j lie in the same column in T_1, then $(i,j)f = -f$. Let I be the collection of indices labeling boxes in T_1. Let $g = \tilde{f}_I^*$. In view of Theorem 6.10(ii), it suffices to prove $g \in \mathrm{id}_2(A \otimes G)$, i.e., $g(a_1 \otimes b_1, \ldots, a_n \otimes b_n) = 0$ where the b_i are odd iff i appears in T_1. But let us see what happens with $(i,j) \bullet g$.

If i, j lie in the same row in T_0, then $(i,j) \bullet g = g$, as before.

If i, j lie in the same column in T_1, then $(i,j) \bullet g = -(-g) = g$.

Let h denote the superpolynomial obtained from g by specializing all indeterminates in the i row of T_0 to y_i, for $1 \le i \le k$, and specializing all indeterminates in the j column of T_1 to z_j, for $1 \le j \le \ell$. Then g is the multilinearization of h (up to a scalar multiple), by Remark 1.30. But h involves only $k\ell$ indeterminates in x, as well as $y_1, \ldots, y_k$ and $z_1, \ldots, z_\ell$. Thus, $h \in \mathrm{id}_2(\tilde{A})$ iff $h \in \mathrm{id}_2(\mathcal{A})$.

Now we put everything together. By hypothesis $g \in \mathrm{id}_2(\tilde{A})$, so $h \in \mathrm{id}_2(\tilde{A})$, implying $h \in \mathrm{id}_2(\mathcal{A}) = \mathrm{id}_2(A \otimes G)$, so its multilinearization $g \in \mathrm{id}_2(A \otimes G)$, as desired. $\square$

To illustrate the power of this theorem, let us present some corollaries.

Corollary 6.21. *Any PI-algebra A satisfies the identities of $M_n(G)$ for suitable n.*

Proof. The theorem shows that $A \sim_{\mathrm{PI}} G(\tilde{A})$ for a suitable affine superalgebra $\tilde{A}$. But Theorem 2.57 says $J(\tilde{A})$ is nilpotent, implying by Lewin's Theorem that $\tilde{A}$ satisfies all identities of $M_n(F)$ for some n, so is a homomorphic image of some generic matrix algebra R. Then $G(\tilde{A}) \subset \tilde{A} \otimes G$ satisfies all identities of $R \otimes G$, which satisfies all identities of $M_n(G)$. $\square$

We shall improve this result in Exercise 4. A stronger result holds in characteristic p, cf. Theorem 7.5.

Corollary 6.22. *Any relatively free algebra has a unique largest nilpotent T-ideal.*

This in fact follows at once from Kemer's solution to Specht's problem, so we leave it as an exercise for the time being.

6.2.1   The structure of finite dimensional superalgebras

Since superalgebras play such a crucial role in the theory, let us review their structure.

Remark 6.23. The center of a superalgebra R is a (commutative) superalgebra. Indeed, if $z = z_0 \oplus z_1 \in \mathrm{Cent}(R)$, then z_0 commutes with any homogeneous component of R and thus $z_0 \in \mathrm{Cent}(R)$; likewise for z_1.

Remark 6.24. Any superalgebra $R = R_0 \oplus R_1$ has an automorphism σ of order 2, given by $r_0 + r_1 \mapsto r_0 - r_1$. (Indeed,

$$\sigma((r_0 + r_1)(s_0 + s_1)) = \sigma(r_0 s_0 + r_1 s_1 + r_0 s_1 + r_1 s_0)$$
$$= r_0 s_0 + r_1 s_1 - r_0 s_1 - r_1 s_0 = (r_0 - r_1)(s_0 - s_1),$$

proving σ is a homomorphism, and clearly $\sigma^2 = 1_R$.)

Conversely, if $\frac{1}{2} \in R$ and R has an automorphism σ of order 2, then we can define a 2-grade on R by putting

$$r_0 = \frac{r + \sigma(r)}{2}, \qquad r_1 = \frac{r - \sigma(r)}{2}, \qquad \forall r \in R.$$

Thus, $r \in R_0$ iff $\sigma(r) = r$, and $r \in R_1$ iff $\sigma(r) = -r$.

The following is a special case of [Row88, Lemma 2.5.39].

Lemma 6.25. *If R is a superalgebra and $\frac{1}{2} \in R$, then $J = J(R)$ is a superideal, and likewise the upper nilradical $N(R)$ is a superideal.*

Proof. Suppose $r_0 + r_1 \in J$. Notation as in Remark 6.24,

$$r_0 - r_1 \in \sigma(J) = J,$$

so $2r_0 \in J$, implying $r_0 \in J$, and hence $r_1 = r - r_0 \in J$. The same argument holds for $N(R)$ since $\sigma(N(R)) \subseteq N(R)$. $\qquad\square$

It follows at once that $R/J(R)$ is also a superalgebra, and we would like to determine its structure in the finite dimensional case. We say R is *supersimple* if R has no proper nonzero superideals.

Remark 6.26. If R is supersimple, then R_0 is simple. (If $I \triangleleft R_0$, then $I + R_1 I R_1 \triangleleft_2 R$.)

Here is an example of a supersimple superalgebra that is not simple.

Example 6.27. $R = M_n(F[c])$, where $R_0 = M_n(F)$, $R_1 = M_n(F)c$, and $c^2 = 1$. Any nonzero superideal I must contain some a or ac, for $a \in M_n(F)$.

If $a \in I$, then, since R_0 is simple, $R_0 = R_0 a R_0 \in I$, implying $1 \in I$; if $ac \in I$, then $R_0 c = (R_0 a R_0)c = R_0 a c R_0 \subseteq I$, implying $c \in I$, so $1 = c^2 \in I$.

Hence $I = R$ in either case, implying R is supersimple. On the other hand, R is not simple, since its center $F[c]$ is not a field.

In one important case, this is the only example.

Proposition 6.28. *Suppose F is an algebraically closed field.*

(i) *The only nontrivial $\mathbb{Z}/2$-gradings on $M_n(F)$ yield superalgebras isomorphic to those in Example 1.34.*

(ii) *Any f.d. supersimple superalgebra is either as in (i) or as in Example 6.27.*

(iii) *Any f.d. semisimple superalgebra R is isomorphic (as superalgebra) to a direct product $\prod_{i=1}^{k} S_i$ of supersimple superalgebras of form (i) or (ii).*

Proof. (i) Write $R = M_n(F) = R_0 \oplus R_1$, and assume $R_1 \neq 0$. $F^{(n)}$ cannot be a simple R_0-module, since otherwise by the density theorem $R_0 \approx \mathrm{End} F_D^{(n)}$ for some finite dimensional division algebra D over F; $D = F$ since F is algebraically closed, and we would conclude $R_0 = M_n(F)$.

So let V_0 be a proper R_0-submodule of $F^{(n)}$, and $V_1 = R_1 V_0$. Then $V_0 + V_1, V_0 \cap V_1$ are R-submodules of $F^{(n)}$, which is simple over R, so

$$V_0 + V_1 = F^{(n)} \qquad \text{and} \qquad V_0 \cap V_1 = 0.$$

Thus,

$$V_0 \oplus V_1 = R_0(V_0 \oplus V_1) \subseteq R_0 V_0 \oplus R_0 V_1 \subseteq V_0 \oplus V_1,$$

implying $R_0 V_0 = V_0$ and $R_0 V_1 = V_1$;

$$V_1 \oplus V_0 = R_1(V_0 \oplus V_1) \subseteq R_1 V_0 \oplus R_1 V_1 \subseteq V_1 \oplus V_0,$$

implying $R_1 V_0 = V_1$ and $R_1 V_1 = V_0$. Writing this in terms of bases of V_0, V_1 yields the desired result.

(ii) If a f.d. supersimple superalgebra R is simple, then it is as in (i), so we assume R is not simple. Let $0 \neq I \triangleleft R$. Since R is supersimple, we have $I \nsubseteq R_1$, so I contains some element $a = a_0 + a_1$ for $a_i \in R_i$, $a_0 \neq 0$. But R_0 is simple by Remark 6.26, so $R_0 a_0 R_0 = R_0$, and replacing a by the appropriate element of $R_0 a R_0$, we may assume $a_0 = 1$. Let $c = a_1 \in R_1$, so $1 + c \in I$. Then

$$c^2 + c = c(1 + c) \in I$$

so

$$c^2 - 1 = (c^2 + c) - (1 + c) \in I \cap R_0 = 0$$

(since R_0 is simple), implying $c^2 = 1$. In particular, c is invertible. Furthermore,

$$[c, R_1] = [1 + c, R_1] \subseteq [R_1, R_1] \cap [I, R_1] \subseteq R_0 \cap I = 0;$$

$$[c, R_0]c = [c, R_0c] \subseteq [c, R_1] = 0,$$

implying $[c, R_0] = 0$. Thus, $c \in \mathrm{Cent}(R)$ and

$$R_1 = R_1 c^2 = (R_1 c)c \subseteq R_0 c \subseteq R_1,$$

implying $R_1 = R_0 c$ and $R = R_0 + R_0 c$; we have recovered Example 6.27.

(iii) First we consider the case where R_0 is simple. We claim R is supersimple. Indeed, any proper superideal I must intersect R_0 trivially (since otherwise, $I \cap R_0 \triangleleft R_0$ implies $1 \in I \cap R_0$). But then $I \subseteq R_1$ so $0 = IR_1 \supseteq I^2$, implying $I = 0$. Thus, we are done by (ii).

In general, $\mathrm{Cent}(R)$ is a commutative superalgebra of the form $Z_0 \oplus Z_1$, where Z_0 must be semisimple (since any nilpotent ideal would generate a nilpotent ideal of R). Writing $Z_0 = Fe_1 \times \cdots \times Fe_k$ for each $Fe_i \approx F$, i.e., the e_i are central idempotents with $\sum e_i = 1$, we see $R = Re_1 \times \cdots \times Re_k$ is a direct product of superalgebras having simple even part, so we are done by the previous paragraph. □

In order to pass up the radical, we need to check that the standard process of lifting idempotents ([Row88, Proposition 1.1.25ff]) respects the grade. Note that any homogeneous idempotent $e \neq 0$ must be even, since $e = e^2$.

Lemma 6.29. *If I is a nil superideal of a superalgebra R, then any even idempotent e of R/I can be lifted to an even idempotent of R.*

Proof. Clearly, $e = r + I$ for some $r \in I_0$. Then $(r^2 - r)^n = 0$ for some n, so r^n is a sum of higher powers of r. Conclude as in [Row88, Proposition 1.1.26], noting the idempotent constructed there is even. □

Proposition 6.30. *If I is a nil superideal of a superalgebra R, then any set of homogeneous matrix units $\{e_{ij} : 1 \leq i, j \leq n\}$ of R/I can be lifted to homogeneous matrix units of R of the same parity.*

Proof. First note that the orthogonal idempotents $e_{11}, \ldots, e_{nn}$ can be lifted to orthogonal (even) idempotents $x_{11}, \ldots, x_{nn}$ of R, since all the elements used in the proof of [Row88, Proposition 1.1.25(i)] are even. But for $i \neq j$ lift e_{ij} to b_{ij} of R of the same parity, and mimic the proof of [Row88, Proposition 1.1.25(iii)] (noting the elements computed there are always homogeneous). □

Theorem 6.31 (Wedderburn Principal supertheorem). *Any f.d. superalgebra R over an algebraically closed field of characteristic $\neq 2$ has a semisimple super-subalgebra $\bar{R}$ such that $R = \bar{R} \oplus J(R)$.*

Proof. We only sketch the proof, since it is completely analogous to the standard ungraded proofs, cf. [Row88, Theorem 2.5.37], for example. The reduction to the case $J^2 = 0$ is standard. In this case, write $\bar{R} = \prod_{i=1}^{k} S_i$, where each S_i is supersimple, i.e., has the form of Proposition 6.28(ii). Write $\bar{e}_i$ for the multiplicative unit of S_i; then $\bar{e}_1, \ldots, \bar{e}_k$ are orthogonal central idempotents of $\bar{R}$ which we lift to orthogonal (but not necessarily central) even idempotents $e_1, \ldots, e_k$ of R. We need to show S_i is isomorphic to a super-subalgebra of $R_i = e_i R e_i$, for $1 \leq i \leq k$. If S_i is simple, then it is $M_{n_i}(F)$ under some grade, and the matrix units can be lifted by Proposition 6.30, so we have a corresponding copy of $M_{n_i}(F)$ in R_i.

Thus, we may assume S_i is not simple, so S_i is as in Example 6.27, i.e., $S_i = M_{n_i}(F[\bar{c}])$ where $\bar{c}^2 = 1$. Again we lift the matrix units to a set of even matrix units $\{e_{uv} : 1 \leq u, v \leq n\}$ of R; we also take odd $c_1 \in R_i$ such that $c_1 + J = \bar{c}$. Thus, $c_1^2 = a + 1$ for some $a \in J$, clearly even. Note that

$$[a, c_1] = [c_1^2 - 1, c_1] = 0.$$

We define

$$c_2 = c_1 - \frac{a}{2}c_1.$$

Since $J^2 = 0$, we see

$$c_2^2 = c_1^2 - ac_1^2 = (1 + a)(1 - a) = 1 - a^2 = 1.$$

Unfortunately c_2 need not be scalar, so we put

$$c = \sum e_{u1} c_2 e_{1u},$$

which clearly commutes with each matrix unit. Furthermore $[c_2, e_{uv}] \in J$ since $\bar{c}_2 = \bar{c}_1$ is scalar, so we see

$$c^2 = \sum_{u,v=1}^{n_i} e_{u1} c_2 e_{1u} e_{v1} c_2 e_{1v} = \sum_{u=1}^{n} e_{u1} c_2 e_{11} c_2 e_{1u}$$

$$= \sum_{u=1}^{n} e_{u1} c_2 c_2 e_{1u} + \sum_{u=1}^{n} e_{u1} [c_2, e_{11}] c_2 e_{1u}$$

$$= 1 + \sum_{u=1}^{n} e_{u1} [c_2, e_{11}]([c_2, e_{1u}] + e_{1u} c_2)$$

$$= 1 + \sum_{u=1}^{n} e_{u1} [c_2, e_{11}] e_{1u} c_2$$

$$= 1 + \sum (e_{u1} c_2 e_{1u} - e_{u1} c_2 e_{1u}) c_2 = 1 + 0 = 1.$$

$\square$

When using the Wedderburn Principal supertheorem, we recall from Proposition 6.28 that the idempotents used in constructing $\bar{R}$ may all be taken to be homogeneous, and clearly any radical substitution is the sum of an odd and even radical substitution. Thus, in checking whether or not a superpolynomial f is an identity of a f.d. superalgebra A, we may confine our attention to homogeneous semisimple and radical substitutions.

We also shall need the super-analog of Theorem J, a way of absorbing traces into superpolynomials. Given a f.d. superalgebra $A = A_0 \oplus A_1$, we define its *superdimension* $\dim_2 A = (t_0, t_1)$ where $t_u = \dim A_u$, $u = 0, 1$.

A superpolynomial f is (t_0, t_1)-*alternating* if, for $t = t_0 + t_1$, f is alternating in a set of t indeterminates, t_0 of which are even and t_1 of which are odd. Such a superpolynomial will clearly be a superidentity of any superalgebra of superdimension (t'_0, t'_1) unless $t'_0 \geq t_0$ and $t'_1 \geq t_1$; this is seen by checking homogeneous bases.

Theorem 6.32. *Suppose A is a finite dimensional superalgebra of superdimension (t_0, t_1), and $a_{iu}, r_j \in A$, for $1 \leq i \leq t_u$, and $1 \leq j \leq m$. Let*

$$V_u = \sum_{i=1}^{t_u} F a_{iu}; \qquad V = V_0 \oplus V_1.$$

For any C-linear homogeneous map $T : V \to V$, letting

$$\lambda^t + \sum_{k=1}^{t} (-1)^k \alpha_k \lambda^{t-k}$$

denote the characteristic polynomial of T (as a $t \times t$ matrix), then for any (t_0, t_1)-alternating superpolynomial f, and any homogeneous $a_1, \ldots, a_t$ in V, t_0 of which are even, we have

$$\alpha_k f(a_1, \ldots, a_t, r_1, \ldots, r_m) = \sum f(T^{k_1} a_1, \ldots, T^{k_t} a_t, r_1, \ldots, r_m),$$

summed over all vectors $(k_1, \ldots, k_t)$ for which $k_1 + \cdots + k_t = k$.

Proof. As in Theorem J.   □

There are some technical difficulties concerning the grading in connection with traces. On the one hand, we want the base field F to have even values, but on the other hand, it is convenient for $\mathrm{tr}(a)$ to have the same parity as a. One cheap way to satisfy both conditions is to build the *trace expediting superalgebra* (in analogy with Definition 4.63) only by adjoining traces of even elements. Since for any odd element a, a^2 is even and thus integral in the trace expediting superalgebra, we see that every element of the trace expediting superalgebra is integral, and thus the trace expediting superalgebra is super-representable.

6.2.2   Proof of Kemer's Superrepresentability Theorem

Having the general preliminaries in hand, let us go back to modify Kemer's proof of Specht's conjecture in the affine case, cf. Chapter 4, and prove Specht's conjecture for arbitrary T-ideals in characteristic 0. In view of Theorem 6.20, it suffices to prove the ACC for T_2-ideals of affine superalgebras. As before, the main step is the following theorem.

Theorem 6.33 (Kemer's PI-Superrepresentability Theorem). *Every affine superalgebra is PI_2-equivalent to a finite dimensional superalgebra.*

Fortunately, the same program we followed in Chapter 4 (section 4.1.2) for proving Theorem 4.1 can be followed here. Unfortunately, everything is more complicated than before because we need to deal with more classes of identities, and the notation doubles (at least). So we go through the same steps, with care. We start with some preliminaries. First we consider the super-version of the T-ideals of matrices.

Definition 6.34. $\mathcal{M}_n = \mathrm{id}(M_n(F))$, viewed as superidentities via Definition 6.3(v). $\mathcal{M}_{k,\ell} = \mathrm{id}_2(M_{k,\ell}(F))$, cf. Example 1.34. $\mathcal{M}'_n = \mathrm{id}_2(M_n(F[c]))$ of Example 6.27. Define

$$\tilde{\mathcal{M}}_n = \left(\bigcap_{k=1}^{n-1} \mathcal{M}_{k,n-k} \right) \cap \mathcal{M}_n \cap \mathcal{M}'_n = \mathrm{id}_2 \left(\prod M_{k,n-k} \times M_n(F) \times M_n(F[c]) \right).$$

Whereas in the ungraded case, any simple PI-algebra of PI-class n has T-ideal $\mathcal{M}_n$, the graded case also includes $\mathcal{M}_{k,\ell}$, for $\ell = n-k$, $1 \leq k \leq n-1$, and $\mathcal{M}'_n$, a total of $n + 1$ possibilities for each n. These are all distinct and have no inclusions among themselves, as seen easily via the *Capelli superpolynomials* $c_{k,\ell}$, which involve k even x_i and ℓ odd x_i; then $c_{k,\ell}$ is not a superidentity of $M_{k,\ell}(F)$ but is a superidentity of all other $M_{k',\ell'}(F)$ where $k' + \ell' = k + \ell$, since obviously, $k > k'$ or $\ell > \ell'$.

Fortunately, we still have the following inclusions:

Remark 6.35.

(i) $\mathcal{M}_{k,\ell} \subseteq \mathcal{M}_{k',\ell}$ for $k \geq k'$, and likewise $\mathcal{M}_{k,\ell} \subseteq \mathcal{M}_{k,\ell'}$ for $\ell \geq \ell'$.

(ii) $\mathcal{M}'_n \subseteq \mathcal{M}'_m$ for $n \geq m$.

Remark 6.36. Suppose a superalgebra A is semiprime (as ungraded algebra) of PI-class n. Then

$$\mathrm{id}_2(A) \supseteq \tilde{\mathcal{M}}_n,$$

because we have natural embeddings of matrices that also respect the grade.

In general, for A a superalgebra satisfying a PI, $J = J(A)$ still is nilpotent by the Razmyslov-Kemer-Braun theorem, but now is a superideal.

Program to prove Theorem 6.33.

Step 1. This requires the super-analog of Lewin's Theorem. The notation always becomes more complicated because it must reflect the grade, but everything else is the same as before, so we shall omit quite a few details. (We still have a long haul.) We may assume F is algebraically closed. Since $J = J(A)$ is graded, we write $J = J_0 \oplus J_1$, and likewise $\bar{A} = \bar{A}_0 \oplus \bar{A}_1$ for $\bar{A} = A/J(A)$, and let $\dim \bar{A}_j = t_j$ for $j = 0, 1$. Hence, by the Wedderburn Principal supertheorem,

$$A = \bar{A}_0 \oplus J_0 \oplus \bar{A}_1 \oplus J_1,$$

so to check a superidentity, we may make all even substitutions in $\bar{A}_0$ or J_0, and all odd substitutions in $\bar{A}_1$ or J_1. Thus, we shall talk of *semisimple* or *radical*, *even* or *odd* substitutions.

Theorem 6.37. *Suppose R is a superalgebra and $I_1, I_2 \lhd_2 R$. Then there is a graded R, R bimodule M, such that $R/(I_1 I_2)$ can be embedded in*

$$\begin{pmatrix} R/I_1 & M \\ 0 & R/I_2 \end{pmatrix}. \tag{6.10}$$

Proof. We use the ungraded version (Theorem 1.72) to embed $R/(I_1 I_2)$ naturally in

$$\tilde{R} = \begin{pmatrix} R/I_1 & M \\ 0 & R/I_2 \end{pmatrix} \tag{6.11}$$

via $r \mapsto \begin{pmatrix} r + I_1 & \delta(r) \\ 0 & r + I_2 \end{pmatrix}$ where $\delta : R \to M$ is a derivation. We replace M by $\{\delta(r) : r \in R\}$ and grade $\tilde{R}$ by matching odd and even parts with R, i.e., if $r \in R$ is even (resp. odd), then so is $r + I_1$, $\delta(r)$, and $r + I_2$. Since

$$\delta(rs) = r\delta(s) + \delta(r)s,$$

we see by definition of grade that this makes $\tilde{R}$ a superalgebra, as desired. $\square$

Corollary 6.38. *For any affine PI-superalgebra R over a field F, there is a finite dimensional F-superalgebra A such that $\mathrm{id}_2(A) \subseteq \mathrm{id}_2(R)$. More precisely, if $\mathrm{id}_2(R/J(R)) \supseteq \tilde{\mathcal{M}}_n$ and $J(R)^k = 0$ then $\tilde{\mathcal{M}}_n^k \subseteq \mathrm{id}_2(R)$.*

Proof. As in the proof of Corollary 4.9. $\square$

Step 2. Given t_0, t_1 we define a superpolynomial f to be *s-fold* (t_0, t_1)-*alternating* if there are sets of homogeneous indeterminates $X_j = Y_j \cup Z_j$, $1 \le j \le s$, with $|Y_j| = t_0$, $|Z_j| = t_1$, (Y_j comprised of even indeterminates, and Z_j of odd indeterminates) such that f is (t_0, t_1)-alternating in each X_j.

Given a f.d. superalgebra A and a T_2-ideal Γ, we define

$$\beta(A) = (\beta_1, \beta_2)$$

where (β_1, β_2) is lexicographically greatest such that for any s, there is an s-fold (β_1, β_2)-alternating superpolynomial $f(X_1, \ldots, X_s; Y) \notin \mathrm{id}_2(A)$.

Now, given $\beta(A) = (t_0, t_1)$, we define

$$\gamma(A) = (\gamma_1, \gamma_2)$$

to be the smallest $s = (s_0, s_1)$ such that every s_0-fold (t_0+1, t_1)-alternating and s_1-fold (t_0, t_1+1)-alternating superpolynomial $f(X_1, \ldots, X_{s-1}; Y)$ that is μ-fold (t_0, t_1)-alternating for large enough μ is in $\mathrm{id}(A)$.

Definition 6.39. The *Kemer superindex* of a PI-superalgebra A, denoted $\mathrm{index}_2 A$, is $(\beta(A), \gamma(A))$, which now has four components (two for β and two for γ). We order this 4-tuple lexicographically.

Clearly, if $\mathrm{id}(A_1) \subseteq \mathrm{id}(A_2)$, then $\mathrm{index}_2(A_1) \ge \mathrm{index}_2(A_2)$.

Step 3. Any f.d. superalgebra is written as a subdirect product of subdirectly irreducible superalgebras, and Definition 4.40 (full) is once again applicable (this time for superalgebras, i.e., requiring the $\bar{e}_{11}^{(k)}$ to be homogeneous), enabling us to prove the superversions of Kemer's two key lemmas (describing β and γ in terms of the structure of a suitable subdirectly irreducible superalgebra).

Remark 6.40 (Kemer's First Superlemma). If A is a full, subdirectly irreducible PI-superalgebra, then $\beta(A) = (\dim_F \bar{A}_0, \dim_F \bar{A}_1)$. In the proof we should recall that the idempotent matrix units $e_{\ell\ell}^{(k)}$ used in Kemer's First Lemma are all even, and thus we can tack on evaluations of Capelli superpolynomials as before, without affecting the grade.

To proceed to Kemer's Second Superlemma, we need to take into account even and odd radical substitutions. Accordingly we write (s_0, s_1) *radical substitutions*, to denote s_0 even radical substitutions and s_1 odd radical substitutions. We say for a superideal N of A that $N^{(s_0, s_1)} = 0$ if any product of elements including s_0 from N_0 and s_1 from N_1 equals 0. The *supernilpotence index* of N is the smallest (s_0, s_1) (lexicographically) such that $N^{(s_0, s_1)} = 0$.

Thus, instead of (t_A, s_A), we have the 4-tuple $(\dim_F \bar{A}_0, \dim_F \bar{A}_1; s_0, s_1)$, where (s_0, s_1) is the supernilpotence index of $J(A)$. Superalgebras A and A' are PI-*equivalent* if $\mathrm{id}_2(A) = \mathrm{id}_2(A')$. A f.d. superalgebra A is PI-*minimal* (among all f.d. superalgebras PI-equivalent to A) if $(\dim_F \bar{A}_0, \dim_F \bar{A}_1; s_0, s_1)$ is minimal (lexicographically). Now we define:

A multilinear superpolynomial f has *Property SuperK* if it vanishes on any (graded) substitution with fewer than (s_0, s_1) radical substitutions, but has a nonvanishing specialization with (s_0, s_1) radical substitutions.

Lemma 6.41. *Any PI-minimal f.d. superalgebra A whose radical has supernilpotence index (s_0, s_1) has a nonidentity satisfying Property SuperK.*

Definition 6.42. A (s_0, s_1, μ)-*Kemer superpolynomial* for W is a superpolynomial f not in $\mathrm{id}(W)$, in $s_0 + s_1 + \mu$ sets of indeterminates X_j, which is s_0-fold $(t_0 + 1, t_1)$-alternating, s_1-fold $(t_0, t_1 + 1)$-alternating, and having another μ folds that are (t_0, t_1)-alternating, where $\beta(A) = (t_0, t_1)$. We call this a *Kemer superpolynomial*, supressing s_0, s_1, and μ in the notation.

Kemer's Second Superlemma now states: If A is PI-minimal of characteristic 0, then $\gamma(A) = (s_0, s_1)$ where (s_0, s_1) is the supernilpotence index of $J = J(A)$. Furthermore, A has a (s_0, s_1, μ)-Kemer superpolynomial, for any (s_0, s_1) less than the supernilpotence index of J (under the lexicographic order), and for any μ. The proof is the same as before: Using f satisfying Property SuperK, we can tack on as many folds of (t_0, t_1)-alternating superpolynomials as we want, and we have s_0 extra even variables and s_1 extra odd variables that we must insert into the folds.

At the same time, just as in Chapter 4, we can verify the phoenix property, that any consequence of a Kemer superpolynomial has a consequence that is a Kemer superpolynomial.

Proposition 6.43. *Suppose $A, A_1, \ldots, A_m$ are f.d. superalgebras, satisfying $\mathrm{index}_2(A_i) < \mathrm{index}_2(A)$ for each i. Then for any $f \notin \mathrm{id}_2(A)$, the T_2-ideal generated by f contains a Kemer superpolynomial for A that is a superidentity of each A_i.*

Proof. This observation follows immediately from the definitions. Let

$$\mathrm{index}_2 A = (t_0, t_1, s_0, s_1) \quad \text{and} \quad \mathrm{index}_2(A_i) = (t_{i0}, t_{i1}, s_{i0}, s_{i1}).$$

If $(t_{i0}, t_{i1}) < (t_0, t_1)$, then by definition, there is some number μ_i such that any μ_i-fold (t_0, t_1)-alternating superpolynomial is in $\mathrm{id}(A_i)$. If $(t_{i0}, t_{i1}) = (t_0, t_1)$, then by definition, any s_{i0}-fold $(t_0 + 1, t_1)$-alternating and s_{i1}-fold $(t_0, t_1 + 1)$-alternating superpolynomial is in $\mathrm{id}(A_i)$. Hence, taking $\omega_2(A)$ analogously to Definition 4.21 and

$$\mu = \max\{\mu_1, \ldots, \mu_m, \omega_2(A_1), \ldots, \omega_2(A_m)\},$$

we see by definition that any μ-Kemer superpolynomial is a superidentity of each A_i. □

Definition 6.44. An arbitrary affine PI-superalgebra A is *supervarietally reducible* if there are affine PI-superalgebras A_1, A_2 satisfying the conditions $\mathrm{id}_2(A_1) \cap \mathrm{id}_2(A_2) = \mathrm{id}_2(A)$ but $\mathrm{id}_2(A_i) \supset \mathrm{id}_2(A)$ for $i = 1, 2$.

Remark 6.45. As in the ungraded case, any f.d. superalgebra is PI_2-equivalent to a direct product of supervarietally irreducible f.d. superalgebras.

Definition 6.46. A PI-*basic* superalgebra is a PI-minimal, full, supervarietally irreducible, f.d. superalgebra.

Step 4. Now that we have the necessary inductive framework to make the proof go through, it remains to obtain a graded analog of the trace ring. This is done as in the proof of Proposition 4.59:

Proposition 6.47. *Any f.d. superalgebra A is PI-equivalent to a direct product of f.d. superalgebras*

$$\tilde{A} \times A_1 \times \cdots \times A_t,$$

where $A_1, \ldots, A_t$ are PI-basic, and where $\tilde{A}$ is a direct product of PI-basic superalgebras $\tilde{A}_1, \ldots, \tilde{A}_u$, such that

$$\begin{aligned}
\mathrm{index}_2(\tilde{A}_1) = \mathrm{index}_2(\tilde{A}_2) = \cdots &= \mathrm{index}_2(\tilde{A}_u) \\
&> \mathrm{index}_2(A_1) \geq \mathrm{index}_2(A_2) \geq \cdots \geq \mathrm{index}_2(A_t).
\end{aligned} \tag{6.12}$$

Furthermore, there exists a Kemer superpolynomial f for each of $\tilde{A}_1, \ldots, \tilde{A}_u$, which is a superidentity for $A_1 \times \cdots \times A_t$.

Proof. Analogous to that of Proposition 4.58. □

Proposition 6.48. *Suppose $\mathcal{A}$ is the relatively free superalgebra of a PI-basic superalgebra A over an algebraically closed field of characteristic 0, and $\mathcal{S}$ be a set of Kemer superpolynomials for A. Let I be the ideal generated by all $\{f(\mathcal{A}) : f \in \mathcal{S}\}$. Then $\mathcal{A}/I$ is superrepresentable (i.e., is embeddable as a superalgebra into a f.d. superalgebra).*

Proof. Construct the relatively free superalgebra $\mathcal{A}$ as in Example 3.26, where we introduce both even and odd commuting indeterminates into Λ, in order to make "generic" even and odd elements; hence, we may assume $\mathcal{A}$ has homogeneous generators. We can extend the trace linearly to $\mathcal{A}$, to take on values in $F[\Lambda]$. Letting $\hat{F}$ denote the commutative affine subalgebra of $F[\Lambda]$ consisting of all the traces of powers ($\leq n_k$) of products (of length up to the Shirshov height) of the generators of $\mathcal{A}$, we let $\hat{A} = \mathcal{A}\hat{F} \subseteq A[\Lambda]$. By Proposition 2.38, $\hat{A}$ is a f.g. module over $\hat{F}$, and in particular, is a Noetherian algebra. By Remark 4.55, I is also closed under multiplication by elements of $\hat{F}$, so in fact, I is a common ideal of $\mathcal{A}$ and $\hat{A}$. Thus, $\mathcal{A}/I \subseteq \hat{A}/I$, which is f.g. over the image of the affine algebra $\hat{F}$, so $\mathcal{A}/I$ is superrepresentable, by the graded version of Proposition 1.92, cf. Exercise 6. $\qquad\square$

Corollary 6.49. *Notation as in Proposition 6.47, let* $\kappa = \mathrm{index}_2(\tilde{A})$, *and let* $\mathcal{S}$ *denote the set of* Kemer superpolynomials of $\tilde{A}$ *that are superidentities of* $A_1 \times \cdots \times A_t$. *Also let* U_i *denote the relatively free superalgebra of* $\tilde{A}_i$, *for* $1 \leq i \leq u$. *Then* $\mathrm{index}_2(U_i/\mathcal{S}(U_i)) < \kappa$.

Proof. Otherwise, some of the $\tilde{A}_j/\mathcal{S}(\tilde{A}_j)$ have Kemer superindex κ. For these j, there is some Kemer superpolynomial f that is a superidentity of all the other factors. But obviously $f \notin \mathrm{id}(A_j)$, implying f is a Kemer superpolynomial of $\tilde{A}$, and thus $f(\tilde{A}) \subseteq \mathcal{S}(\tilde{A})$, implying $f \in \mathrm{id}_2(\tilde{A}/\mathcal{S}(\tilde{A}))$, contrary to choice of f. $\qquad\square$

Lemma 6.50. *Suppose* Γ *is a* T_2-*ideal properly containing* $\mathrm{id}_2(A)$ *for some f.d. superalgebra* A. *Then there is a f.d. superalgebra* A' *such that* $\Gamma \supseteq \mathrm{id}_2(A')$ *but* $\mathrm{index}_2(A') = \mathrm{index}_2(\Gamma)$. *Furthermore, taking* A' *to be a direct product of PI-basic superalgebras* $\prod A_i$, *then for each component* A_i' *of highest Kemer superindex, and for large enough* μ, *no* (s_0, s_1, μ)-*superKemer polynomial of* A_i' *is in* $\Gamma + \mathrm{id}_2(A_i')$.

Proof. The proof is virtually identitical to that of Proposition 4.61. $\qquad\square$

We are ready, finally, to conclude the proof of Theorem 6.33, analogously to the proof of Theorem 4.66:

Proof. Let $\kappa = \mathrm{index}_2(\Gamma)$. $\Gamma \supseteq \mathrm{id}_2(A)$ for suitable f.d. A, cf. Corollary 6.38. By Lemma 6.50, we may assume $\mathrm{index}_2(A) = \kappa$, and furthermore, for μ large enough, Γ and A satisfy precisely the same (s_0, s_1, μ)-superKemer polynomials (since no (s_0, s_1, μ)-Kemer superpolynomial of one can be an identity of the other).

Let $\mathcal{S}$ be the set of (s_0, s_1, μ)-Kemer superpolynomials of Γ, and $\Gamma' = \Gamma + \mathcal{S}$. By definition,

$$\mathrm{index}_2(\Gamma') < \kappa,$$

so by induction there is some f.d. superalgebra A' with $\mathrm{id}_2(A') = \Gamma'$.

Now write $A = \tilde{A} \times A_1 \times \cdots \times A_m$ where $A_1, \ldots, A_u$ have Kemer superindex κ and the other A_i have lower Kemer superindex. Let $\mathcal{S}_i$ be T_2-ideal generated by (s_0, s_1, μ)-Kemer superpolynomials of A_i; then $\cup \mathcal{S}_i = \mathcal{S}$.

For each $i \leq u$, let $\hat{U}_i$ denote the trace expediting superalgebra of A_i, and let $\mathcal{A}'_i = \hat{U}_i / \Gamma(U_i)$. Thus, $\mathrm{id}_2(\mathcal{A}'_i) = \Gamma + \mathrm{id}_2(A_i)$. Using the phoenix property as in the proof of Theorem 4.66, and using the $\mathcal{A}'_i$ and a trace argument, we can prove that $\Gamma(\hat{U}_i) \cap \bar{\mathcal{S}}_i = 0$.

Embedding $\mathcal{A}'_i$ into a f.d. superalgebra A''_i, for $1 \leq i \leq u$, we repeat the argument at the end of Theorem 4.66 to conclude

$$\Gamma = (\Gamma + \mathcal{S}) \cap \mathrm{id}_2(A''_1) \cap \cdots \cap \mathrm{id}_2(A''_u) = \mathrm{id}_2(A' \times A''_1 \times \cdots \times A''_u),$$

as desired. $\square$

6.2.3   Kemer's Main Theorem completed

Theorem 6.51 (Kemer). *The ACC holds for T_2-ideals of affine superalgebras (over a field of characteristic 0).*

Proof. We want to show that any chain

$$\mathcal{I}_1 \subseteq \mathcal{I}_2 \subseteq \mathcal{I}_3 \subseteq \dots \tag{6.13}$$

of T_2-ideals terminates. Since ordered 4-tuples of natural numbers satisfy the descending chain condition, we may start far enough along our chain of T-ideals and assume that all the A_j in (4.13) have the same Kemer superindex κ. We do this for κ minimal possible, and induct on κ.

By Proposition 6.47, we can write each

$$A_j = \tilde{A}_j \times A_{j,1} \times \cdots \times A_{j,t_j},$$

where each $\mathrm{index}_2(A_{j,i}) < \kappa$. But by the superversion of Remark 4.68, we may assume $\tilde{A}_j = \tilde{A}$ for some superalgebra $\tilde{A}$ that is a direct product of PI_2-basic components of Kemer superindex κ.

Let $\mathcal{I}$ be the T-ideal generated by all Kemer superpolynomials of $\tilde{A}$ that are superidentities of $A_{1,1} \times \cdots \times A_{1,t}$. As we argued in Chapter 4, $\tilde{A}/\mathcal{I}(\tilde{A})$ cannot contain any Kemer superpolynomials (for κ) that are superidentities of $A_{1,1} \times \cdots \times A_{1,t}$, so its Kemer superindex must decrease and the chain

$$\mathcal{I}_1 + \mathcal{I} \subseteq \mathcal{I}_2 + \mathcal{I} \subseteq \mathcal{I}_3 + \mathcal{I} \dots \tag{6.14}$$

must stabilize. On the other hand, for each j,

$$\mathcal{I} \cap \mathrm{id}_2(\tilde{A}) = \mathcal{I} \cap \mathrm{id}_2(A_1) = \mathcal{I} \cap \mathcal{I}_1 \subseteq \mathcal{I} \cap \mathcal{I}_j = \mathcal{I} \cap \mathrm{id}_2(A_j) \subseteq \mathcal{I} \cap \mathrm{id}_2(\tilde{A}),$$

so equality must hold at each stage, i.e., $\mathcal{I} \cap \mathcal{I}_1 = \mathcal{I} \cap \mathcal{I}_j$ for each j. But coupled with (6.14), this means the chain (6.13) stabilizes, contradiction.

$\square$

We are finally ready for Kemer's solution to Specht's problem.

Theorem 6.52 (Kemer). *The ACC holds for T-ideals, over any field of characteristic 0.*

Proof. Combine Theorem 6.20 with Theorem 6.51. □

As we shall see in Chapter 7, Theorem 6.52 is true only in characteristic 0, although Theorem 6.51 can be proved more generally in arbitrary characteristic.

6.3   Consequences of Kemer's Theory—The Verbal Structure of an Algebra

Kemer's theorem opens the door to a new structure theory of PI-algebras. We shall present the outline here, and give an application of Berele to GK dimension in Chapter 9. We want to transfer the theory of T-ideals of $F\{X\}$ to an arbitrary algebra A over F, by using the theory of Section 1.8.

6.3.1 T-ideals of relatively free algebras

Our starting point is for relatively free algebras. Kemer's theorem implies the ACC for T-ideals of any relatively free algebra $\mathcal{A}$. This means using Theorem 1.98:

Theorem 6.53. *For any T-ideal $\mathcal{I}$ of a relatively free algebra, the radical $\sqrt{\mathcal{I}}$ is a finite intersection of prime T-ideals, and $\sqrt{\mathcal{I}}^k \subseteq \mathcal{I}$ for some k.*

Viewed another way, any relatively free algebra $\mathcal{A}$ has a nilpotent T-deal $\mathcal{I}$, such that $A/\mathcal{I}$ is the finite subdirect product of relatively free algebras whose identities constitute prime T-ideals.

Since prime T-ideals are so fundamental from this point of view, let us define them explicitly (as a special case of Definition 1.96):

Definition 6.54. A T-ideal $\mathcal{P}$ is *prime* if, for all T-ideals $\mathcal{I}_1, \mathcal{I}_2$ properly containing $\mathcal{P}$, we have $\mathcal{I}_1 \mathcal{I}_2 \not\subseteq \mathcal{P}$. We call a relatively free algebra T-*prime* is 0 is prime as a T-ideal.

The first question is to determine the T-prime relatively free algebras, or equivalently, "What are the prime T-ideals of $F\{X\}$?" The answer lies in the translation to T_2-ideals via the Grassman envelope $G(A)$.

Remark 6.55. $\mathcal{P}$ is a prime T-ideal of $F\{X\}$ iff $\mathcal{P}^*$ is a prime T_2-ideal of the free superalgebra $F\{Y, Z\}$. (Indeed, $(\mathcal{I}_1 \mathcal{I}_2)^* = \mathcal{I}_1^* \mathcal{I}_2^* \supseteq \mathcal{P}^*$ iff $\mathcal{I}_1 \mathcal{I}_2 \subseteq \mathcal{P}$.)

Thus, by Kemer's theory, the determination of the prime T-ideals of $F\{X\}$ (for char$(F) = 0$) is precisely the determination of the prime T_2-ideals of affine superalgebras, and thus of finite dimensional superalgebras. We need some more structure theory for PI-superalgebras, along the lines of Lemma 6.25. We always assume our algebras contain the element $\frac{1}{2}$.

Remark 6.56. If $M_n(F)$ is graded, then any multilinear central polynomial must have some nonzero evaluation of homogeneous elements; designating the corresponding indeterminates as even or odd (according to this substitution) gives a central superpolynomial.

Remark 6.57. Recall Lemma 6.24, which attaches to any superalgebra $R = R_0 \oplus R_1$ the automorphism $\sigma : r_0 + r_1 \mapsto r_0 - r_1$. Then $P \cap \sigma(P) \triangleleft_2 R$ for any $P \triangleleft R$.

Proposition 6.58. *Suppose R is a superalgebra, notation as in Remark 6.57.*

(i) *If P is a prime ideal of R, then $P \cap \sigma(P)$ is a prime superideal, in the sense that if $I_1, I_2 \triangleleft_2 R$ with $I_1 I_2 \subseteq P \cap \sigma(P)$, then I_1 or I_2 is contained in $P \cap \sigma(P)$.*

(ii) *Conversely, given a prime superideal P, take $P' \triangleleft R$ maximal with respect to $P' \cap \sigma(P') \subseteq P$, which exists by Zorn's lemma; then P' is a prime ideal of R.*

Proof. (i) $I_1 I_2 \subseteq P$, so we may assume $I_1 \subseteq P$; hence, $I_1 = I_1 \cap \sigma(I_1) \subseteq P \cap \sigma(P)$.

(ii) If on the contrary $I_1 I_2 \subseteq P'$ with $I_1, I_2 \supset P'$, then $I_1 \cap \sigma(I_1) \not\subseteq P$ and $I_2 \cap \sigma(I_2) \not\subseteq P$, so $(I_1 \cap \sigma(I_1))(I_2 \cap \sigma(I_2)) \not\subseteq P$, contradiction. $\square$

Hence, we can grade the basic theorems in the PI-structure theory, and prove for example:

Proposition 6.59. *If R is superprime PI with center $C = C_0 \oplus C_1$, and $S = C_0 \setminus \{0\}$, then the localization $S^{-1}R$ is supersimple and finite dimensional over $F = S^{-1}C_0$, the field of fractions of C_0. If $S^{-1}R$ is not simple, then it is the direct product of two isomorphic simple algebras.*

Proof. R must be torsion-free over C_0. Indeed, more generally, no homogeneous central element c can be a zero divisor. Otherwise, if $rc = 0$ for $r \neq 0$, then the homogeneous parts of r also annihilate c, implying c is annihilated by a superideal I, and then $I(Rc) = 0$, contrary to R superprime. Hence one can construct $S^{-1}R$, which is a superalgebra whose even component is the field F. Replacing R by $S^{-1}R$, we may assume $C_0 = F$ and we want to prove R is supersimple, of finite dimension over F.

Since $R/P \approx R/\sigma(P)$, they each have a central polynomial and we can apply Theorem C to show that every ideal intersects the center nontrivially. But any proper superideal I intersects the field C_0 at 0, implying I must have a nonzero odd element c, and by the previous paragraph $0 \neq c^2 \in I \cap F$, contradiction.

Thus, R is supersimple, and taking a maximal ideal P we must have $P \cap \sigma(P) = 0$, implying

$$R \approx R/P \times R/\sigma(P),$$

by the Chinese Remainder Theorem. But R/P and $R/\sigma(P)$ are f.d. by Kaplansky's (ungraded) theorem, so R is f.d. $\square$

Corollary 6.60. *Any prime superalgebra R is superPI-equivalent to one of the algebras in Proposition 6.28.*

Proof. In view of the proposition, we may assume R is supersimple and f.d.; tensoring by the algebraic closure of the even part F_0 of the center we may assume F_0 is algebraically closed, so we have reduced to Proposition 6.28. $\square$

Proposition 6.61. *Any prime T_2-ideal $\mathrm{id}_2(A)$ can be written as $\mathrm{id}_2(A)$ for a supersimple algebra A.*

Proof. As noted, we may assume A is the relatively free superalgebra of a finite dimensional superalgebra, and thus is superrepresentable inside some f.d. superalgebra $\hat{A}$. But $J(A)$ is a nilpotent T_2-ideal, cf. Lemma 6.25 and Proposition 3.32, so this reduces us to the case where $J(A) = 0$.

Next, $J(\hat{A}) \cap A \subseteq J(A) = 0$, implying A is superrepresentable inside the super-semisimple superalgebra $\hat{A}/J(\hat{A})$. It follows that $\mathrm{id}_2(A) = \mathrm{id}_2(\hat{A})$.

Write $\hat{A}/J(\hat{A}) = \prod_{i=1}^{q} A_i$ for A_i supersimple. Then $\mathrm{id}_2(A) = \cap \mathrm{id}_2(A_i)$, so, being prime, $\mathrm{id}_2(A) = \mathrm{id}_2(A_i)$ for some i. $\square$

Corollary 6.62. *The prime T_2-ideals are precisely $\mathcal{M}_n$, $\mathcal{M}_{k,\ell}$, and $\mathcal{M}'_n$.*

Proof. The T_2-ideal of a supersimple PI-algebra must be that of one of the algebras in Proposition 6.28(ii). $\square$

Remark 6.63. Suppose Γ is a prime T-ideal. Writing $\Gamma = \mathrm{id}(G(R))$ for supersimple PI-algebra R over an algebraically closed field, we see $\mathrm{id}_2(R)$ must be a prime T_2-ideal, and thus one of $\mathcal{M}_n$, $\mathcal{M}_{k,\ell}$, and $\mathcal{M}'_n$. It remains to translate these back to A. Let V denote the underlying vector space of G. Thus, $G = G_0 \oplus G_1 V$.

Case 1. $R = M_n(F)$ with the trivial grade. Then $A = M_n(F)$.

Case 2. $R = M_{k,\ell}(F)$. Then (for $n = k + \ell$) A is the subalgebra of $M_n(G)$ spanned by $G_0 e_{ij}$ for $1 \le i, j \le k$ or $k + 1 \le i, j \le n$, and by $G_1 e_{ij}$ for the other i, j ($i > k$ and $j \le \ell$, or $i \le k$ and $j > \ell$).

Case 3. $R = M_n(F) \oplus M_n(F)c$, where $c_2 = 1$. Then $A = G_0 M_n(F) \oplus G_1 M_n(F)c$, which is clearly PI-equivalent to $M_n(G)$.

Putting everything together, we can classify the prime T-ideals in characteristic 0.

Theorem 6.64. *The following list is a complete list of the prime T-ideals of $F\{X\}$, for any field F of characteristic 0:*

(i) $\mathrm{id}(M_n(F))$;

(ii) $\mathrm{id}(M_{k,\ell})$;

(iii) $\mathrm{id}(M_n(G))$.

(Note that whereas (i) is a prime ideal of $F\{X\}$, (ii) and (iii) are prime only as T-ideals.)

Remark 6.65. Although Theorem 6.64 is very explicit and is proved directly by passing to T_2-ideals, the transition between T-ideals and T_2-ideals often is quite delicate, involving taking all the different p_I^* of an identity p, cf. Proposition 6.10. This makes it difficult to answer such basic questions as: "What are the minimal identities of $M_n(G)$?"

Having determined the prime T-ideals in characteristic 0, we are led to try to find the analogs of basic structure theorems related to the classic Nullstellensatz of Hilbert. Any prime T-ideal is the intersection of maximal T-ideals, cf. Exercise 7. Unfortunately, the Jacobson radical of a relatively free algebra, although a T-ideal, need not necessarily be nilpotent, as evidenced by the relatively free algebra of the Grassman algebra (whose radical is nil but not nilpotent). Nevertheless, one has the following positive result, following immediately from Kemer's Theorem.

Remark 6.66. If the image of a T-ideal $\mathcal{I}$ of a relatively free algebra $\mathcal{A}$ is zero in every T-prime homomorphic image of $\mathcal{A}$, then $\mathcal{I}$ is nilpotent, by Corollary 1.101.

The situation in characteristic p is much harder, as we shall discuss in Chapter 7. Even though much goes wrong, the analog to Remark 6.66 remains open, and in fact is conjectured by Kemer to be true.

6.3.2   Verbal ideals

What about applying ACC on T-ideals to arbitrary algebras?

Unfortunately, unless A is relatively free, there could be too few homomorphisms from A to itself for Definition 3.5 to be meaningful. To proceed further, we need another definition.

Definition 6.67. A *verbal ideal* of an F-algebra A is the set of evaluations in A of a T-ideal $\mathcal{I}$ of $F\{X\}$, where X is a countably infinite set of indeterminates.

In other words, the verbal ideal of A corresponding to $\mathcal{I}$ is its total image in

$$\{f(a_1,\ldots,a_m) : f(x_1,\ldots,x_m) \in \mathcal{I}, \qquad a_i \in A\}.$$

The key to our discussion lies in the following easy observation:

Remark 6.68. Given any $r_j = f_j(a_1,\ldots,a_{m_j})$, there is a homomorphism $\Phi : F\{X\} \to A$ such that

$$r_1 = \Phi(f_1(x_1,\ldots,x_{m_1})), \quad r_2 = \Phi(f_2(x_{m_1+1},\ldots,x_{m_1+m_2})), \cdots;$$

the point here is that different indeterminates x_i appear in the different positions.

We write $\mathcal{I}(A)$ for the verbal ideal of A corresponding to a T-ideal $\mathcal{I}$.

Lemma 6.69.

(i) $\mathcal{I}(F\{X\}) = \mathcal{I}.$

(ii) *Any verbal ideal is an ideal.*

(iii) $(\mathcal{I}_1 + \mathcal{I}_2)(A) = \mathcal{I}_1(A) + \mathcal{I}_2(A).$

(iv) $(\mathcal{I}_1\mathcal{I}_2)(A) = \mathcal{I}_1(A)\mathcal{I}_2(A).$

(v) *If I is a verbal ideal of A, then* $\mathcal{I} = \{f \in F\{X\} : f(A) \subseteq I\} \triangleleft_T F\{X\}.$

(vi) *For char $F = 0$, any F-algebra satisfies the ACC for verbal ideals.*

Proof. (i) is clear.

(ii) Follows easily from Remark 6.68: For example, if $r = f(a_1,\ldots,a_m)$ and $a \in A$, then take a homomorphism Φ sending $x_i \mapsto a_i$ for $1 \le i \le m$ and $x_{m+1} \mapsto a$, and conclude $ar = \Phi(x_{m+1}f(x_1,\ldots,x_m)) \subseteq \mathcal{I}$. The same argument proves (iii) and (iv).

(v) $f(h_1, \ldots, h_m)(a_1, \ldots, a_n) = f(h_1(a_1, \ldots, a_n), \ldots, h_m(a_1, \ldots, a_n)) \in I$, for all $a_i \in A$.

(vi) Given verbal ideals $I_1 \subset I_2 \subset \ldots$ let

$$\mathcal{I}_j = \{f \in F\{X\} : f(A) \subseteq I_j\} \lhd_T F\{X\},$$

so we conclude using Kemer's theorem. $\qquad\qquad\square$

Thus, the program of Chapter 1 is available:

Theorem 6.70. *Every verbal ideal I has only a finite number of verbal primes $P_1, \ldots, P_t$ minimal over I, and some finite product of the P_i is contained in I.*

Proof. By Theorem 1.98. $\qquad\qquad\square$

We are led to characterize the prime verbal ideals.

Proposition 6.71. *Each prime verbal ideal of A is the image of a prime T-ideal.*

Proof. Suppose P is a prime verbal ideal of A, and take $\mathcal{P}$ as in Lemma 6.69(v). If $\mathcal{I}_1 \mathcal{I}_2 \subseteq \mathcal{P}$, then

$$\mathcal{I}_1(A)\mathcal{I}_2(A) \subseteq \mathcal{P}(A) = P,$$

so some $\mathcal{I}_j \subseteq \mathcal{P}$. This proves $\mathcal{P}$ is a prime T-ideal. $\qquad\qquad\square$

Here is a related result, inspired by Kemer's theory.

Definition 6.72. A verbal ideal I is *verbally irreducible* if we cannot write $I = J_1 \cap J_2$ for verbal ideals $J_1, J_2 \supset I$.

Proposition 6.73. *Every verbal ideal I is a finite intersection of verbally irreducible verbal ideals.*

Proof. Otherwise, take a maximal counterexample I. Then I is verbally reducible, so write $I = I_1 \cap I_2$. By hypothesis, I_1, I_2 both are finite intersection of verbally irreducible verbal ideals, so I also is such an intersection. $\qquad\qquad\square$

Corollary 6.74. *Every relatively free affine PI-algebra U is a finite subdirect product of relatively free PI-algebras of PI-basic algebras.*

Proof. By the Proposition, we may assume $\mathrm{id}(U)$ is verbally irreducible. But $U = U(A)$ for some f.d. algebra A, which, by Proposition 4.58, is PI-equivalent to the finite direct sum of PI-basic algebras. Thus, we can replace A by one of these components. $\qquad\qquad\square$

6.3.3   Standard identities versus Capelli identities

Since so much of the theory depends on sparse identities, and more specifically Capelli identities, let us conclude this chapter with some results about Capelli identities. The material uses some of Kemer's ideas, although it predated (and provided motivation for) his solution to Specht's problem. In Example 3.34 we saw that the Capelli identity c_n implies the standard identity s_n, but s_n does not imply c_n. Furthermore, G does not satisfy any Capelli identity at all. This raises the following question: Does s_n imply $c_{n'}$ for suitable n'? The solution, due to Kemer [Kem78], although not used in the remainder of this book, is quite pretty, and has historical interest since it is the first time Grassman techniques were applied to the Capelli polynomial. Our presentation incorporates a slight modification in Berele [Ber85a].

We prove:

Theorem 6.75 (Kemer). *Any algebra of characteristic 0 satisfying s_n satisfies $c_{n'}$ for some n' that depends only on n.*

The rest of this subsection is dedicated to the proof of this theorem.

Example 6.76. Let us construct the relatively free superalgebra $\tilde{G}$ with respect (only) to the graded identities of (6.1) of Remark 6.5. Thus, we consider sets of usual indeterminates $\bar{X} = \{\bar{x}_i : i \in I\}$ and odd indeterminates $\bar{Z} = \{\bar{z}_i : i \in I\}$, where the $\bar{z}_j$ must strictly anticommute with each other. (We do not bother with the even indeterminates.) Let $\tilde{G} = F\{\bar{X}, \bar{Z}\}$. Since all the identities involve the $\bar{z}_i$, we see $F\{\bar{X}\} \approx F\{X\}$, the free algebra. Let $\tilde{G}_m$ denote all expressions in $\tilde{G}$ that are linear in each of $\bar{z}_1, \ldots, \bar{z}_m$. Since the identities are homogeneous, $\tilde{G}_m$ is well-defined, and Remark 1.28 has a more precise formulation.

Remark 6.77. Write $F\{X; Z\}$ for the free algebra in (usual) indeterminates $x_1, x_2, \ldots$ and $z_1, z_2, \ldots$. Let F_m denote the polynomials that are alternating in $z_1, z_2, \ldots, z_m$. There is a 1:1 correspondence $F_m \mapsto \tilde{G}_m$ given by

$$\sum_{\sigma \in S_n} (\mathrm{sgn}\sigma) h_1 z_{\sigma(1)} \cdots h_m, z_{\sigma(m)} h_{m+1} \mapsto h_1 \bar{z}_1 \cdots h_m \bar{z}_m h_{m+1}.$$

Let W_m denote the set of words of length m in the $\bar{x}_j$. Also let $\langle W_m \rangle$ denote the ideal of $F\{\bar{X}; \bar{Z}\}$ generated by the images of the elements of W_m.

Lemma 6.78. *If $w_1, \ldots, w_n \in W_m$, then*

$$w_{\sigma(1)} a_1 \ldots a_{n-1} w_{\sigma(n)} = \epsilon w_1 a_1 \ldots a_{n-1} w_n, \quad \forall a \in \tilde{G},$$

where $\epsilon = 1$ for m even and $\epsilon = \mathrm{sgn}(\sigma)$ for m odd.

Proof. Since every permutation is a product of transpositions, it is enough to check this for $n = 2$, in which case, this follows at once from the definition, since we are making m^2 switches of odd elements, and $m^2 \equiv m$ (mod 2). $\qquad\square$

Let $\mathcal{I}_m$ be the T-ideal of s_m, and let $\mathcal{J}_n$ be the T-ideal of c_n. Given a T-ideal $\mathcal{T}$, we let $\mathcal{T}(\tilde{G})$ denote the evaluations of $\mathcal{T}$ on $\tilde{G}$, under $x_i \mapsto \bar{x}_i$ and $z_i \mapsto \bar{z}_i$.

Lemma 6.79. $W_m \subseteq \mathcal{I}_m(\tilde{G})$.

Proof. $z_{i_1} \ldots z_{i_m} = \frac{1}{m!} s_n(z_{i_1}, \ldots, z_{i_m})$. $\qquad\square$

Lemma 6.80. *For any T-ideal $\mathcal{T}$ of $F\{X\}$, $\mathcal{J}_m \subseteq \mathcal{T}$ iff $\langle W_1 \rangle^m \subseteq \mathcal{T}(\tilde{G})$.*

Proof. We take $a_i \in \tilde{G}$ arbitrary and $w_i \in W_1$.

($\Rightarrow$) $m! a_0 w_1 a_1 \ldots w_m a_{m+1} = a_0 c_m(w_1, \ldots, w_m; a_1, \ldots, a_m) \in \tilde{G}\mathcal{T}(\tilde{G}) \subseteq \mathcal{T}(\tilde{G})$.

($\Leftarrow$) We need to show $c_m \in \mathcal{I}$. But

$$c_m(\bar{z}_1, \ldots \bar{z}_m; \bar{x}_1, \ldots \bar{x}_m) = m! \bar{z}_1 \bar{x}_1 \ldots \bar{z}_m \bar{x}_m \bar{z}_{m+1} \in \langle W_m \rangle^m \subseteq \mathcal{T}(\tilde{G}).$$

Thus, there is $f \in \mathcal{T}$ such that

$$f(\bar{z}_1, \ldots, \bar{z}_{m+1}; \bar{x}_1, \ldots, \bar{x}_m) = c_m(\bar{z}_1, \ldots, \bar{z}_m; \bar{x}_1, \ldots, \bar{x}_m).$$

Taking the alternator

$$\tilde{f} = \sum_{\sigma \in S_n} \text{sgn}(\sigma) f(z_{\sigma(1)}, \ldots, z_{\sigma(m)}; x_1, \ldots, x_m),$$

which is m-alternating in the z_i, we see that

$$\tilde{f}(\bar{z}_1, \ldots, \bar{z}_m; \bar{x}) = m! c_m((\bar{z}_1, \ldots, \bar{z}_m; \bar{x})),$$

and thus Remark 6.77 shows $c_m = \frac{1}{m!} f \in \mathcal{T}$. $\qquad\square$

We recall $\tilde{s}_n(x_1, \ldots, x_n) = \sum_{\pi \in S_n} x_{\pi(1)} \ldots x_{\pi(t)}$, cf. (1.6) of Chapter 1.

Proposition 6.81. *If $a_1, \ldots, a_{t+1} \in W_m \tilde{G}$, then*

$$\tilde{s}_t(a_1, \ldots, a_t) a_{t+1} \in \begin{cases} \langle W_{2m} \rangle + \mathcal{I}_{2t}(G), & m \text{ odd} \\ \langle W_{2m-1} \rangle + \mathcal{I}_{2t}(G), & m \text{ even.} \end{cases}$$

Proof. For m odd, write $a_i = w_i g_i$ for $w_i \in W_m$ and $g_i \in \tilde{G}$; then Lemma 6.78 yields

$$\tilde{s}_t(a_1, \ldots, a_t) a_{t+1} = \sum_\sigma w_{\sigma(1)} g_{\sigma(1)} \cdots w_{\sigma(t)} g_{\sigma(t)} w_{t+1} g_{t+1}$$

$$= \sum_\sigma \text{sgn}(\sigma) w_1 g_{\sigma(1)} \cdots w_t g_{\sigma(t)} w_{t+1} g_{t+1}$$

$$= \frac{1}{t!} \sum_{\sigma, \tau} \text{sgn}(\sigma\tau) w_{\tau(1)} g_{\sigma(1)} \cdots w_{\tau(t)} g_{\sigma(t)} w_{t+1} g_{t+1}.$$

But the last sum differs from $s_{2t}(w_1, g_1, \ldots, w_t, g_t)$ only by terms where two ws are adjacent, i.e., in $\langle W_{2m} \rangle$, so we conclude the right side is in $\langle W_{2m} \rangle + \mathcal{I}_{2t}(\tilde{G})$.

For m odd, write $a_i = w_i g_i$ for $w_i \in W_{m-1}$ and $g_i \in W_1 \tilde{G}$, and apply the same reasoning. (We need the subscript of W to be odd, in order for Lemma 6.78 to be applicable.) $\qquad\square$

The continuation of the proof is of the flavor of the proof we have given of the Amitsur-Levitzki theorem, relying heavily on the Dubnov-Ivanov-Nagata-Higman Theorem 3.40. Take β as in Theorem 3.40.

Corollary 6.82. $\langle W_m \rangle^{\beta(t)} \subseteq \langle W_{2m} \rangle^{2t} + \mathcal{I}_{2t}(G)$ *for m odd, and* $\langle W_m \rangle^{\beta(t)} \subseteq \langle W_{2m-1} \rangle^{2t} + \mathcal{I}_{2t}(G)$ *for m even.*

Proof. Theorem 3.40 says $(W_m \tilde{G})^{\beta(t)}$ is contained in the T-ideal without 1 generated by $\tilde{s}_t$, which by the Proposition is contained in $\langle W_{2m} \rangle + \mathcal{I}_{2t}(G)$ or $\langle W_{2m-1} \rangle + \mathcal{I}_{2t}(G)$. $\qquad\square$

Now define the function $\gamma : \mathbb{N} \to \mathbb{N}$ inductively by $\gamma(1) = 1$ and

$$\gamma(m+1) = \begin{cases} 2\gamma(m) & \text{for } m \text{ odd}; \\ 2\gamma(m) - 1 & \text{for } m \text{ even}. \end{cases}$$

Corollary 6.83. *Suppose $n = 2t$, and m satisfies $\gamma(m) > n$. Then*

$$\langle W_1 \rangle^{\beta(t)^m} \subseteq W_{\gamma(m)} + \mathcal{I}_n(\tilde{G}) \subseteq \mathcal{I}_n(\tilde{G}).$$

Proof. Clear, since $W_{\gamma(m)} \subseteq W_n \subseteq \mathcal{I}_n(\tilde{G})$ by Remark 6.79. $\qquad\square$

Confronting this corollary with Lemma 6.80, we see for n even and $m' = \beta(m)^t$ that $\mathcal{J}_{m'} \subseteq \mathcal{I}_n$, i.e., s_n implies $c_{m'}$. Since s_n implies s_{n+1}, we may always assume n is even, and thus the corollary is applicable and we have proved the theorem. Furthermore, the proof is constructive, depending on the function β of Theorem 3.40, which can be improved, cf. Exercise 12.4.

Now that we have Kemer's solution to Specht's problem and the subsequent structure theory of T-ideals, we can improve Thoerem 6.75 (albeit without an explicit bound).

Proposition 6.84. *If an F-algebra A of characteristic 0 satisfies an identity f that is not a consequence of the Grassman identity $[[x_1, x_2], x_3]$, then A satisfies all identities of $M_n(F)$, for some n.*

Proof. We sketch the easy proof. We may assume that A is relatively free. By the structure theory of T-ideals, A has a nilpotent T-ideal N such that A/N is a subdirect product of T-prime algebras. In view of Lewin's Theorem 1.72, we may pass to A/N and thereby assume $N = 0$; thus we may assume that A is T-prime. In view of Theorem 6.64, we are done unless $A \sim_{\mathrm{PI}} M_n(G)$. But then f is an identity of $M_n(G)$ and thus of G, contrary to hypothesis.   □

In particular we can take f to be c_m or s_m, and thereby recover Theorem 6.75.

6.3.4   Specht's problem for T-spaces.

Recall the definition of T-space from Definition 3.10(i). Shchigolev [Shch01] strengthened Kemer's theorem and proved T-spaces are finitely based in characteristic 0. The proof is based on the following

Theorem 6.85 (V. V. Shchigolev). *Let S be a T-space in the free algebra $A = F\{X\}$ over a field of characteristic 0.*

(i) If $S \backslash [A, A] \neq \emptyset$, then S includes a nonzero T-ideal.

(ii) If $S \subseteq [A, A]$, then there exists a T-ideal Γ such that $[\Gamma, A] \subseteq S$.

Using this theorem, Shchigolev can then apply Kemer's theory to prove that T-spaces are finitely based. In Chapter 7, we consider T-spaces from the opposite point of view; in characteristic $p > 0$ we shall construct T-spaces that are not finitely based, and use them as the jumping board to construct T-ideals that are not finitely based.

Chapter 7

PI-Algebras in Characteristic p

Since some of the theory developed in Chapters 4 through 6 relied on the assumption of characteristic 0 (especially that portion in Chapters 5 and 6 using the Young-Frobenius theory of representations of S_n), the theory in characteristic $p \neq 0$ requires a separate discussion. The most striking feature is that without the characteristic 0 assumption, Specht's problem now has a negative solution, and most of this chapter will be devoted to various counterexamples.

7.1 Positive Results

We start by discussing those aspects of the PI-theory that carry over in characteristic $p > 0$.

We already proved in Chapter 2 that every affine PI-algebra satisfies a Capelli identity, thereby yielding the Razmyslov-Kemer-Braun theorem in all characteristics.

The Kemer index is a useful tool for affine PI-algebras in any characteristic, which we shall need in Chapter 8.

Remark 7.1. The most direct approach to Specht's conjecture for affine PI-algebras in characteristic p would be to try to apply the techniques of Chapter 4 and 6. The methods given in Chapter 4 yield representability up to multilinear polynomials. One must also use characteristic coefficients rather than the trace alone, but this is a technical difficulty that already was surmounted in Proposition 2.39. Otherwise, the theory goes through without modification, and one can prove that for any ascending chain of verbal ideals in an affine algebra, the multilinear part stabilizes.

Most of the theory for affine algebras goes over in general, provided that the base field is infinite. In this case, all identities can be recovered from homogeneous identities, whereas this cannot be done over a finite field. Our (rather intricate) proof of the Second Kemer Lemma 4.54 required the radical substitutions to be in variables that are linear; since a priori we do not know which substitutions are to be radical, we must then deal with multilinear polynomials. However, a careful treatment of the linearization process in characteristic p enables one to bypass this difficulty. Kemer [Kem91b] proved Theorem 4.66 (the PI-representability of affine PI-algebras) for arbitrary characteristics over an arbitrary infinite field.

Actually, Belov verified the ACC for T-ideals in $C\{x_1, \ldots, x_\ell\}$ over an arbitrary commutative Noetherian ring C, including the case C finite, but his proof [Bel2002] (which, so far, only exists in Russian) is too long and technical for the scope of this book.

One can also ask for an affine algebra A whether $\mathrm{id}(A)$, viewed as a T-ideal of $F\{X\}$ for an infinite set of indeterminates X, is finitely based. This question follows at once from Kemer's theorem when F has characteristic 0, but, for characteristic p, remains unsolved. Also, there are many more prime T-ideals in characteristic p, and their classification remains perhaps the most important open question in characteristic p.

7.1.1 De-multilinearization

The major obstacle in characteristic p is how to reverse the multilinearization process. Here is our main technique.

Remark 7.2. Suppose we want to verify that a given homogeneous polynomial $f(x_1, \ldots, x_d)$ is an identity of A. Let $n_i = \deg_i f$. If each $n_i < p$, i.e., p does not divide $n_i!$ for any i, then it is enough to check the multilinearization of f, cf. Remark 1.12(ii).

In general, we define a *word specialization* to be a specialization to words in the generators of A. To verify f is an identity, it is enough to check that every partial multilinearization of f is an identity, and also that f vanishes under all word specializations, seen by working backwards on the definition of Δ_1 in Definition 1.11, e.g.,

$$f(x_1 + x_2; y_1, \ldots, y_m) = f(x_1; y_1, \ldots, y_m) + f(x_2; y_1, \ldots, y_m) \\ + \Delta_1 f(x_1, x_2; y_1, \ldots, y_m). \tag{7.1}$$

But then continuing inductively, it is enough to check that the partial multilinearizations (as well as f) vanish under all word specializations.

This remark enables us to employ the combinatorics of the previous chapters even in characteristic p. We shall see several instances of this in the sequel, and start now with a trivial example, which is just a restatement of the Frobenius automorphism.

Remark 7.3. Suppose $\text{char}(F) = p$, and $A = F\{a_1, \ldots, a_\ell\}$ is a commutative algebra without 1, and $a_i^p = 0$ for $1 \le i \le \ell$. Then $x^p \in \text{id}(A)$. Indeed, this is true for products of the generators, whereas all coefficients of $(a + b)^p - a^p - b^p$ are multiples of p and thus 0.

Similarly, we shall see in Exercise 7 that the Grassman algebra in characteristic $p > 2$ satisfies the identity $(x + y)^p - x^p - y^p$. (For $p = 2$, one gets $(x + y)^4 - x^4 - y^4$, cf. Exercise 3.)

Remark 7.4. We continue the logic of Remark 7.2. Suppose $\deg_i f = n_i \le p$. We claim that if the linearization $\tilde{f}_i$ of f in x_i is an identity of A, then every partial multilinearization g of f in x_i is also an identity of A.

Indeed, by induction it is enough to take $g = \Delta_i f$, notation as in (1.2) of Chapter 1. Write $g = \sum_{j=1}^{n_i-1} f_j$ where f_j has degree j in x_i (and thus degree $n_i - j$ in x_i'). As in (1.4) of Chapter 1, we can recover $j!(n_i - j)!f_j$ from $\tilde{f}_i$. But p does not divide $j!(n_i - j)!$, so we see $\tilde{f}_i \in \text{id}(A)$ iff each $f_j \in \text{id}(A)$, which clearly implies $g \in \text{id}(A)$.

Repeating this argument for each original indeterminate x_i, we see that if $\deg_i f \le p$ for each p and the total multilinearization of f is an identity of A, then each partial multilineariation is also an identity of A.

7.1.2 Kemer's matrix identity theorem

A much stronger version of Lewin's theorem holds in characteristic p:

Theorem 7.5. [Kem95], [Bel99] *Any PI-algebra A in characteristic $p > 0$ satisfies the identities of $M_m(\mathbb{Z}/p)$ for suitable m.*

The proof is out of the scope of our book, but let us give an overview of the proof. First, Kemer [Kem93] showed (cf. Exercise 7.21) that every PI-algebra in characteristic > 0 satisfies a standard identity, although the bounds are very large. Applying the representation theory in characteristic p described briefly at the end of Chapter 5 to [Kem80], Kemer proved in [Kem95]:

Any PI-algebra over a field of characteristic p satisfies all *multilinear* identities of some matrix algebra, and in particular, the appropriate Capelli identity.

This is already a striking result which is clearly false for characteristic 0, when we recall that the Grassman algebra of an infinite dimensional vector space over $\mathbb{Q}$ does not satisfy a Capelli identity.

As a consequence, the Razmyslov-Zubrillin theory expounded in Chapter 2 and the appendix to Chapter 4 (and Exercise 4.7) become available to all PI-algebras of characteristic p. Furthermore, Donkin [Do92] proved the following generalization of a theorem of Procesi and Razmyslov to be discussed in Chapter 12:

Theorem 7.6. *Every invariant in several matrices is a consequence of the coefficients of the characteristic polynomial.*

Zubkov [Zu96] showed relations among these invariants follow from the Hamilton-Cayley polynomial. Putting the pieces together yields the proof of Theorem 7.5.

7.2 The Extended Grassman Algebra

As in characteristic 0, the key to much of our theory is a close study of the Grassman algebra G, whose identities were seen in Exercise 3.5 to be consequences of the Grassman identity $[[x_1, x_2], x_3]$. Now assume for the remainder of this section that $\operatorname{char}(F) = 2$. (Thus, $[a, b] = ab + ba$.) Unfortunately, the Grassman algebra is now commutative, so we need a replacement algebra for G.

Definition 7.7. Define $F[\varepsilon_i : i \in \mathbb{N}]$ to be the commutative polynomial algebra modulo the relations $\varepsilon_i^2 = 0$. We denote this algebra as $F[\varepsilon]$.

Thus, any element $c \in F[\varepsilon]$ inherits the natural degree function, and if f has constant term 0, then $c^2 = 0$. (This is seen at once from Remark 7.3, since $F[\varepsilon]$ is commutative.) This provides us the perfect setting for our analog of the Grassman algebra in characteristic 2.

Definition 7.8. The *extended Grassman algebra* G^+ is defined over $F[\varepsilon]$ by means of the quadratic relations

$$[e_i, e_j] = \varepsilon_i \varepsilon_j e_i e_j, \ \forall i \neq j.$$

Thus, G^+ is a superalgebra, and in contrast with G, we now have $e_i^2 \neq 0$. Nevertheless, we do have useful positive information. G^+ clearly has a base B consisting of words $e_{i_1} \dots e_{i_t}$ with $i_1 \leq i_2 \leq \cdots \leq i_t$. Given a word $b = e_{i_1} \dots e_{i_t} \in B$, let us write $\varepsilon_b = \sum_{j=1}^{t} \varepsilon_{j_u}$.

Remark 7.9. $\varepsilon_b^2 = 0$ for any word b, since

$$\left(\sum_{u=1}^{t} \varepsilon_{j_u} \right)^2 = \sum_{u=1}^{t} \varepsilon_{j_u}^2 = \sum_{u=1}^{t} 0 = 0.$$

Definition 7.10. There is a natural F-linear map $\varphi : F[\varepsilon] \to F$ given by each $\varepsilon_i \mapsto 1$. We call an element of $F[\varepsilon]$ *even* if it is in $\ker \varphi$, and *odd* otherwise. Likewise, an element of G^+ is *even* if, written as a linear combination of elements of B, each coefficient is even.

Lemma 7.11. G^+ *satisfies the following properties, for* $b = e_{j_1} \dots e_{j_t}$:

(i) $[e_i, b] = \varepsilon_i \varepsilon_b e_i b$;

(ii) $\varepsilon_b b \in \text{Cent}(G^+)$.

Proof. First we show $\varepsilon_i e_i \in \text{Cent}(G^+)$, by proving it commutes with each e_j :

$$[\varepsilon_i e_i, e_j] = \varepsilon_i [e_i, e_j] = \varepsilon_i^2 \varepsilon_j e_i e_j = 0.$$

Now

$$[e_i, b] = \sum_{u=1}^{t} e_{j_1} \dots e_{j_{u-1}} [e_i, e_{j_u}] e_{j_{u+1}} \dots e_{j_t}$$

$$= \sum_{u=1}^{t} \varepsilon_i \varepsilon_{j_u} e_{j_1} \dots e_{j_{u-1}} e_i e_{j_u} e_{j_{u+1}} \dots e_{j_t}$$

$$= \sum_{u=1}^{t} \varepsilon_{j_u} e_{j_1} \dots e_{j_{u-1}} (\varepsilon_i e_i) e_{j_u} e_{j_{u+1}} \dots e_{j_t}$$

$$= \varepsilon_i e_i \sum_{u=1}^{t} \varepsilon_{j_u} e_{j_1} \dots e_{j_{u-1}} e_{j_u} e_{j_{u+1}} \dots e_{j_t}$$

$$= \varepsilon_i (\sum_{u=1}^{t} \varepsilon_{j_u}) e_i e_{j_1} \dots e_{j_{u-1}} e_{j_u} e_{j_{u+1}} \dots e_{j_t},$$

proving (i).

To prove (ii), it is enough to commute with any e_i:

$$\left[e_i, \left(\sum_{u=1}^{t} \varepsilon_{j_u} \right) b \right] = \sum_{u=1}^{t} \varepsilon_{j_u} [e_i, b]$$

$$= \sum_{u=1}^{t} \varepsilon_{j_u} \varepsilon_i \left(\sum_{u=1}^{t} \varepsilon_{j_u} \right) e_i e_{j_1} \dots e_{j_t}$$

$$= \varepsilon_i \left(\sum_{u=1}^{t} \varepsilon_{j_u} \right)^2 e_i e_{j_1} \dots e_{j_t} = 0.$$

$\square$

This is generalized in Exercise 1. However, we take another tack. We say a sum is *homogeneous* if each summand has the same degree in each indeterminate. For example, the sum in Remark 3.42 is homogeneous.

Remark 7.12. In the free associative algebra $C\{X\}$, any commutator $[w_1, w_2]$ of words in the x_i can be rewritten as a homogeneous sum $\sum_i [w_{3i}, x_i]$ for suitable words w_{3i}. This is seen immediately by applying induction to Remark 3.42(iii).

Proposition 7.13. G^+ *satisfies the Grassman identity. (Equivalently, $[x_1, x_2]$ is G^+-central.)*

Proof. It suffices to prove $[w_1, w_2]$ is central for any words w_1, w_2 in G^+; by Remark 7.12, we may assume w_2 is a letter e_i, so we are done by Lemma 7.11. $\qquad\square$

Corollary 7.14. $[x_1^2, x_2] \in \mathrm{id}(G^+)$. *(Equivalently, x_1^2 is G^+-central.)*

Proof. For any $a, b \in G^+$, we have

$$[a^2, b] = a[a, b] + [a, b]a = a[a, b] + a[a, b] = 2a[a, b] = 0.$$

$\qquad\square$

We also are interested in squares in G^+.

Corollary 7.15. $[x_1, x_2]^2 \in \mathrm{id}(G^+)$.

Proof. This is by Corollary 3.44. $\qquad\square$

Proposition 7.16. *Suppose $a \in G^+$ is homogeneous of degree 1 in each of in $e_1, \ldots, e_t$. Then for some $\gamma \in F$,*

$$a^2 = \gamma^2 (e_1 \ldots e_t)^2.$$

Proof. Noting

$$e_j e_i = e_i e_j - [e_i, e_j] = e_i e_j + \varepsilon_i \varepsilon_j e_i e_j = (1 + \varepsilon_1 \varepsilon_j) e_i e_j,$$

we can rewrite a as

$$c e_1 \ldots e_t,$$

where $c \in F[\varepsilon]$. Hence

$$a^2 = c^2 (e_1 \ldots e_t)^2;$$

writing $c = \gamma + c'$ where $\gamma \in F$ and $c' \in F[\varepsilon]$ has constant term 0, we have

$$c^2 = \gamma^2 + (c')^2 = \gamma^2 + 0 = \gamma^2,$$

as desired. $\qquad\square$

It remains to compute squares of these words:

Proposition 7.17. *Let*

$$\bar{\varepsilon}_t = \sum_{j=0}^{[t/2]} \sum_{i_1 < i_2 \cdots < i_{2j}} \varepsilon_{i_1} \ldots \varepsilon_{i_{2j}},$$

i.e., the sum of all even products of the ε_i, where the term for $j = 0$ is 1. Then

$$(e_1 \ldots e_t)^2 = \bar{\varepsilon}_t e_1^2 \ldots e_t^2.$$

Proof. By induction on t, the assertion being obvious for $t = 1$. But then by induction

$$(e_1 \ldots e_{t-1})^2 = \bar{\varepsilon}_{t-1} e_1^2 \ldots e_{t-1}^2,$$

and letting $b = e_1 \ldots e_{t-1}$, we have

$$(e_1 \ldots e_t)^2 = (be_t)^2 = be_t be_t = b(be_t + [b, e_t])e_t = b(1 + \varepsilon_t \varepsilon_b)be_t e_t$$
$$= (1 + \varepsilon_t \varepsilon_b)b^2 e_t^2 = (1 + \varepsilon_t \varepsilon_b)\bar{\varepsilon}_{t-1} e_1^2 \ldots e_{t-1}^2 e_t^2$$

so it remains to show $(1 + \varepsilon_t \varepsilon_b)\bar{\varepsilon}_{t-1} = \bar{\varepsilon}_t$, i.e., we need

$$\varepsilon_t \varepsilon_b \bar{\varepsilon}_{t-1} = \bar{\varepsilon}_t - \bar{\varepsilon}_{t-1},$$

which is clearly $\sum_j \sum_{i_1 < \cdots < i_{2j-1}} \varepsilon_{i_1} \ldots \varepsilon_{i_{2j-1}} \varepsilon_t$. In other words, we want to show

$$\varepsilon_b \bar{\varepsilon}_{t-1} = \sum_j \sum_{i_1 < \cdots < i_{2j-1}} \varepsilon_{i_1} \ldots \varepsilon_{i_{2j-1}}.$$

But multiplying ε_i by $\varepsilon_{i_1} \ldots \varepsilon_{i_{2j-2}}$ yields 0 unless $i \neq i_k$ for each $1 \leq k \leq 2j - 2$, and there are $2j - 1$ ways to get a product of the given form $\varepsilon_{i_1} \ldots \varepsilon_{i_{2j-1}}$. Hence

$$\varepsilon_b \bar{\varepsilon}_{t-1} = \sum_j \sum_{i_1 < \cdots < i_{2j-1}} (2j - 1)\varepsilon_{i_1} \ldots \varepsilon_{i_{2j-1}}$$
$$= \sum_j \sum_{i_1 < \cdots < i_{2j-1}} \varepsilon_{i_1} \ldots \varepsilon_{i_{2j-1}},$$

since $2j \equiv 0 \pmod{2}$. $\qquad\square$

The important thing for us will be that the number of terms is

$$1 + \binom{t}{2} + \binom{t}{4} + \cdots = 2^{t-1},$$

which is even.

Corollary 7.18. *The product of squares of words involving distinct letters is even, provided at least one of the factors has length > 1.*

Proof. The products have coefficients whose number of terms are a power of 2, without cancellation, and thus are even. $\qquad\square$

For example, $(e_1e_2)^2(e_3e_4)^2 = (1 + \varepsilon_1\varepsilon_2)(1 + \varepsilon_3\varepsilon_4)e_1^2e_2^2e_3^2e_4^2$. Note on the other hand that

$$(e_1e_2)^2(e_1e_3)^2 = (1 + \varepsilon_1\varepsilon_2)(1 + \varepsilon_1\varepsilon_3)e_1^2e_2^2e_1^2e_3^2 = (1 + \varepsilon_1\varepsilon_2 + \varepsilon_1\varepsilon_3)e_1^4e_2^2e_3^2,$$

which is odd.

7.2.1 Computing in the Grassman and extended Grassman algebras

We now turn to the heart of our examples—some of the combinatorial properties of the Grassman identity $[[x_1, x_2], x_3]$ which make it possible for us to perform complicated computations involving polynomials. The key role is played by commutators, which by hypothesis are central. We start with some formal (and very useful) consequences of Latyshev's Lemma 3.43:

Lemma 7.19. *The following identities are consequences of the Grassman identity:*

(i) $[x_1, x_2]u[x_3, x_4] = -[x_1, x_4]u[x_2, x_3]$.

(ii) $[x_1, x_2]u[x_1, x_3] = 0$.

(iii) *The identity* $[x_1, x_2]w[x_3, x_4] = 0$ *holds in any 2-generated subalgebra.*

(iv) *In a 2-generated subalgebra, any single expression (formed by taking regular products and Lie products only) involving two sets of commutators is 0. (For example, $[a_1a_2, [a_2, a_1a_2^2]a_1] = 0$.)*

Proof. (i) follows immediately from Lemma 3.43, since u commutes with all commutators. One gets (ii) by specializing x_4 to x_1.

To prove (iii), one may assume x_i are words in the e_i, and proceed by an easy induction on their lengths: If all have length 1, then two of them are equal, and the only nontrivial case is $[e_i, e_j]u[e_i, e_k]$, which is 0 by (ii). Finally, the inductive step: If $x_1 = rs$, then

$$[rs, x_2]w[x_3, x_4] = r[s, x_2]w[x_3, x_4] + [r, x_2]sw[x_3, x_4],$$

both of which are 0 by induction.

(iv) Any single commutator is central, and so could be extracted from the expression; thus, we are done by (iii). $\qquad\square$

We say that a product of m commutators in G or G^+ is in *standard form* if it can be written as a homogeneous sum of the form

$$[e_{j_1}, e_{j_2}][e_{j_3}, e_{j_4}] \ldots [e_{j_{2m-3}}, e_{j_{2m-2}}][v_m, e_{j_{2m}}], \qquad (7.2)$$

where v_m is an arbitrary word.

Lemma 7.20. *If f is a product of m commutators, then any specialization of f in G or in G^+ can be put in standard form.*

Proof. We need to rewrite $[v_1, w_2] \ldots [v_m, w_m]$ (for any words v_i, w_i) as a sum of the form (7.2).

Step 1. By Remark 7.12, we may reduce to the case $w_i = e_{j_i}$ (since our expression is sums of expressions of this form).

Step 2. In view of Lemma 7.19(ii), we may assume the e_{j_i} are distinct.

Step 3. $[v_1, e_{j_1}][v_2, e_{j_2}] = -[e_{j_1}, v_1][v_2, e_{j_2}] = [e_{j_1}, e_{j_2}][v_2, v_1]$, by Lemma 7.19(i). Thus, if $m \geq 2$, we may assume the first commutator is $[e_{j_1}, e_{j_2}]$, say $[e_1, e_2]$. By Step 1 again, we may replace v_1 by some new e_{j_2}. So now we have

$$[e_1, e_2][v_2, e_{j_2}] \ldots [v_m, e_{j_m}].$$

Step 4. Repeating the argument of Step 3, we reduce to

$$[e_1, e_2][e_3, e_4] \ldots [e_{2m-3}, e_{2m-2}][v_m, e_{j_m}],$$

which is the desired form. $\qquad \square$

7.3 Non-Finitely Based T-Ideals in Characteristic 2

There is a rather simple strategy for constructing a non-finitely based T-ideal: Construct a series of polynomials $f_1, f_2, \ldots$ and algebras $A_1, A_2, \ldots$ such that for each n, $f_1, \ldots, f_n \in \mathrm{id}(A_n)$ but $f_m \notin \mathrm{id}(A_n)$ for some $m > n$. This means that the T-ideal generated by all $f_1, f_2, \ldots$ cannot be a consequence of a finite number of f_i, and thus cannot be finitely based. Although somehow this approach is (perhaps surprisingly) defeated in characteristic 0 by the Kemer polynomials, there are enough new identities in characteristic p to carry it through.

We start with the simplest counterexample, which is in characteristic 2. The example is due to Belov and Grishin, with proof shortened by using a construction of Gupta and Krasilnikov. We utilize the extended Grassman algebra G^+, cf. Definition 7.8.

Lemma 7.21. *Let I_n be the $\mathbb{Z}/2$-subspace of G^+ spanned by all $\bar{x}_{i_1}^2 \dots \bar{x}_{i_m}^2$, where $m < n$, for $\bar{x}_i \in G^+$. Then*

$$e_{i_1}^2 e_{i_2}^2 \dots e_{i_n}^2 \notin I_n,$$

for any n. (We require $i_1, i_2, \dots i_n$ distinct.)

Proof. Taking homogeneous parts in $e_{i_1}, e_{i_2}, \dots e_{i_n}$, it suffices to show that

$$e_{i_1}^2 e_{i_2}^2 \dots e_{i_n}^2$$

cannot be written as a sum of products of fewer than n squares. But each such product is even, by Corollary 7.18, so the sum must be even, whereas $e_{i_1}^2 e_{i_2}^2 \dots e_{i_n}^2$ has coefficient 1. $\square$

Example 7.22. Define

$$f_n = [y_1^2, y_2] x_1^2 x_2^2 \dots x_n^2 [y_1^2, y_2]^3.$$

Over a field F of characteristic 2, the T-ideal generated by $\{f_n : n \in \mathbb{N}\}$ is not finitely based.

Proof. We work in the matrix algebra $M_5(G^+)$ with its standard set of matrix units e_{ij}, not to be confused with the $e_i \in G^+$ used above. Let R be the subalgebra of $M_5(G^+)$ consisting of upper triangular matrices, and let A be the F-subalgebra of R generated by 1, $v = \sum_{i=1}^4 e_{i,i+1}$ and all $\{b(e_{22} + e_{44}) : b \in G^+\}$. Thus, A consists of elements of the form

$$w_j = \begin{pmatrix} \alpha_j & b_j' & * & * & * \\ 0 & b_j & b_j'' & * & * \\ 0 & 0 & \alpha_j & b_j' & * \\ 0 & 0 & 0 & b_j & b_j'' \\ 0 & 0 & 0 & 0 & \alpha_j \end{pmatrix},$$

where $\alpha_j, \beta_j \in F$ and $b_j, b_j', b_j'' \in G^+$. Note that $G^+ e_{15} \subseteq A$ since

$$b e_{15} = v b (e_{22} + e_{44}) v^3.$$

On the other hand, $b(e_{22} + e_{44})$ and v both annihilate $G^+ e_{15}$ from each side, so any F-subspace of $G^+ e_{15}$ is an ideal of A; in particular, $I_n e_{15} \triangleleft A$, taking I_n as in Lemma 7.21. The continuation is an easy exercise in computing upper triangular matrices.

Let $A_n = A / I_n e_{15}$. In view of Corollary 7.14, $b_j^2 \in \text{Cent}(G^+)$, and certainly $\alpha_j^2 \in \text{Cent}(G^+)$, so any evaluation of $[w_1^2, w_2]$ in A_n has diagonal 0 thus is of the form

$$s = \begin{pmatrix} 0 & b' & * & * & * \\ 0 & 0 & b'' & * & * \\ 0 & 0 & 0 & b' & * \\ 0 & 0 & 0 & 0 & b'' \\ 0 & 0 & 0 & 0 & 0 \end{pmatrix}.$$

If $r = w_1^2 \ldots w_m^2 \in A$, the $2,2$ entry of r is $b_1^2 \ldots b_m^2$. Then, taking s as above, the only possible nonzero entry of srs^3 in A comes from

$$(b'e_{12})re_{22}(b''e_{23})(b'e_{34})(b''e_{45}),$$

so is $(b'b'')^2b_1^2 \ldots b_m^2 e_{15}$. This is in $I_n e_{15}$ whenever $m < n-2$, implying $f_1, \ldots, f_{n-3} \in \mathrm{id}(A_n)$.

On the other hand,

$$[(e_{22}+e_{44})^2, s] = [e_{22}+e_{44}, s] = (b'e_{23}+b'e_{45}) - (b''e_{12}+e_{34}) = s$$

since $+1 = -1$, so taking $\bar{x}_i = e_i$ and $b' = 1$ and $b'' = e_n$ in G^+, we see f_{n-1} takes on the value $e_1^2 \ldots e_{n-1}^2 e_n^2$, which is not in I_n, so $f_{n-1} \notin \mathrm{id}(A_n)$.

Thus, the chain of T-ideals generated by $\{f_1, \ldots, f_n : n \in \mathbb{N}\}$ never stabilizes, as desired. $\hspace{2cm} \square$

7.3.1 T-spaces evaluated on the extended Grassman algebra

Unfortunately, Shchigolev [Shch00] showed that the natural analog to Example 7.22 in odd characteristic is finitely based; the T-space generated by the polynomials $Sh_n = x_1^p \cdots x_n^p$ is finitely based (even in the free algebra!), by the set $Sh_1, \ldots, Sh_{(p+1)/2}$.

Accordingly, we shall take a different approach to constructing infinitely based T-ideals, which although more intricate, leads naturally to examples in all positive characteristics. The idea is to build a T-space (cf. Definition 3.10) that is not finitely based, and then modify the construction for T-ideals. Thus, we divide the problem into easier parts, but the down side is that we can only use one test algebra at a time. In characteristic 2, it is natural to try the test algebra G^+.

Finding a non-finitely based T-space means we want to find a sequence $f_1, f_2, \ldots$ of polynomials such that, for each n, some value of f_n cannot be spanned over F by specializations of $f_m, m < n$. It turns out that these polynomials are not linear, so we have to worry about linearizing the f_m.

Towards this end, by Remark 7.2, it suffices to check this for word specializations and for partial linearizations. It turns out in all our examples that the contribution of the partial linearizations *must be* 0! Let us make the idea explicit by means of a definition.

Definition 7.23. A p-*monomial* is a monomial h for which $\deg_i h \in \{0, p\}$ for each i. A polynomial is called p-*admissible* if it is a linear combination of p-monomials. A specialization of a homogeneous polynomial f in G (resp. G^+ for $p = 2$) to words v_i in the e_j to be p-*admissible* if each e_j occurs of degree a multiple of p, for $1 \leq i \leq 2n$.

We shall choose $f_n(x_1, \ldots, x_{2n})$ with $\deg_i f_n = p$ for all i. Assuming $\text{char}(F) = p$, we see by Remark 7.4 that to check the partial linearizations it is enough to show in the multilinearization of f_m that the homogeneous parts of degree p in each e_i of any specialization of f_m is 0. Thus, we shall show for each $m < n$ that each p-admissible word specialization of a multilinearization of f_m is 0, and also show the p-admissible word specializations of the f_m cannot cancel f_n.

Let us see how this works for $p = 2$.

Lemma 7.24. *If f is a product of m commutators, then any 2-admissible specialization of f in G^+ is 0.*

Proof. In view of Lemma 7.20, we need show (7.2) is 0 if it is 2-admissible. We argue by induction on the degree ℓ of v_m, and then on m. If $\ell = 1$ and $m = 1$, then our specialization is $[e_1, e_1] = 0$.

If $\ell = 1$ and $m > 1$, then v_m is some e_i, say $v_m = e_i$. We consider the general situation after dealing with two important cases.

Case 1. Suppose e_i appears in v_m with degree 1, where $i \neq m$. By definition of 2-admissibility, e_i must appear somewhere else; let us assume for notational convenience that $i = 1$. Then $v_m = e_1 v'$ and, since commutators are central,

$$[e_1, e_2][e_3, e_4] \ldots [v_m, e_{j_m}] = \ldots [e_1 v', e_{j_m}]$$
$$= [e_3, e_4] \ldots [e_1[e_1, e_2]v', e_{j_m}]$$
$$= [e_3, e_4] \ldots [\varepsilon_1 \varepsilon_2 e_1^2 e_2 v', e_{j_m}]$$
$$= [e_3, e_4] \ldots [e_2 v', e_{j_m}] \varepsilon_1 \varepsilon_2 e_1^2.$$

But $[e_3, e_4] \ldots [e_2 v', e_{j_m}] = 0$ since it is a 2-admissible specialization of a product of commutators with smaller m and the same ℓ, implying (7.2) is 0.

Case 2. Suppose $v_m = v' e_i e_j v''$. Then we can replace v_m by $v' e_j e_i v''$. Indeed, it suffices to prove the difference,

$$[e_1, e_2][e_3, e_4] \ldots [e_{2m-3}, e_{2m-2}][v'[e_i, e_j]v'', e_{j_m}],$$

is 0. But $[e_i, e_j]$ is central, so we get

$$[e_1, e_2][e_3, e_4] \ldots [e_{2m-3}, e_{2m-2}][e_i, e_j][v'v'', e_{j_m}].$$

But the length of $v'v''$ is $\ell - 2$, so this is 0 by induction, as desired.

Now we handle the general (7.2) in general. In view of Case 2, we can rearrange the e_i in v_m in ascending order. But any e_i^2 is central, so can be removed from the commutator. Disregarding this e_i^2 still produces a 2-admissible specialization of f, which by induction on ℓ is 0 (since we have decreased ℓ by 2). Hence, we are done unless all the the e_i appearing in v_m are distinct, but then we are done by Case 1 unless $v_m = e_m$, in which case we have $\ldots [e_m, e_m] = 0$. $\qquad \square$

Now define

$$P_n = x_1[x_1, x_2]x_2x_3[x_3, x_4]x_4 \ldots x_{2n-1}[x_{2n-1}, x_{2n}]x_{2n}. \qquad (7.3)$$

This definition is only valid for characteristic 2; we give the version for characteristic p below.

Lemma 7.25. *Any 2-admissible specialization of any partial linearization P_n in G^+ is 0.*

Proof. In view of Remark 7.4, it suffices to check the multilinearization. Let us linearize first at x_1. Then $x_1[x_1, x_2]x_2$ becomes

$$x_1[x_1', x_2]x_2 + x_1'[x_1, x_2]x_2$$

which (recalling that we are in characteristic 2) is

$$[x_1, x_1']x_2^2 + [x_1x_2, x_1'x_2].$$

Now linearizing at x_2 yields

$$[x_1, x_1'][x_2, x_2'] + [x_1x_2, x_1'x_2'] + [x_1x_2', x_1'x_2].$$

Applying this argument to each pair gives us a sum of products of commutators, each of which is 0 by Lemma 7.24. $\qquad \square$

Let us now check word specializations.

Lemma 7.26. *For any words $v_1, \ldots, v_{2m}$ such that $\deg_i v_1 \ldots v_{2m} \leq 2$ for each i, the term*

$$v_1[v_1, v_2]v_2v_3[v_3, v_4]v_4 \ldots v_{2m-1}[v_{2m-1}, v_{2m}]v_{2m}$$

is even, cf. Definition 7.10, unless all the v_i are single letters.

Proof. Suppose

$$q_j = v_{2j-1}[v_{2j-1}, v_{2j}]v_{2j} \neq 0.$$

If e_i occurs in v_1, it already occurs twice and cannot occur in any other v_j. Likewise with v_2, so the ε_i occurring in the calculation of q_1 cannot occur in the calculation of any other q_j. It follows at once that the parity of $q_1 \ldots q_m$ is the product of the parities of $p_1, p_2, \ldots, p_m$ individually. So it suffices to prove that some q_j is even. Now

$$v_{2j-1}[v_{2j-1}, v_{2j}]v_{2j} = v_{2j-1}^2 v_{2j}^2 - (v_{2j-1}v_{2j})^2.$$

These are even, by Corollary 7.18, unless both v_{2j-1} and v_{2j} are single letters. So we are done unless $v_1, \ldots, v_{2m}$ are all single letters, as claimed. $\qquad\square$

Theorem 7.27. *The T-space generated by $\{P_n : n \in \mathbb{N}\}$ is not finitely based.*

Proof. $P_n(e_1, \ldots, e_n) = \varepsilon_1 \ldots \varepsilon_{2n} e_1^2 \ldots e_{2n}^2$, which is 2-admissible and odd (nonzero). Lemma 7.25 shows that any multilinearization of any P_m cannot cancel this term, so it suffices to check word specializations.

But Lemma 7.26 says for $m < n$ that any word specialization of P_m provides an even contribution to $e_1^2 \ldots e_{2n}^2$, so we cannot obtain 0 by adding word specializations. $\qquad\square$

7.4   Non-Finitely Based T-Ideals in Odd Characteristic

Our next task is to construct a non-finitely based T-ideal in characteristic $p > 2$. It turns out that the natural analog of the polynomials P_n in characteristic p plays the key role, but the analysis is rather intricate. First we shall find a non-finitely based T-space containing the consequences of the Grassman identity. Then we modify this to a non-finitely based T-ideal. However, we must be careful, for we already saw in Exercise 3.5 that any T-ideal containing the Grassman identity is finitely based!

Since the Grassman algebra is 2-graded, we are led to utilize superidentities in our treatment. We shall use w_i, x_i to denote ordinary indeterminates, y_i for even indeterminates, and z_i for odd indeterminates.

Throughout, we work over an infinite field F of characteristic $p > 2$. We start with the Grassman algebra G over a countably infinite dimensional F-vector space V. As usual, we write $\{e_i : i \in \mathbb{N}\}$ for the standard base of V.

7.4.1 Superidentities of the Grassman algebras

Let us obtain some superidentities of G that lie at the core of our investigation.

Remark 7.28. We recall Remark 6.5, which says $zxz \in \mathrm{id}_2(G)$, for all z odd.

Lemma 7.29. *Suppose x_i are indeterminates, y_i are even indeterminates, and z_i are odd indeterminates. Then G satisfies the following superidentities:*

(i) $(y + z)^n = y^n + n y^{n-1} z$;

(ii) $x_0(y + z)x_1(y + z)\ldots x_{n-1}(y + z)x_n$
$$= y^n x_0 \ldots x_n + y^{n-1} \sum_{k=0}^{n-1} x_0 \ldots x_k z x_{k+1} \ldots x_n;$$

(iii) $zx(y + z)^n = zxy^n$;

(iv) $[y_1 + z_1, y_2 + z_2] = 2z_1 z_2$.

Proof. (i) In the expansion of $(y + z)^n$, any monomial containing two occurrences of z is 0. But this only leaves the monomials of degree at most 1 in z, and since y is central, we get (i) by the binomial expansion.

(ii) The same argument as in (i), although here we note that z does not necessarily commute with the x_i.

(iii) The same idea as before, noting that z already occurs at the beginning so the only monomial that survives in the expansion of $(y + z)^n$ is y^n.

(iv) The y_i are central, so the only nonzero part of the commutator is $[z_1, z_2]$, which is $2z_1 z_2$ since the z_i anticommute. □

This already lays the groundwork for an example by Shchigolev [Shch00] of a non-finitely based T-space, cf. Exercise 16, utilizing techniques that we need in the sequel. However, since Belov proved Specht's conjecture for affine algebras in characteristic $p > 0$, the T-ideal generated by Exercise 16 must be finitely based, so instead we proceed directly toward a counterexample to Specht's problem.

7.4.2 The test algebra A

We shall assume henceforth that $\mathrm{char}(F) = p > 2$. In order to get a non-finitely based T-ideal, we need an example that intrinsically requires an infinite number of indeterminates and we use a natural generalization of the P_n used earlier in characteristic 2. The ensuing computations require a sophisticated analysis that occupies the remainder of this chapter. Our choice of test algebra is crucial, since the Grassman algebra G does not do the trick.

Rather than working over G itself, we take the relatively free superalgebra $\mathcal{G}$ of G, which has generic even elements $\bar{y}_i$ and generic odd elements $\bar{z}_i$ satisfying the various superidentities of G. Let $A = \mathcal{G}/I$, where I is the ideal generated by all words h on the $\bar{y}_i, \bar{z}_i$, for which $\deg_i h \geq p$ for some i. Since the $\bar{y}_i$ and $\bar{z}_j$ either commute or anticommute, clearly the p-th power of any word in A is 0. Clearly, A is a superalgebra, where a word is odd or even depending on how many $\bar{z}_i$ appear in it. Furthermore, A inherits the natural degree from $\mathcal{G}$.

Lemma 7.30. $\bar{y}^p = 0$ for any even element $\bar{y} \in A$ without constant term.

Proof. By Remark 7.3, since y is spanned by even elements whose p power is 0. $\qquad\square$

Proposition 7.31. A satisfies the identity

$$w x_1 w x_2 \cdots x_{p+1} w$$

for indeterminates w, x_i.

Proof. Consider a specialization $x_i \mapsto \bar{x}_i$, and $w \mapsto \bar{y} + \bar{z}$ where $\bar{y}$ is even (and thus central) and $\bar{z}$ is odd. Then $w x_1 w x_2 \cdots x_{p+1} w$ specializes to a sum of terms where each w is replaced by $\bar{y}$ or $\bar{z}$. But if $\bar{z}$ appears twice, then the term is 0 by Remark 6.5. Thus, we only are concerned when at most one of the w are replaced by $\bar{z}$. Then $\bar{y}$ appears p times and is central, so our term equals $\bar{y}^p$ times the remaining part. But $\bar{y}^p = 0$ by Lemma 7.30. $\qquad\square$

The key to our analysis is the polynomial

$$P(x_1, x_2) = x_1^{p-1}[x_1, x_2]x_2^{p-1}.$$

Clearly, $P(x_1, x_2)$ is a p-admissible polynomial, cf. Definition 7.23.

Lemma 7.32 (Shchigolev).

(i) *Given any monomial h, write h_i for the monomial obtained by substituting 1 for x_i. Then*

$$[w, h] = \sum d_i[w, x_i]x_i^{d_i} h_i \qquad (7.4)$$

holds identically in A, where $d_i = \deg_i h$.

(ii) Any p-admissible polynomial is central or an identity of A.

(iii) The polynomial $P(x_1, x_2)$ is A-central.

Proof. (i) Lemma 7.19(ii) yields the identity

$$[w, x_i]vx_i = [w, x_i]x_iv - [w, x_i][x_i, v] = [w, x_i]x_iv,$$

which enables us to move the x_i to the left; applying this to each occurrence of x_i yields

$$[w, x_i]h = [w, x_i]x_i^{d_i}h_i.$$

But we know that

$$[w, x_{i_1} \ldots x_{i_n}] = \sum_{j=1}^{n} x_{i_1} \ldots x_{i_j-1}[w, x_{i_j}]x_{i_j+1} \ldots x_{i_n}. \tag{7.5}$$

Let $h = x_{i_1} \ldots x_{i_n}$. Since commutators in A are central, we can move the commutator in (7.5) to the left, and get

$$[w, h] = \sum_{j}[w, x_{i_j}]x_{i_1} \ldots x_{i_j-1}x_{i_j+1} \ldots x_{i_n}$$

and noting each x_i occurs exactly d_i times, we can combine repetitions and get (7.4).

(ii) By (i), $[w, h] \in \mathrm{id}(A)$ if h is a p-monomial.

(iii) Substituting

$$e_{ip+1} + e_{ip+2} + \cdots + e_{(i+1)p}$$

for x_i gives a nonzero value in A. $\qquad\square$

Recall Definition 2.62, of *cyclic shift* and *cyclically conjugate*.

Proposition 7.33.

(i) Suppose h is a p-monomial, and h' is cyclically conjugate to h. Then h and h' have the same value in A under any word specialization.

(ii) Suppose g is any central polynomial in A, and h, h' are two cyclically conjugate words. If gh is p-admissible, then gh−gh' always specializes to 0 under word specializations.

Proof. (i) It suffices to consider the case of a cyclic shift. Write $d_j = \deg_j h$. We need to prove that $[x_j, h] \in \mathrm{id}(A)$ whenever $x_j h$ is a p-monomial, i.e., $d_j = p - 1$ and $d_i \equiv 0 \pmod p$ for all $i \neq j$. But this follows from Lemma 7.32(i), since $[x_j, h]$ can be replaced by

$$\sum_j d_i[x_i, x_j] x_i^{d_i} h_i;$$

each summand is 0, because $d_i \equiv 0$ for $j \neq i$, and because $[x_i, x_i] = 0$ for $j = i$.

(ii) Immediate from (i). If $h = uv$ and $h' = vu$, then gh takes on the same value as ugv, which is cyclically conjugate to gvu. $\qquad\square$

Corollary 7.34. *Suppose g is a central polynomial in A, and words $v = v_1 v_2$, $v' = v_3 v_4$ are cyclically conjugate, with gv p-admissible. Then*

$$v_1 g v_2 = gv = gv' = v_3 g v_4$$

holds identically under all word specializations in A.

7.4.3 Shchigolev's non-finitely based T-space

Here is the correct version of our polynomial P_n in arbitrary characteristic $p > 0$, cf. (7.3):

$$P_n = \prod_{i=1}^{n} P(x_{2i-1}, x_{2i}) = x_1^{p-1}[x_1, x_2] x_2^{p-1} \dots x_{2n-1}^{p-1}[x_{2n-1}, x_{2n}] x_{2n}^{p-1}.$$

We formally write the indeterminates $x_i = y_i + z_i$ for y_i even and z_i odd. The bulk of the proof will involve considering in turn all the even or odd replacements for the x_i.

Lemma 7.35. *The polynomials $x_1^{p-1}[x_1, x_2] x_2^{p-1}$ and $2z_1 z_2 y_1^{p^n-1} y_2^{p^n-1}$ are equivalent on G.*

Proof. By Lemma 7.29 (iv) and then (iii),

$$x_1^{p^n-1}[x_1, x_2] x_2^{p^n-1} = 2x_1^{p^n-1} z_1 z_2 x_2^{p^n-1} = 2z_1 z_2 y_1^{p^n-1} y_2^{p^n-1}.$$

$\qquad\square$

Remark 7.36. By Lemma 7.35, P_n is equivalent over G to

$$2z_1 z_2 y_1^{p-1} y_2^{p-1} \dots 2z_{2n-1} z_{2n} y_{2n-1}^{p-1} y_{2n}^{p-1} = 2^n z_1 y_1^{p-1} \dots z_{2n} y_{2n}^{p-1},$$

since the y_i are central.

The supermonomial on the RHS is linear in the z_i and has degree $p-1$ in each y_i. For notational convenience, we drop the coefficient 2^n since it is prime to p.

The first step of our argument will be to show that the T-space generated by the P_n, $n \in \mathbb{N}$, is not finitely based. Intuitively, one might try to demonstrate this on A, but the assertion is then false! Indeed,

$$P(y_1 y_2 y_3 + z_1 z_2 z_3, y_4 + z_4) = 2z_1 z_2 z_3 z_4 (y_1 y_2 y_3)^p y_4^p$$
$$= 2z_1 y_1^p z_2 y_2^p z_3 y_3^p z_4 y_4^p$$
$$= P_2(y_1 + z_1, y_2 + z_2, y_3 + z_3, y_4 + z_4).$$

So we must modify our test space, and let $\tilde{A}$ be the subalgebra generated by all $\bar{x}_i = \bar{y}_i + \bar{z}_i$. Thus, $\bar{y}_i$ is not an element of $\tilde{A}$, and so this counterexample is not valid. We shall need to deal with the multilinearization of $P(x_1, x_2)$, which is

$$\sum_{\sigma_1, \sigma_2 \in S_p} x_{1,\sigma_1(2)} \cdots x_{1,\sigma_1(p)} [x_{1,\sigma_1(1)}, x_{2,\sigma_2(1)}] x_{2,\sigma_2(2)} \cdots x_{2,\sigma_2(p)}$$

and clearly equivalent to

$$\sum_{\sigma_1, \sigma_2 \in S_p} x_{1,\sigma_1(2)} \cdots x_{1,\sigma_1(p)} z_{1,\sigma_1(1)} z_{2,\sigma_2(1)} x_{2,\sigma_2(2)} \cdots x_{2,\sigma_2(p)}.$$

Thus, writing

$$\hat{P}' = \sum_{\sigma \in S_p} x_{1,\sigma(2)} \cdots x_{1,\sigma(p)} z_{1,\sigma(1)}, \qquad \hat{P}'' = \sum_{\sigma \in S_p} z_{2,\sigma(1)} x_{2,\sigma(2)} \cdots x_{2,\sigma(p)},$$

we see $\hat{P}' \hat{P}''$ is the multilinearization of P. We need a more subtle argument to switch the other occurrences of x with y as in Lemma 7.35, since the odd parts do not repeat so conveniently (and thus Lemma 7.29 is not available).

Lemma 7.37. *Writing $x_{1,i} = y_{1,i} + z_{1,i}$ for $2 \le i \le p$, we have*

$$\hat{P}' = \sum_{\sigma \in S_p} y_{1,\sigma(2)} \cdots y_{1,\sigma(p)} z_{1,\sigma(1)}.$$

Proof. We could see this by multilinearizing Lemma 7.35, but let us see it directly. Substituting $y_{1,i} + z_{1,i}$ for $x_{1,i}$ and opening $\hat{P}'$ accordingly we consider terms involving some $z_{1,i}$ other than $z_{1,\sigma(1)}$. Suppose $i = \sigma(j)$. Then switching $z_{1,\sigma(1)}$ and $z_{1,\sigma(j)}$ yields

$$\cdots z_{1,\sigma(j)} \cdots z_{1,\sigma(1)} = - \cdots z_{1,\sigma(1)} \cdots z_{1,\sigma(j)},$$

which cancels the term corresponding to $\sigma\tau$ where τ is the transposition $(1j)$. Thus, all these terms pair off with opposite signs and cancel, leaving only the terms in the assertion. $\square$

Since the $y_{1,i}$ are central, we can rewrite

$$\hat{P}'(x_{11},\ldots,x_{1p}) = \sum_{\sigma_1 \in S_p} z_{1,\sigma(1)} y_{1,\sigma(2)} \cdots y_{1,\sigma(p)}, \qquad (7.6)$$

which has the same form as $\hat{P}''$. Thus, we write $\hat{P}$ for $\hat{P}'$ and see that the multilinearization of P_m is

$$\hat{P}(x_{1,1},\ldots,x_{1,p})\hat{P}(x_{2,1},\ldots,x_{2,p})\ldots\hat{P}(x_{2m,1},\ldots,x_{2m,p}). \qquad (7.7)$$

The next theorem is a more complicated version of Exercise 16.

Theorem 7.38. (char $F = p > 2$.) P_n does not belong to the T-space generated by the $P_m, \forall m < n$.

Proof. Since $P_n(\bar{x}_1, \bar{x}_2, \ldots x_{2n}) = 2^n \bar{y}_1^{p-1} \bar{z}_1 \bar{y}_2^{p-1} \bar{z}_2 \ldots \bar{y}_{2n}^{p-1} \bar{z}_{2n}$, we shall show that for any $m < n$, the part of any evaluation $P_m(u_1, \ldots, u_{2m})$ of P_m which has degree precisely $p-1$ in each $\bar{y}_i, i \leq n$, and degree 1 in each $\bar{z}_i, i \leq n$, must be 0. This will show that P_n cannot be a consequence of P_m for $m < n$, and thus yields the desired result.

As in Remark 7.2, this can be checked in two steps: first when the u_i is a word in $\tilde{A}$, and second to show each partial linearization of P_m in the desired terms is 0. For any word v, write v' for the odd part and v'' for the even part. Thus, $\bar{x}_i' = \bar{z}_i$ and $\bar{x}_i'' = \bar{y}_i$.

Step 1. We check the assertion for any substitutions of words. Suppose some u_j is a word of degee > 1, i.e., $u_j = \bar{x}_i v_j$. We claim P_m then has value 0. For notational simplicity, we assume j is odd, i.e., $j = 2k - 1$, and see

$$P(u_j, u_{j+1}) = P(\bar{x}_i v_j, u_{j+1}) = 2(\bar{x}_i v_j)' u_{j+1}'((\bar{x}_i v_j)'')^{p-1}(u_{j+1}'')^{p-1}.$$

Then $(\bar{x}_i v_j)' = \bar{x}_i' v_j'' + \bar{x}_i'' v_j' = \bar{z}_i v_j'' + \bar{y}_i v_j'$, and $(\bar{x}_i v_j)'' = \bar{y}_i v_j'' + \bar{x}_i' v_j'$, so we get

$$2(\bar{z}_i v_j'' + \bar{y}_i v_j')u_{j+1}'(\bar{y}_i v_j'' + \bar{z}_i v_j')^{p-1}(u_{j+1}'')^{p-1}.$$

But

$$(\bar{z}_i v_j'' + \bar{y}_i v_j')u_{j+1}'\bar{z}_i v_j' = 0$$

since it is a sum of terms in which either the odd element $\bar{z}_i$ or v_j' repeats, so we are left with

$$2(\bar{z}_i v_j'' + \bar{y}_i v_j')u_{j+1}'(\bar{y}_i v_j'')^{p-1}(u_{j+1}'')^{p-1},$$

or

$$2(\bar{z}_i \bar{y}_i^{\,p-1} v_j''^{\,p} + \bar{y}_i^{\,p} v_j''^{\,p-1} v_j')u_{j+1}'(u_{j+1}'')^{p-1},$$

each term of which is 0 by Lemma 7.30. (This argument could be stream-lined a bit using Lemma 7.20.)

In other words, for word specializations, the only way to get a nonzero value in P_m is for our word specialization to be linear in the $\bar{z}$, i.e., $\bar{z}_i \bar{y}_i^{\,p-1}$. In this case, the degree is $2m < 2n$, and we cannot cancel P_n; in general, we can delete any linear substitution and continue by induction. Thus, we may assume there are no word specializations.

So it remains to prove:

Step 2. Every p-admissible word specialization in $\tilde{A}$ of any partial linearization (cf. Definition 1.11) of P_m is 0. Since we have introduced the 2-graded structure into our considerations, we consider a partial linearization f of P_m in the supersense, i.e., take f to be a homogeneous component of

$$\ldots (z_{i,1} + z_{i,2})(y_{i,1} + y_{i,2})^{p-1} - z_{i,1}y_{i,1}^{p-1} - z_{i,2}y_{i,2}^{p-1} \ldots,$$

i.e., the i, 1-degree (in both y_i and z_i together) is some k_i and the i, 2-degree is $p - k_i$, where $0 < k_i < p$, and $1 \le i \le 2m$. We consider substitutions (from $\tilde{A}$) in f in total degree p in each $\bar{x}_i$, and need to show, in any such substitution the part of degree $p - 1$ in each $\bar{y}_i, i \le n$, and degree 1 in each $\bar{z}_i, i \le n$, must be 0. In order to avoid repeating this condition again, we call this the "prescribed part" of the substitution.

In view of Remark 1.12(iv), since all indeterminates have degree $\le p$, it is enough to check all (total) linearizations of f in the indeterminates in which we do not make word specializations. But we already ruled out word specializations, so we may assume f is the multilinearization of P_m; in other words, we may assume f has the form (7.6), which we write more concisely as

$$\prod_{j=1}^{2n} \hat{P}(x_{j1}, \ldots, x_{jp}).$$

It suffices to check that all word specializations of each $x_{j1}, \ldots, x_{jp}$ have prescribed part 0. We shall prove this for all m (even including $m \ge n$), and argue by induction on n.

So we consider the various substitutions

$$x_{jk} \mapsto w_{jk} = \bar{x}_{jk1} \ldots \bar{x}_{jkt},$$

where $t = t(j, k)$ depends on j and k. Now each $\bar{x}_{jk\ell} = \bar{y}_{jk\ell} + \bar{z}_{jk\ell}$, the sum of its even and odd parts, and we have to consider all terms arising from one or the other. Towards this end, we define the *choice vector* $\mathbf{m} = (m_{jk\ell})$, where $m_{jk\ell}$ is 0 if we are to choose $\bar{y}_{jk\ell}$ (the even part) and $m_{jk\ell}$ is 1 if we are to choose $\bar{z}_{jk\ell}$ (the odd part) taken in v_{jk}.

We define $\bar{x}_{jk\ell}^{\mathbf{m}}$ to be $\bar{y}_{jk\ell}$ if $m_{jk\ell} = 0$, and $\bar{z}_{jk\ell}$ if $m_{jk\ell} = 1$. Also write $w_{jk}^{\mathbf{m}}$ for

$$\bar{x}_{jk1}^{\mathbf{m}} \ldots \bar{x}_{jkt}^{\mathbf{m}}.$$

Thus, we have

$$w_{jk} = \sum_{\mathbf{m}} w_{jk}^{\mathbf{m}},$$

summed over 2^t possible choices for $\mathbf{m}$. Of course, many of these automatically give 0 in f. We consider only those substitutions for which

$$\sum_{\sigma} w_{j,\sigma(1)}^{\mathbf{m}} w_{j,\sigma(2)}^{\mathbf{m}} \cdots w_{j,\sigma(p)}^{\mathbf{m}}$$

could be nonzero.

For example, by Lemma 7.37, when $k = \sigma(1)$, we must take $\bar{z}_{jk\ell}$ an odd number of times, and for all k, we must take $\bar{z}_{jk\ell}$ an even number of times. In other words, to evaluate, we first designate the special k_0 which will go to the first position of (7.6), pick $\bar{z}_{jk_0\ell}$ an odd number of times, pick $\bar{z}_{jk\ell}$ an even number of times for all other k, sum over all such choices, and sum over those σ such that $\sigma(1) = k_0$.

We need to evaluate

$$w_j = \sum_{\sigma \in S_p} \sum_{\mathbf{m}} w_{j,\sigma(1)}^{\mathbf{m}} w_{j,\sigma(2)}^{\mathbf{m}} \cdots w_{j,\sigma(p)}^{\mathbf{m}}, \qquad (7.8)$$

and show $w_1 w_2 \ldots w_n = 0$. Note that each summand of $w_1 w_2 \ldots w_n$ is p-admissible; we should bear this in mind always, and this is why we have to take the product (since an individual w_j may not have length a multiple of p, and then the proof would fail.)

We subdivide (7.8) as $S_1 + S_2$, where S_1 is taken over all $\mathbf{m}$ such that $\sum_{\ell=1}^{t(j,k)} m_{jk\ell} \leq 1$ for all j, k, and S_2 is taken over all other $\mathbf{m}$. In other words, when we evaluate our specialization of f, which is

$$\hat{P}(w_{1,1}, \ldots, w_{1,p}) \hat{P}(w_{2,1}, \ldots, w_{2,p}) \ldots \hat{P}(w_{2m,1}, \ldots, w_{2m,p}), \qquad (7.9)$$

S_1 consists of all terms where from *each* $\hat{P}(w_{j,1}, \ldots, w_{j,p})$, we only take those summands with at most one z, and S_2 consists of all terms containing a summand of *some* $\hat{P}(w_{j,1}, \ldots, w_{j,p})$ that has at most two zs.

We conclude by showing each of S_1 and S_2 are 0.

Proof that S_1 is 0. Induction on n. We may assume each w_{jk}^m has fewer than two odd terms, and thus, as shown above, w_{jk}^m has no odd terms unless $k = \sigma(1)$. Thus, the only nonzero contribution to S_1 comes from

$$x_{jk} \mapsto \bar{y}_{jk1} \dots \bar{y}_{jkt}, \quad k \neq \sigma(1);$$

$$x_{jk} \mapsto \bar{z}_{jk1}\bar{y}_{jk2} \dots \bar{y}_{jkt} + \bar{y}_{jk1}\bar{z}_{jk2} \dots \bar{y}_{jkt} + \dots + \bar{y}_{jk1}\bar{y}_{jk2} \dots \bar{z}_{jkt}, \quad k = \sigma(1).$$

Since all $\bar{y}_{jk\ell}$ are central, we can rearrange the letters. Writing d_{ijk} for the degree of $\bar{x}_i$ in the substitution of x_{jk}, and $d_{i,j} = \sum_k d_{ijk}$, we see $\hat{P}(x_{j1}, \dots, x_{jp})$ specializes to

$$\sum_{i=1}^{2n} \sum_{k=1}^{p} \sum_{\sigma \in S_p} d_{ijk} \bar{z}_i \bar{y}_1^{d_{1,j}} \dots \bar{y}_i^{d_{i,j}-1} \dots \bar{y}_{2n}^{d_{2n,j}}$$

$$= (p-1)! \sum_{i=1}^{2n} \sum_{k=1}^{p} d_{ijk} \bar{z}_i \bar{y}_1^{d_{1,j}} \dots \bar{y}_i^{d_{i,j}-1} \dots \bar{y}_{2n}^{d_{2n,j}}$$

$$= - \sum_{i=1}^{2n} d_{i,j} \bar{z}_i \bar{y}_1^{d_{1,j}} \dots \bar{y}_i^{d_{i,j}-1} \dots \bar{y}_{2n}^{d_{2n,j}}.$$

(We have $(p-1)!$ because we sum over the $(p-1)!$ permutations σ for which $\sigma(1) = k$, but $(p-1)! \equiv -1 \pmod{p}$.)

At this stage, if $m < n$, then in multiplying these to get our specialization of f, we see that in each summand a suitable $\bar{z}_i$ does not appear, so we could factor out y_i^p. Writing $S_1(\hat{x}_i)$ to denote the sum corresponding to the further specialization of the w_{jk} where we erased all occurrences of $\bar{x}_i$ (i.e., substitute 1 for $\bar{x}_i$), we see that

$$S_1 = \sum_{i_1=1}^{n} \bar{y}_{i_1}^p S_1(\hat{x}_{i_1}) - \sum_{i_1,i_2=1}^{n} \bar{y}_{i_1}^p \bar{y}_{i_2}^p S_1(\hat{x}_{i_1}, \hat{x}_{i_2})$$

$$+ \dots \pm \sum_{i_1,i_2,\dots i_{n-m}=1}^{n} \bar{y}_{i_1}^p \bar{y}_{i_2}^p \dots \bar{y}_{i_{n-m}}^p S_1(\hat{x}_{i_1}, \dots, \hat{x}_{i_{m-n}}).$$

Each of the summands is 0 by induction on n, so it remains to prove $S_1 = 0$ when $m = n$. Taking the product over j, we see f specializes to

$$- \sum d_{i_1,1} \dots d_{i_{2n},2n} \bar{z}_{i_1} \dots \bar{z}_{i_{2n}} \bar{y}_1^{p-1} \dots \bar{y}_{2n}^{p-1}. \qquad (7.10)$$

If i repeats, then the same $\bar{z}_i$ occurs twice and the term is 0. Thus, $i_1, \dots, i_{2n}$ are distinct, and

$$\bar{z}_{i_1} \dots \bar{z}_{i_{2n}} = \text{sgn}(\pi) \bar{z}_1 \dots \bar{z}_{2n}$$

where $\pi \in S_{2n}$ is the permuation sending $\pi(j) = i_j$. Thus, (7.10) equals

$$-\sum \mathrm{sgn}(\pi)d_{i_1,1}\ldots d_{i_{2n},2n}\bar{z}_1 \ldots \bar{z}_{2n}\bar{y}_1^{p-1}\ldots \bar{y}_{2n}^{p-1}$$
$$= -\det(d_{i,j})\bar{z}_1 \ldots \bar{z}_{2n}\bar{y}_1^{p-1}\ldots \bar{y}_{2n}^{p-1}.$$

But $\sum d_{i,j} = 0 \pmod{p}$, so the determinant is 0 (since char$(F) = p$), yielding the desired result.

In summary, the proof of this case boils down to showing that a certain determinant that calculates the coefficients of S_1 is always 0 modulo p.

Proof that S_2 is 0. The idea here is to pull out double occurrences of z, and thus reduce to the previous case via induction. Unfortunately, the argument is extremely delicate, since our reduction involves counting certain terms more than once; we need to show the extraneous terms that we counted also total 0, by means of a separate argument.

By assumption, we are only counting terms where some w_{jk} involves at least two odd substitutions. Fixing this j, k, we make the following claim.

Claim 1. For any $t' \geq 2$, the sum of terms in which

$$m_{j,k,\ell_1} = m_{j,k,\ell_1+1} = 1$$

for specified ℓ_1 (i.e., in which v_{jk} has two odd terms in the specified positions $(\ell_1, \ell_1 + 1)$) is 0.

Indeed, written explicitly

$$v_{jk} = \bar{x}_{jk1}^{\mathbf{m}} \ldots \bar{x}_{jk\ell_1-1}^{\mathbf{m}}\bar{z}_{jk\ell_1}\bar{z}_{jk\ell_1+1}\bar{x}_{jk\ell_1+2}^{\mathbf{m}} \ldots \bar{x}_{jk\ell_t}^{\mathbf{m}}.$$

But $\bar{z}_{jk\ell_1}\bar{z}_{jk\ell_1+1}$ is even and thus central so we can write

$$w_{jk}^{\mathbf{m}} = \bar{z}_{jk\ell_1}\bar{z}_{jk\ell_1+1}w_{jk}^{\mathbf{m}'}$$

where

$$w_{jk}^{\mathbf{m}'} = \bar{x}_{jk1}^{\mathbf{m}} \ldots \bar{x}_{jk\ell_1-1}^{\mathbf{m}} \ldots \bar{x}_{jk\ell_2-1}^{\mathbf{m}}\bar{x}_{jk\ell_2}^{\mathbf{m}} \ldots \bar{x}_{jk\ell_{t'}}^{\mathbf{m}}, \ldots.$$

Since $\bar{z}_{jk\ell_1}\bar{z}_{jk\ell_{t'}}$ is central, so we could pull it out of the evaluation of f, i.e.,

$$\hat{P}(w_{j1}^{\mathbf{m}},\ldots,w_{jp}^{\mathbf{m}}) = \bar{z}_{jk\ell_1}\bar{z}_{jk\ell_2}\hat{P}(w_{j1}^{\mathbf{m}},\ldots,v_{jk}',\ldots,w_{jp}^{\mathbf{m}}).$$

But $\bar{x}_{j,k,\ell_1}$ is some $\bar{x}_{i_1}$, and $\bar{z}_{j,k,\ell_1+1}$ is some $\bar{x}_{i_2}$. Having used up the odd occurrences of $\bar{x}_{i_1}, \bar{x}_{i_2}$, we only have even occurrences left in the prescribed part, which can also be pulled out of those vs in which they appear. Writing

v_{jk} for that part of w_{jk}^m obtained by deleting all occurrences of $\bar{x}_{i_1}, \bar{x}_{i_2}$, we see that

$$\hat{P}(w_{j1}^m, \ldots, w_{jp}^m) = \bar{z}_{i_1}\bar{y}_{i_1}^{p-1}\bar{z}_{i_2}\bar{y}_{i_2}^{p-1}\hat{P}(v_{j1}, \ldots, v_{jp}), \qquad (7.11)$$

which involves two fewer indeterminates. Thus, the prescribed part in (7.11) is 0, proving the claim.

Next, recall that $[x_i, x_j] = 2z_iz_j$, so Claim 1 implies that we get 0 when we take the sum in (7.8) over all terms containing a commutator $[\bar{x}_{i,j,\ell}^m, \bar{x}_{i,j,\ell+1}^m]$. But

$$\bar{x}_{i,j,\ell}^m\bar{x}_{i,j,\ell+1}^m = [\bar{x}_{i,j,\ell}^m, \bar{x}_{i,j,\ell+1}^m] + \bar{x}_{i,j,\ell+1}^m\bar{x}_{i,j,\ell}^m,$$

so this proves the sum in (7.8) is still the same if we interchange two positions. This proves:

Claim 2. For any ℓ_1, ℓ_2, the sum of terms in which $m_{j,k,\ell_1} = m_{j,k,\ell_2} = 1$ for specified ℓ_1, ℓ_2 (i.e., in which w_{jk}^m has odd letters in two specified positions ℓ_1, ℓ_2) is 0.

Since any term in S_2 has some $m_{j,k,\ell_1} = m_{j,k,\ell_2} = 1$, by definition of S_2, we would like to conclude the proof by summing over all such possible pairs (ℓ_1, ℓ_2). Unfortunately, letting

$$\bar{S}_2 = \sum_{\sigma \in S_p} \sum_{m_{j,k,\ell_1} = m_{j,k,\ell_2} = 1} w_{j,\sigma(1)}^m \cdots w_{j,\sigma(p)}^m,$$

we see that $\bar{S}_2 - S_2$ consists of terms which are doubly counted, and we want to show this discrepancy also is 0. Towards this end, for any $(\ell_1, \ldots, \ell_q)$, we define $S_{j,k,(\ell_1,\ldots,\ell_q)}$ to be the sum of terms in S_2 for which

$$m_{j,k,\ell_1} = \cdots = m_{j,k,\ell_q} = 1,$$

since this is the source of our double counting for $q \geq 3$.

Claim 3. $S_{j,k,(\ell_1,\ldots,\ell_q)} = 0$ whenever $q \geq 2$ is even.

Claim 3 is proved analogusly to Claim 1, this time drawing out $\frac{q}{2}$ pairs of odd letters at a time from w_{jk}^m instead of a single pair.

Now define $\bar{S}_{j,k,q} = \sum_{(\ell_1,\ldots,\ell_q)} S_{j,k,(\ell_1,\ldots,\ell_q)}$. Clearly, $S_2 = \sum c_{j,k,q}\bar{S}_{j,k,q}$ for suitable integers $c_{j,k,q}$. Thus, it remains to prove:

Claim 4. $c_{j,k,q} = 0$ for q odd.

Given Claim 4, we see S_2 is summed over $\sum c_{j,k,q}\bar{S}_{j,k,q}$ for q even, so is $\sum c_{j,k,q}0 = 0$, proving the theorem. So it remains to prove Claim 4, which we do by showing that for q even, the computation of

$$c_{j,k,2}\bar{S}_{j,k,2} + c_{j,k,4}\bar{S}_{j,k,4} + \cdots + c_{j,k,q}\bar{S}_{j,k,q}$$

(which is 0) counts each $2m + 1$-tuple exactly once.

For example, consider $\bar{S}_{j,k,3}$, i.e., suppose w_{jk}^m has (at least) three odd letters, which we may assume are for positions $\ell = 1, 2, 3$. Then we counted $z_{j,k,1}z_{j,k,2}z_{j,k,3}$ three times (the number of times we can choose two places of three). But when we moved $z_{j,k,1}z_{j,k,3}$ out first, we had to move $z_{j,k,3}$ past $z_{j,k,2}$, which reverses the sign, so in counting the duplication, we have twice with sign $+1$ and once with sign -1, so the total is $1 + 1 - 1 = 1$, and there is no extra duplication.

When we try to apply this argument in case w_{jk}^m has (at least) four odd letters, there is extra duplication, as seen in considering $z_{j,k,1}z_{j,k,2}z_{j,k,3}z_{j,k,4}$. Here we choose two of four, yielding six possibilities, four of which can be pulled out of w_{jk}^m without changing sign (odd positions in $(1,2)$, $(2,3)$, $(3,4)$, or $(1,4)$) and two of which reverse the sign (odd positions in $(1,3)$ or $(2,4)$). Hence, the sums are counted twice.

Next consider $q = 5$. Now we have to worry about the duplication arising from counting $\overline{S_{j,k,2}}$ and counting $\overline{S_{j,k,4}}$. We need to prove the duplication for each is the same as for $t(j,k) = 4$, for all the duplications to cancel. This is not too difficult to see directly. The contribution from counting $\overline{S_{j,k,2}}$ is the number of pairs of form $(i, i+1)$ or $(i, i+3)$ minus the number of pairs of form $(i, i+2)$ or $(i, i+4)$, which is

$$(4 + 2) - (3 + 1) = 6 - 4 = 2,$$

the same as the contribution from counting $\overline{S_{j,k,2}}$ for $q = 4$. Counting $\overline{S_4}$ for $q = 5$ gives $+$ for (1,2,3,4), (1,2,4,5), and (2,3,4,5) and $-$ for (1,2,3,5) and (1,3,4,5), for a total of $3 - 2 = 1$, the same as for $q = 4$.

To conclude the proof of Claim 4 in general, we need to show that for any numbers t, u the contribution arising in counting $\overline{S_{2u}}$ for $q = 2m$ is the same as the contribution arising in counting $\overline{S_{2u}}$ for $q = 2m + 1$.

As before, the contribution arises in the number of switches we need to make to move all the odd terms to the left side; we multiply by (-1) raised to the number of such switches. Given odd positions in $\ell_1, \ell_2, \ldots, \ell_{2u}$, listed in ascending order, we could compute the number of such switches (to arrive at position $(1, 2, \ldots, 2u)$) to be

$$\ell_1 - 1 + \ell_2 - 2 + \cdots + \ell_{2u} - 2u = \sum_{i=1}^{2u} \ell_i - u(2u+1).$$

Since the crucial matter is whether this is even or odd, we can replace $u(2u + 1)$ by u and define:

$\gamma(u, q, 0)$ is the number of $(\ell_1, \ell_2, \ldots, \ell_{2u})$ (written in ascending order) with $\ell_{2u} \leq q$, such that $\sum_{i=1}^{2u} \ell_i - u$ is even.

$\gamma(u, q, 1)$ is the number of $(\ell_1, \ell_2, \ldots, \ell_{2u})$ (written in ascending order) with $\ell_{2u} \leq q$, such that $\sum_{i=1}^{2u} \ell_i - u$ is odd.

$\delta(u, q) = \gamma(u, q, 0) - \gamma(u, q, 1)$.

Then $\delta(u, q)$ is the number of times we are counting terms in $\overline{S_{2u}}$ with q odd letters. We showed above that $\delta(2, 2) = \delta(2, 3)$, $\delta(2, 4) = \delta(2, 5)$, and $\delta(4, 4) = \delta(4, 5)$; we need to prove $\delta(u, 2m) = \delta(u, 2m + 1)$ for each m.

Note that the difference between $\gamma(u, 2m, 0)$ and $\gamma(u, 2m + 1, 0)$ arises precisely from those sequences $(\ell_1, \ell_2, \ldots, \ell_{2u})$ for which $\ell_{2u} = 2m + 1$; likewise for $\gamma(u, 2m, 1)$ and $\gamma(u, 2m + 1, 1)$.

Thus, we need only consider sequences $(\ell_1, \ell_2, \ldots, \ell_{2u-1}, \ell_{2m+1})$ used in the computation of $\delta(u, 2m+1)$. Since $2m+1-u$ is a constant that appears in all the sums, we need to prove:

Claim 5. Let $\beta(u, m, 0) = $ the number of sequences $(\ell_1, \ell_2, \ldots, \ell_{2u-1})$ with $\ell_{2u-1} \leq 2m$ and $\sum \ell_i$ even, and $\beta(u, m, 1) = $ the number of sequences $(\ell_1, \ell_2, \ldots, \ell_{2u-1})$ with $\ell_{2u-1} \leq 2m$ and $\sum \ell_i$ odd.

Then $\beta(u, m, 0) = \beta(u, m, 1)$.

(Note that for this claim to be nonvacuous we must have $u \leq m$.) We prove this by induction on m. For $m = 1$ it is trivial, since then $u = 1$ and there are two numbers to choose from, namely 2, which is even, and 1, which is odd.

We say $\ell_1, \ell_2, \ldots, \ell_{2u-1}$ has:

type 1 if $\ell_{2u-1} \leq 2(m - 1)$;

type 2 if $\ell_{2u-2} \leq 2(m - 1)$ but $\ell_{2u-1} \geq 2m - 1$;

type 3 if $\ell_{2u-2} = 2m - 1$ and so $\ell_{2u-1} = 2m$.

In type 1 we are counting $\beta(u, m - 1, 0)$ and $\beta(u, m - 1, 1)$, which are equal by induction on m.

In type 3 we are counting $\beta(u-1, m-1, 0)$ and $\beta(u-1, m-1, 1)$, which also are equal by induction on m.

So it remains to consider type 2. In this case there is a 1:1 correspondence between the sequences with $\ell_{2u-1} = 2m - 1$ and the sequences with $\ell_{2u-1} = 2m$ (obtained by switching ℓ_{2u-1} from $2m - 1$ to $2m$), which obviously have opposite parities, and so their contributions cancel. This concludes the proof of Claim 5 and thus of Theorem 7.38. $\qquad \square$

Corollary 7.39. *The T-space generated by $\{P_n : n \in \mathbb{N}\}$ is not finitely based.*

7.4.4 The test algebra $\widehat{A}$

Unfortunately, as noted earlier, in order to pass to T-ideals, we need to pass to another test algebra which does not satisfy the Grassman identity. For convenience we introduce notation for $\tilde{A}$, which we shall carry to the end of this section. We write the image of x_{2i-1} as a_i and the image of x_{2i} as b_i. Thus, $\tilde{A}$ is generated by $\{a_i, b_i : i \in \mathbb{N}\}$, and any word in these in which either a_i or b_i repeats p times is 0 in $\tilde{A}$.

We extend the algebra $\tilde{A}$ to an algebra $\widehat{A}$ by adjoining new elements e, f, t to A, satisfying the following relations where $q = p^2$:

$$t^{4q+1} = 0, \quad [t, a_i] = [t, b_i] = 0,$$
$$[t, fe] = [fe, a_i] = [fe, b_i] = 0,$$
$$a_i e = b_i e = te = e^2 = 0, \quad fa_i = fb_i = ft = f^2 = 0;$$

any word with more than q occurrences of e or more than q occurrences of f is 0.

As before, we shall exhibit a countably infinite set of polynomials whose T-ideal in $\widehat{A}$ is not finitely based, and so the T-ideal itself is not finitely based. Before defining these polynomials, let us examine properties of $\widehat{A}$, especially in connection with the Grassman polynomial.

Remark 7.40.

(i) t and fe centralize $\tilde{A}$, and e, f, t are nilpotent.

(ii) For any $c \in \{a_i, b_j, e, f, t\}$, $ce = 0$ unless $c = f$, and $fc = 0$ unless $c = e$.

(iii) Each nonzero word in $\widehat{A}$ can be written as

$$w = e^{\varepsilon_1} s v f^{\varepsilon_2}$$

where $\varepsilon_1, \varepsilon_2 \in \{0, 1\}$, $s = t^{k_1}(fe)^{k_2}$ for suitable k_1, k_2, and v is a word from $\tilde{A}$; that is, in the a_i and b_i. (This follows at once from (i) and (ii).)

(iv) In (iii), if w has respective degree d_1, d_2 in e, f, then letting $d = \min\{d_1, d_2\}$, we see $k_2 = d$ and $\varepsilon_i = d_i - d$ for $i = 1, 2$.

Definition 7.41. The word w in $\widehat{A}$ is called a p-word if v (as in(iii)) is a p-monomial in $\tilde{A}$. A p-element of $\widehat{A}$ is a sum of p-words.

Remark 7.42. In view of Remark 7.40(ii) and (iii), every commutator of elements of $\widehat{A}$ has one of the forms (up to reversing the order of the commutator, which means multiplying by -1):

$$[es_1v_1, es_2v_2] = 0;$$

$$[s_1v_1f, s_2v_2f] = 0;$$

$$[es_1v_1, s_2v_2] = es_1s_2v_1v_2;$$

$$[s_1v_1f, s_2v_2] = -s_1s_2v_2v_1f;$$

$$[es_1v_1f, es_2v_2f] = es_1v_1fes_2v_2f - es_2v_2fes_1v_1f = efes_1s_2[v_1, v_2]f;$$

$$[es_1v_1, s_2v_2f] = es_1v_1s_2v_2f - s_2v_2fes_1v_1;$$

$$[es_1v_1, es_2v_2f] = -es_2v_2fes_1v_1 = es_1s_2fev_2v_1;$$

$$[s_1v_1f, es_2v_2f] = s_1v_1fes_2v_2f = s_1s_2fev_1v_2f.$$

We shall see that the last four equations are irrelevant, since their degrees in e and f are too high. Take new indeterminates $\tilde{e}, \tilde{f}, \tilde{t}$. We want to see when $[[\tilde{e}, \tilde{t}], \tilde{t}]$ can have a nonzero specialization on $\widehat{A}$. Let us linearize this to

$$[[\tilde{e}, \tilde{t}_1], \tilde{t}_2] + [[\tilde{e}, \tilde{t}_2], \tilde{t}_1].$$

Lemma 7.43. *Any nonzero word specialization of*

$$[[\tilde{e}, \tilde{t}_1], \tilde{t}_2]$$

to $\widehat{A}$ must have some occurrence of e or f.

Proof. $[[x_1, x_2], x_3]$ is an identity of $\tilde{A}$ and thus of $\tilde{A}[t]$, which is the subalgebra of $\widehat{A}$ without the e or f. □

7.4.5 The counterexample

Definition 7.44. Let $q = p^2$, and

$$Q_n = [[\tilde{e}, \tilde{t}], \tilde{t}]P_n(x_1, \ldots, x_{2n})\left([\tilde{t}, [\tilde{t}, \tilde{f}]]\,[[\tilde{e}, \tilde{t}], \tilde{t}]\right)^{q-1}[\tilde{t}, [\tilde{t}, \tilde{f}]],$$

where $P_n = P(x_1, x_2) \cdots P(x_{2n-1}, x_{2n})$, with $P(x_1, x_2) = x_1^{p-1}[x_1, x_2]x_2^{p-1}$.

Theorem 7.45. *The T-ideal generated by the polynomials $\{Q_n : n \in \mathbb{N}\}$ is not finitely based on $\widehat{A}$ (and so is not a finitely based T-ideal).*

Lacking a shortcut generalizing the proof of Example 7.22, we return to the generic approach based on Remark 7.2, which is to verify it first for words and then for partial multilinearizations.

7.4.6   Specializations to words

Let us fix an arbitrary specialization $\hat{e}, \hat{f}, \hat{t}_i$ in $\widehat{A}$ of $\tilde{e}$, $\tilde{f}$, and $\tilde{t}_i$.

Lemma 7.46.

(i) *In any nonzero word specialization in $\widehat{A}$ from partial linearizations of Q_n, the specializations $\bar{x}_i$ of x_i (in P_n) are in $\tilde{A}[t]$, and the evaluation c of P_n is a sum of terms each of which contains suitable a_j or b_j; furthermore, the following conditions must hold:*

e has degree 1 in $[[\hat{e}, \hat{t}_1], \hat{t}_2] + [[\hat{e}, \hat{t}_2], \hat{t}_1]$.

f has degree 1 in $[\hat{t}_1, [\hat{t}_2, \hat{f}]] + [\hat{t}_2, [\hat{t}_1, \hat{f}]]$.

(ii) *In each substitution, any word containing a_i or b_i has multiplicity $< p$.*

Proof. (i) By Lemma 7.43, e and f must have a total of at least $2q$ occurrences, and by the last line defining the relations of $\widehat{A}$, this must mean e and f each appear exactly q times, i.e., each $[[\hat{e}, \hat{t}], \hat{t}]$ has degree exactly 1 in e (or f) in any term contributing to the evaluation.

This does not leave any room for e, f in the $\bar{x}_i$, so these are in $\tilde{A}[t]$. If a_j or b_j does not appear in $\bar{x}_{2i-1}$, then $\bar{x}_{2i-1} \in F[t]$, so $[\bar{x}_{2i-1}, \bar{x}_{2i}] = 0$, yielding the first assertion.

We also need $[\hat{t}_{i_1}, [\hat{t}_{i_2}, \hat{f}]][[\hat{e}, \hat{t}_{i_3}], \hat{t}_{i_4}] \neq 0$. This means $[\hat{t}_{i_1}, [\hat{t}_{i_2}, \hat{f}]]$ either starts with e or ends with f. But if it starts with e, it is annihilated by the evaluation of c that precedes it. Thus, the only relevant contribution of $[\hat{t}_{i_1}, [\hat{t}_{i_2}, \hat{f}]]$ ends with f. The same argument shows that the only relevant contribution of $[[\hat{e}, \hat{t}_{i_3}], \hat{t}_{i_4}]$ starts with e.

(ii) By Remark 7.40(iii), all the appearances come together in some subword $v \in \tilde{A}$, so we are done by the defining relations of $\tilde{A}$.   □

The specialization $\tilde{e} \mapsto \hat{e}$, $\tilde{f} \mapsto \hat{f}$, $\tilde{t}_i \mapsto \hat{t}_i$ is *good* if no a_i, b_i appear in $\hat{e}$, $\hat{f}$ or $\hat{t}_i$, and the specialization is *bad* otherwise.

Lemma 7.47. *Any bad word specialization in Q_n is 0.*

Proof. Since $\tilde{A}$ satisfies the identity $z v_1 z v_2 \ldots v_q z = 0$, cf. Proposition 7.31, any term vanishes if it is obtained by a specialization of $\tilde{t}$ or $\tilde{e}$ or $\tilde{f}$ containing a_i or b_i in q.   □

Thus, we may consider only the good word specializations.

Remark 7.48. In any good word specialization, if $[[\hat{e}, \hat{t}_1], \hat{t}_2]] \neq 0$, then two of these terms are powers of t and the third is et^j for some j. If $\hat{e}$ starts with e, then $[[\hat{e}, \hat{t}_1], \hat{t}_2] = \hat{e}\hat{t}_1\hat{t}_2$. (Similarly if $\hat{t}_1$ starts with $\hat{e}$, so then we switch $\hat{e}$ and $\hat{t}_1$ in the notation.) If $\hat{t}_2$ starts with e, then we get 0, since both $\hat{e}$ and $\hat{t}_1$ would have to be in polynomials t, and thus commute.)

Thus, we may assume $\hat{e}$ starts with e, and have $\hat{t}_i = t^{j_i}$ for $j = 1, 2$ and $\hat{e} = et^{j_3}$. Then

$$[[\hat{e}, \hat{t}_1], \hat{t}_2] = et^{j_1+j_2+j_3} = t^{j_1+j_2+j_3-2}[[e, t], t].$$

Likewise, if $\hat{f} = t^{j_4}f$, then

$$[\hat{t}_1, [\hat{t}_2, \hat{f}]] = t^{j_1+j_2+j_4}f.$$

Consequently, any good word specialization from $Q_n(x_i, y_i, \tilde{e}, \tilde{f}, \tilde{t})$ to $\widehat{A}$ will be a multiple of $Q_n(\bar{x}_i, \bar{y}_i, e, f, t)$ by a polynomial in t, where $\bar{x}_i, \bar{y}_i$ are the images of x_i and y_i.

Remark 7.49. Multiplication of any good word specialization of Q_n (or of any of its partial linearizations) by any element of $\widehat{A}$ yields zero. Indeed, the evaluation starts with e, so the only question for left multiplication is by f, and we saw this is 0 by the last line of the defining relations for $\widehat{A}$.

Combining Remarks 7.48 and 7.49, we see that any word specialization in the T-ideal of Q_n has the form

$$t^k e P_n(\bar{x}_1, \ldots, \bar{x}_{2n})(fe)^{q-1}f.$$

These cannot cancel $t^k e P_m(\bar{x}_1, \ldots, \bar{x}_{2m})(fe)^{q-1}f$ in view of Theorem 7.38. In particular, the T-ideal generated by $\{Q_n : n \in \mathbb{N}\}$ cannot be finitely generated via good word specializations.

7.4.7 Partial linearizations

Having disposed of word specializations, we turn to the partial linearizations.

Lemma 7.50 (Basic Lemma). *Any p-admissible specialization of any partial linearization of Q_n is 0.*

Proof. Recall the cyclic shift δ from Remark 2.62. We define an analogous operation σ that acts on the q appearances of the term $[\tilde{t}, [\tilde{t}, \tilde{f}]][\tilde{e}, \tilde{t}], \tilde{t}]$ in

$$Q_n = [[\tilde{e}, \tilde{t}], \tilde{t}]P_n(x_1, \ldots, x_n)([\tilde{t}, [\tilde{t}, \tilde{f}]][[\tilde{e}, \tilde{t}], \tilde{t}])^{q-1}[\tilde{t}, [\tilde{t}, \tilde{f}]],$$

(the last commutator $[\tilde{t}, [\tilde{t}, \tilde{f}]]$ is paired with the first $[[\hat{e}, \hat{t}], \hat{t}]$). More precisely, we write

$$Q_n = \tilde{u} P_n(x_1, \dots, x_n)(\tilde{v}\tilde{u})^{q-1}\tilde{v},$$

where $\tilde{u} = [\tilde{e}, [\tilde{t}, \hat{t}]]$ and $\tilde{v} = [\tilde{t}, [\tilde{t}, \tilde{f}]]$, and let $\sigma = \delta^2$ applied to any partial linearization, where δ now is the usual cyclic shift with respect to the alphabet $\{\tilde{u}, \tilde{v}\}$.

The terms of the specialization are partitioned into orbits of the operator σ. Since $q = p^2$, the orbit of a specialization is of order p^j for $0 \le j \le 2$.

Case 1. $j = 0$, i.e., the specialization is invariant. This is a word specialization so is 0 by Lemma 7.47.

Case 2. $j \ne 0$. The length of the orbit is a multiple of p. By Lemma 7.32 combined with Corollary 7.34, applied to the cyclic shift, if any two p-admissible specializations belong to the same orbit, then they are equal in $\widehat{A}$; hence, their sum is a multiple of p and is thus zero. $\square$

7.4.8   Verification of the counterexample

Having proved the word specialization case and the partial multilinearization case, we can now finish the proof of Theorem 7.45.

Proof of Theorem 7.45. To show that Q_n is not in the T-ideal generated by the $\{Q_i\}_{i<n}$, we consider the ideal generated by its specializations in $\widehat{A}$; since Q_n is admissible, and we can take homogeneous parts, we only consider admissible specializations.

By Lemma 7.50, any specialization of $\tilde{e}, \tilde{f}$, or $\tilde{t}$ in a partial linearization of Q_n on $\tilde{A}$ is zero. Thus, we may consider good specializations of Q_n. By Remarks 7.49 and 7.48, the specialization may be taken to be $\hat{t} \mapsto t$, $\hat{e} \mapsto e$, $\hat{f} \mapsto f$ (since other good specializations are proportional) and substitution of words for x_i, y_i. This reduces us to substitutions as a T-space, which is covered by Theorem 7.38. $\square$

Note 7.51.

(i) We can unify our examples by noting that the same proof holds for $p = 2$, virtually word for word, if we go back to Section 7.4.2 and define the test algebra $A = \mathcal{G}/I$, where I is the ideal generated by all words h on the $\bar{y}_i, \bar{z}_i$, for which $\deg_i h \ge 4$ for some i.

(ii) In the case $p > 2$, we could lower the degrees, and actually take $q = p$. Of course, bad nonzero invariant specializations could arise but, as Shchigolev has noted, their action produces exactly one pth power

v^p. So we could replace each polynomial by its product with v^p. On the other hand, the T-ideals could be constructed using products of polynomials Q_m for each $m \geq 1$. The double commutators $[[e, t], t]$ and $[t, [t, f]]$ could be replaced by other polynomials implied by the Grassman identity. The key is to use a test algebra in which the polynomials from Lemma 7.32 are central.

Chapter 8

Recent Structural Results

This short chapter is devoted to some recent structural results of a combinatorial nature.

8.1 Left Noetherian PI-algebras Are Finitely Presented

Rather than review the full theory of left Noetherian PI-algebras, cf. [Row80, Chapter 5] and [GoWa89] for example, we jump straight to a recent theorem of Belov that answers a question of Irving, Bergman, and others. It is a lovely example of Kemer's theory developed in Chapter 4, applied to a question which, at the outset, seems totally unrelated to this theory.

Definition 8.1. An algebra A is *finitely presented* (as an algebra) if A can be written in the form $F\{X\}/I$ where I is finitely generated as a two-sided ideal.

Clearly there are only countably many such I when F is a countable field, so the same argument as quoted in Example 1.81 shows that there are uncountably many non-finitely presented affine algebras. In fact, these can be taken to be weakly Noetherian, cf. Exercise 2.

Our aim in this section is to prove the following theorem.

Theorem 8.2. *Every affine Noetherian PI-algebra A (over any commutative Noetherian ring) is finitely presented.*

(Note that the theorem can be false if R is not affine. For example, the field of rational functions $F(\lambda)$ in one indeterminate is commutative Artinian but is not finitely presented. On the other hand, Resco and Small [ReS93] produced an example to show the theorem is not true without the PI hypothesis, at least in positive characteristic.)

We introduce the following notation. Write $\mathcal{F} = F\{x_1, \ldots, x_\ell\}$. Consider a left Noetherian affine algebra

$$A = F\{a_1, \ldots, a_\ell\} = \mathcal{F}/\mathcal{I} \qquad (8.1)$$

for suitable $\mathcal{I} \lhd \mathcal{F}$. We need to show that $\mathcal{I}$ is f.g. as an ideal. Let $\mathcal{I}_n$ be the ideal generated by elements of $\mathcal{I}$ of degree $\leq n$, which is certainly f.g. (at most the number of words of length $\leq n$), and let $A_n = \mathcal{F}/\mathcal{I}_n$. Thus, there is a natural surjection $\psi_n : A_n \to A$, given by $a + \mathcal{I}_n \mapsto a + \mathcal{I}$. We aim to show ψ_n is an isomorphism for all large enough n.

Let us also define $\psi_{m,n}$ to be the natural surjection $A_m \to A_n$, for $m < n$. Let $K_{m,n} = \ker \psi_{m,n}$. If $n_1 \leq n_2$, then $K_{m,n_1} \subseteq K_{m,n_2}$, and

$$\bigcup_{n \in \mathbb{N}} K_{m,n} = \ker \psi_n = \mathcal{I}/\mathcal{I}_n.$$

Recall the definition I^{cl} of k-*closed*, from Definition 2.53. By Proposition 2.54, the chain $\{K_{m,n}^{cl} : n \in \mathbb{N}\}$ stabilizes, so given m, we have n such that $K_{m,n}^{cl} = \ker \psi_n^{cl} = \ker \psi_n$, by Theorem 1.91.

The main technique is a way to use the Noetherian property to encode information into a finite set of relations.

Lemma 8.3. *Suppose we are given $b \in A$ arbitrarily. There is some $t \in \mathbb{N}$ (depending on b) and words $w_1, \ldots, w_t$ in the a_i satisfying the property that for any $a \in A$, there are $r_j \in A$ such that*

$$ba = \sum_{j=1}^{t} r_j b w_j, \qquad (8.2)$$

where the $r_j = r_j(a, b)$ depend on a and b.

Proof. $M = AbA \subset A$ is a Noetherian A-module, so we can write $M = \sum_{j=1}^{t} Abw_j$ for finitely many words $w_1, \ldots, w_t$ in the a_i, for suitable t. This means for any $a \in A$ that $ba = \sum r_j b w_j$ for suitable $r_j \in A$, yielding (8.2). $\qquad \square$

We can push this farther, using the same proof.

Lemma 8.4. *For any k, there is some t (depending on k) and words $w_1, \ldots, w_t$ in the a_i satisfying the property that for any $a \in A$, there are $r_j \in A$ such that*

$$va = \sum r_j v w_j \qquad (8.3)$$

for each word v of length $\leq k$, where the $r_j = r_j(a)$ depend on a but are independent of the choice of v. Furthermore, the relations (8.3) are all a consequence of a finite number of relations (viewing $A = \mathcal{F}/\mathcal{I}$).

Proof. Consider the direct sum $M = \oplus_v AvA$, where v runs over all words in $a_1, \ldots, a_\ell$ of length $\leq k$. There are a finite number $(\leq (\ell+1)^k)$ of these, and each AvA is an ideal of A and thus f.g. as A-module, implying M is a f.g. A-module. Let $\hat{v}$ be the vector whose v-component is v. Then $M = \sum_w A\hat{v}w$, summed over all words w in $a_1, \ldots, a_\ell$, and since M is Noetherian we can write $M = \sum_{j=1}^t A\hat{v}w_j$ for finitely many words $w_1, \ldots, w_t$ in the a_i, for suitable t. This means for any $a \in A$ that $\hat{v}a = \sum r_j \hat{v}w_j$ for suitable $r_j \in A$, and checking components we get (8.3).

It remains to show (8.3) is the consequence of a finite number of relations. Let k' be the maximal length of the w_j. We have a finite number of relations (8.3) where a runs over words in the a_i of length $\leq k' + 1$, since there are only a finite number of such words a. From now on, we assume only these relations, but claim they suffice to prove (8.3) for all $a \in A$.

Indeed, by linearity, we need check the assertion only for words a in the a_i, and we proceed by induction on $t = |a|$. The assertion is given for $t \leq k' + 1$. In general, suppose $t > k' + 1$ and write $a = a'a''$ where $|a'| = k' + 1$ and thus, $|a''| = t - (k' + 1)$. By hypothesis, (8.3) holds for a', so we can write

$$va' = \sum_{k=1}^t r_k(a')vw_k,$$

and thus,

$$va = va'a'' = \sum_{k=1}^t r_k(a')vw_k a''.$$

But $|w_k a''| \leq k' + (t - (k' + 1)) \leq t - 1$, so by induction

$$r_k(a')v(w_k a'') = \sum_j r_k(a')r_j(w_k a'')vw_j.$$

Thus,

$$va = \sum_{j=1}^t \left(\sum_{k=1}^t r_k(a')r_j(w_k a'') \right) vw_j,$$

as desired, proving the claim, taking

$$r_j(a) = \left(\sum_k r_k(a')r_j(w_k a'') \right),$$

which is independent of v. $\qquad\square$

Recall from Definition 4.14 the definition of "folds" of a polynomial, and the designated indeterminates (which alternate in the folds); also recall Definition 4.29 of a "monotonic" polynomial, i.e., the undesignated indeterminates occur in a fixed order. We denote the designated indeterminates as x_i and the undesignated indeterminates as y_j. Also recall Definition 4.86, of d-identity for a polynomial with folds.

Lemma 8.5. *Suppose f is a monotonic identity of A. For any $d > 0$, there is n such that f is a d-identity of A_n. Furthermore, if A_n is spanned over its center by words of length $\leq d$, then f is an identity of A_n.*

Proof. Consider $f(\bar{x}_1, \ldots, \bar{x}_m; \bar{y}_1, \ldots, \bar{y}_q)$ for suitable m, q. Taking $\hat{v}$ to be the vector of words $v_1, \ldots, v_m$ in $a_1, \ldots, a_\ell$ of length $\leq d$, we write each $\bar{x}_i$ as some v_{j_i} and apply Lemma 8.4 (where $a = \bar{y}_m$) to replace $\hat{v}\bar{y}_m$ by terms ending in the w_j, and next apply Lemma 8.4 to replace the $\hat{v}\bar{y}_{m-1}r_k(\bar{y}_m)$ by terms ending in the w_j, and so forth. After passing through each $\bar{y}_i$, we get a sum of terms of the form

$$r f(\bar{x}_1, \ldots, \bar{x}_m; w_{j_1}, \ldots, w_{j_m}).$$

Note that Lemma 8.4 required only a finite number of relations. Since the $\bar{x}_i$ are words in the generators of bounded length, and there are a finite number of w_j, we require only a finite number of extra relations $f(\bar{x}_1, \ldots, \bar{x}_m; w_{j_1}, \ldots, w_{j_m}) = 0$, so for n_0 suitably large (the maximum length of these relations) we see the f is a d-identity of A_n whenever $n \geq n_0$.

The last assertion is clear.   □

Proof of Theorem 8.2 concluded. Our goal is to show $A_n = A$ for suitably large n. We shall rely heavily on Lemma 8.3.

Suppose I is any graded ideal of $\mathcal{F}$ not contained in $\mathcal{I}$. Then

$$0 \neq (I + \mathcal{I})/\mathcal{I} \triangleleft A = \mathcal{F}/\mathcal{I};$$

write $\bar{A} = A/((I + \mathcal{I})/\mathcal{I})$. By Noetherian induction, we may assume that the theorem holds for

$$\bar{A} \approx \mathcal{F}/(I + \mathcal{I}).$$

Thus, taking I_n to be the ideal generated by elements of I of degree $\leq n$, and

$$\bar{A}_n = \mathcal{F}/(I_n + \mathcal{I}_n),$$

we have $\bar{A}_n = \bar{A}$ for large enough n, i.e., $I + \mathcal{I} = I_n + \mathcal{I}_n$.

Our goal is to find such an ideal I such that $I \cap \mathcal{I} \subseteq \mathcal{I}_n$ for all large enough n. Indeed, then we could consider the exact sequences

$$0 \to (I_n + \mathcal{I}_n)/\mathcal{I}_n \to A_n \to \bar{A}_n \to 0; \tag{8.4}$$

$$0 \to (I + \mathcal{I})/\mathcal{I} \to A \to \bar{A} \to 0; \tag{8.5}$$

Applying ψ_n to (8.4) yields isomorphisms $(I_n + \mathcal{I}_n)/\mathcal{I}_n \approx (I + \mathcal{I})/\mathcal{I}$ and $\bar{A}_n \approx \bar{A}$ at the two ends, so also an isomorphism at the middle, i.e., $A_n = A$, as desired.

So our goal is to find such an I. If $I = \mathcal{F}h\mathcal{F}$ for some $h \in \mathcal{F}$, then $I_n = A_n \bar{h} A_n$ for $\bar{h} = h + \mathcal{I}_n$, which is f.g. as left ideal by Lemma 8.3.

Recall by Kemer's Capelli Theorem 2.56 that A satisfies some Capelli identity c_m. Hence, Lemma 8.5 shows c_m is an m-identity for A_n for all large enough n. In view of Proposition 4.89, A_n is a PI-algebra (satisfying c_{2m-1}), for all suitable large n. Thus,

$$\text{id}(A_n) \subseteq \text{id}(A_{n+1}) \subseteq \text{id}(A_{n+2}) \cdots,$$

so letting $\kappa_n = (t_n, s_n)$ be the Kemer index of A_n, we have $\kappa_n \geq \kappa_{n+1} \geq \ldots$, and so this must reach a minimum at some n, i.e., at this stage

$$\kappa_n = \kappa_{n+1} = \cdots = \kappa$$

for some $\kappa = (t, s)$.

Claim. $\kappa = \text{index}(A)$.

This requires some machinery from Chapter 4. Otherwise, in view of Proposition 4.25, for any large enough μ, A_n has a Kemer μ-polynomial f such that all of its monotonic components are identities of A. In view of Proposition 4.92, we may assume f is not a $(m + 1)$-identity of A_n for all n. But, by Lemma 8.5, f is a $(m + 1)$-identity of A_n for large enough n. This contradiction shows that $\text{index}(A) = \kappa$, as desired.

Now take a Kemer polynomial f of A which is also a Kemer polynomial of A_n. Thus, there is some multilinear $h \in f(\mathcal{F})$ whose image in A is not 0. Let $I = \langle h \rangle$. Write

$$\bar{I} = (I + \mathcal{I})/\mathcal{I} \lhd A, \qquad \bar{I}_n = (I + \mathcal{I}_n)/\mathcal{I}_n \lhd A.$$

Note that if $s > 1$, then we may choose I such that $\bar{I}_n^2 = 0$, in view of Proposition 4.64, and we conclude easily by the above argument. Thus, we may assume $s = 1$, and again we could use Proposition 4.64 to get central polynomials and conclude fairly easily. However, this argument has only been laid out in characteristic 0, so we provide a characteristic-free argument based on the obstruction to integrality (Definition 2.46).

Let $\bar{L}_m$ denote the obstruction to integrality for A_m of degree t (recalling $\kappa = (t, s)$), cf. Definition 2.51, and likewise $\bar{L}$ denotes the obstruction to integrality for A of degree t. (One can reduce easily to the case $\bar{L} = 0$, but this is not relevant.) For large enough m, $\bar{I}_m = (I + \mathcal{I}_m)/\mathcal{I}_m$ is f.g. as a left ideal of A_m, in view of Lemma 8.3 (which enables us to express the

the relations with respect to a finite number of terms). By Theorem I, $\bar{I}$ annihilates the obstruction to integrality (to degree t) of A_m, and thus of $A_n = A_m/K_{m,n}$ for all $n > m$. Hence $\bar{I}$ is a module over $A_n/\bar{L}_n$ for all $n > m$. In view of Proposition 2.54, taking n large enough, these stabilize, i.e., $A_n/\bar{L}_n \approx A/\bar{L}$. But now $\bar{I}$ is f.g. as a module over the Noetherian ring $A/\bar{L}$, and thus, is a Noetherian module.

Let $K'_n = \bar{I} \cap K_{m,n}$. Then $\{K'_n : n \in \mathbb{N}\}$ is an ascending chain of submodules of the Noetherian module $\bar{I}$, so has a maximal member K'_n; now for any $n' \geq n$, we have $\bar{I} \cap K_{n,n'} = 0$ since otherwise its preimage in $\mathcal{F}/I_{m,n}$ would be more than L_n, contrary to choice of n. It follows that $\bar{I} \cap \ker \psi_n = 0$, which is what we needed to show. $\square$

8.2   Identities of Group Algebras and Enveloping Algebras

Representability plays a special role in the PI-theory of group algebras.

Example 8.6. Suppose G is any group that contains an Abelian group A of finite index. Then $F[G]$ is a f.g. module over the the commutative algebra $F[A]$, so is representable in the weak sense of Remark 1.70.

This easy observation led group theorists to wonder if there are other examples of group algebras satisfying a PI. Isaacs and Passman [IsPa64] proved the striking result that these are the only examples in characteristic 0, although the situation in characteristic $p > 0$ is somewhat more complicated. Their proof was very long, but now in characteristic 0 there is a short, simple proof found in Passman [Pas89], which we present here. Let us start by quoting some well-known facts from the theory of group algebras (to be found in [Pas77] or [Row88]), which we shall use repeatedly without citation. We start with some characteristic free observations.

One of the key tools is $\Delta = \Delta(G)$, the subgroup of G consisting of elements having only a finite number of conjugates. Note that an element $g \in G$ is in Δ iff its centralizer $C_G(g)$ is of finite index in G.

Fact I. Given $g \in \Delta$, the (finite) sum $\sum h$ of conjugates of g is an element of $\text{Cent}(F[G])$, and conversely $\text{Cent}(F[G])$ has a base consisting of such elements. Furthermore, if $\sum_{i=1}^{t} h_i \in \text{Cent}(F[G])$, then the set $\{h_1, \ldots, h_t\}$ includes all conjugates of h_1, implying $h_1 \in \Delta$. In particular, $\text{Cent}(F[G]) \subseteq F[\Delta]$.

Fact II. [Pas77, Lemma 4.1.6]. For any f.g. subgroup H of Δ, the commutator subgroup H' is torsion.

Fact III. [Pas77, Theorem 4.2.12], [Row88]: $F[G]$ is semiprime, for any group G and any field F of characteristic 0.

Lemma 8.7. *Suppose* $R = F[G]$ *has PI-class* n, *with* $\operatorname{char}(F) = 0$. *Then* $[G : \Delta] \le n^2$.

Proof. Take an n^2-alternating multilinear central polynomial h. By Remark G, any n^2 elements of R are dependent over $h(R)$. But if h is a sum of say m monomials, any element of $h(G)$ is a sum of at most m elements, each of which thus lies in Δ. Hence, $h(R) \subseteq F[\Delta]$. Thus, $n^2 \ge [G : \Delta]$. $\square$

Lemma 8.8. *Suppose* $G = \Delta(G)$ *and* $F[G]$ *satisfies a multilinear identity* $f = f(x_1, \ldots, x_d)$ *with coefficients in* $F[Z]$, *where* Z *is a torsion subgroup of* $Z(G)$. *Then* G *has a subgroup* A *of finite index with* A' *finite.*

Proof. By induction on the number of monomials of the identity f. Write $f = f_1 + f_2$ where f_1 is the sum of monomials in which x_i appears before x_j, and f_2 is the sum of the monomials where x_i appears after x_j. Clearly, we could choose i, j such that f_1, f_2 are both nonzero, and we assume for convenience that $i = 1$ and $j = 2$. Let Z_0 be the subgroup of Z generated by the support of the coefficients of f. Then Z_0 is a f.g. torsion Abelian group and thus is finite. If $G' \subseteq Z_0$ is Abelian we are done, so we take $v, w \in G$ with $vw \notin Z_0 wv$. Then $H = C_G(v) \cap C_G(w)$ has finite index in G, and for all $h_i \in H$, we have

$$vw f_1(h_1, \ldots, h_d) + wv f_2(h_1, \ldots, h_d) = f(vh_1, wh_2, h_3, \ldots, h_d) = 0,$$

but also

$$vw(f_1(h_1, \ldots, h_d) + f_2(h_1, \ldots, h_d)) = vw f(h_1, \ldots, h_d) = 0,$$

yielding $(vw - wv)f_2(h_1, \ldots, h_d) = 0$, i.e., letting $s = vwv^{-1}w^{-1} \notin Z_0$,

$$(s - 1)f_2(h_1, \ldots, h_d)wv = (s - 1)wv f_2(h_1, \ldots, h_d) = 0, \quad \forall h_i \in H,$$

implying $(s - 1)f_2(h_1, \ldots, h_d) = 0, \forall h_i \in H$.

If $s \notin H$, then matching components in $F[H]$ shows

$$vw f_2(h_1, \ldots, h_d) = 0, \quad \forall h_i \in H,$$

i.e., f_2 is an identity of $F[H]$, and we are done by induction (using H instead of G).

Thus, we may assume $s \in H$, and so $s \in Z(H)$ since v, w both centralize H. But $s \in H$ is torsion, so we replace Z by $Z\langle s \rangle$, a torsion central subgroup of H, and again are done by induction, since $(s - 1)f_2$ is an identity of $F[H]$. (Note that $(s - 1)f_2 \ne 0$ since $s \notin Z_0$.) $\square$

We are ready for the characteristic 0 case.

Theorem 8.9. *If* $\operatorname{char}(F) = 0$ *and* $F[G]$ *satisfies a PI, then* G *has a normal Abelian subgroup of finite index.*

Proof. By the Lemma, G has a subgroup $A \subseteq \Delta(G)$ of finite index, such that the commutator subgroup A' is finite; we take A such that $|A'|$ is minimal. We shall prove $A' = \{1\}$, i.e., that A is Abelian.

Since $F[A]$ is semiprime, we need to show the image of each element of A' in any prime homomorphic image $R = F[A]/P$ of $F[A]$ is 1. By Corollary D, taking $C = \operatorname{Cent}(R)$ and $S = C \setminus \{0\}$, the central localization $S^{-1}R$ is finite dimensional over $S^{-1}C$, and thus, is generated by the images of a finite number of elements of A; since these each have only a finite number of conjugates, their centralizer in A is a subgroup H of finite index, whose image in R/P is central. Clearly $A' \supseteq H'$ so, by hypothesis, $A' = H'$, whose image in R/P is 1.

Any subgroup A of index n contains a normal subgroup of index $\leq n!$, cf. Exercise 3. □

The situation for the enveloping algebra $U(L)$ of a Lie algebra L turns out to be trivial when $\operatorname{char}(F) = 0$. In this case, we define

$$\Delta = \Delta(L) = \{a \in L : [L : C_L(a)] < \infty\},$$

and have the following information in characteristic 0:

Fact IV. $U(L)$ is a domain (as seen at once from the Poincaré-Birkhoff-Witt Theorem).

Fact V. [BeP92] $\operatorname{Cent}(U(L)) \subseteq U(\Delta)$.

Lemma 8.10. *Suppose the enveloping algebra* $U(L)$ *of a Lie algebra* L *satisfies a PI. Then*

(i) L *has an Abelian Lie ideal of finite codimension.*

(ii) *Any Abelian Lie ideal* A *of* L *is central.*

Proof. Remark G implies $[L : \Delta] < \infty$, and also, letting $C = \operatorname{Cent}(U(L))$ and $S = C \setminus \{0\}$, $S^{-1}U(\Delta)$ is finite dimensional over $S^{-1}C$, spanned, say, by $a_1, \ldots, a_m$ in $U(\Delta)$.

(i) Since each element of Δ has centralizer in L of finite codimension, this implies Δ has a subspace A of finite codimension centralizing all a_i, and thus, A is in the center of Δ, and is certainly Abelian.

(ii) $K = S^{-1}U(A)$ is commutative, and thus a finite field extension of $S^{-1}C$. But for any $b \in L$, ad_b induces a derivation δ on K over F, which must be 0 since K/F is separable [Jac80]. This means A is central. □

Proposition 8.11. (cf. Lichtman [Lic89]) *The enveloping algebra $U(L)$ of a Lie algebra L over a field F of characteristic 0 satisfies a PI iff L is Abelian.*

Proof. ($\Leftarrow$) $U(L)$ is commutative, so certainly PI.

($\Rightarrow$) Tensoring by the algebraic closure of F, we may assume F is algebraically closed. Take an Abelian Lie ideal A of L, of minimal (finite) codimension. We claim L/A is Abelian. Indeed, otherwise there is some $a \in L/A$, such that ad_a has a nontrivial eigenvector $b \in L/A$; thus, $Fa + Fb$ is a 2-dimensional Lie algebra whose Lie ideal Fb is noncentral, contrary to Lemma 8.10 (ii).

Now for any $b \in L$, $A + Lb \lhd L$ since L/A is Abelian, and is Abelian by Lemma 8.10 (ii), contrary to the maximality of A unless L is Abelian. $\square$

A similar argument gives a result of Bakhturin, cf. Exercise 5.

Chapter 9

Poincare-Hilbert Series and Gelfand-Kirillov Dimension

Our aim in this chapter is to study growth of PI-algebras by means of the techniques developed so far. To do this, we shall define the *Hilbert series* of an algebra, otherwise called the Poincaré-Hilbert series, and the important invariant known as the *Gelfand-Kirillov* dimension. The Gelfand-Kirillov dimension of an affine PI-algebra A always exists (and is bounded by the Shirshov height), and is an integer when A is representable. (However, there are examples of affine PI-algebras with non-integral Gelfand-Kirillov dimension.) These reasonably decisive results motivate us to return to the Hilbert series, especially in determining when it is a rational function. We shall discuss (in brief) some important ties of the Hilbert series to the codimension sequence.

9.1   The Hilbert Series of an Algebra

Definition 9.1. An algebra A has a *filtration* with respect to $\mathbb{N}$ if A has f.d. subspaces $\tilde{A}_0 \subseteq \tilde{A}_1 \subseteq \ldots$ with $\cup_n \tilde{A}_n = A$ and $\tilde{A}_m \tilde{A}_n \subseteq \tilde{A}_{m+n}$. Given a ring A with a filtration, define the *Hilbert series* or *Poincaré series* to be

$$H_A = 1 + \sum_{n \geq 1} d_n \lambda^n, \qquad (9.1)$$

where $d_n = \dim_F(\tilde{A}_n / \tilde{A}_{n-1})$ and λ is a commuting indeterminate. When A is ambiguous, we write $d_n(A)$ for d_n.

Likewise, an A-module M is *filtered* if M has f.d. subspaces $\widetilde{M}_n$ with $\cup \widetilde{M}_n = M$ and $\tilde{A}_n \widetilde{M}_m \subseteq \widetilde{M}_{m+n}$. We define

$$H_M = \sum_{n \geq 0} d_n \lambda^n,$$

where $d_0 = 1$ and $d_n = \dim_F(\widetilde{M}^n / \widetilde{M}^{n-1})$. (Note that we are assuming $d_n < \infty$.)

Any f.g. module $M = \sum Aw_j$ over a filtered algebra becomes filtered, by putting $\widetilde{M}_n = \sum_j \tilde{A}_n w_j$. However, at times, it is preferable to choose different filtrations. For example, if $M \triangleleft A$, one might prefer to take the filtration of M given by $\widetilde{M}_n = M \cap \tilde{A}_n$.

Remark 9.2. A is finite dimensional, iff $\tilde{A}_n = A$ for some n, iff H_n is a polynomial of degree $\geq n$.

Thus, the theory is of interest mainly for infinite dimensional algebras.

Remark 9.3.

(i) Any graded algebra $A = \oplus A_n$ is filtered, where $\tilde{A}_n = \sum_{m \leq n} A_m$.

(ii) If A, B are graded algebras over a field F, then $A \otimes_F B$ can be graded via

$$(A \otimes B)_n = \sum_{j=0}^{n} A_j \otimes B_{n-j}.$$

In this case, $d_n(A \otimes B) = \sum_{j=0}^{n} d_j(A) d_{n-j}(B)$, so

$$H_{A \otimes B} = \sum_{j=0}^{n} d_j(A) d_{n-j}(B) \lambda^n = \sum_j d_j(A) \lambda^j \sum_k d_k(B) \lambda^k = H_A H_B.$$

This gives rise to several standard examples.

Example 9.4.

(i) $A = F[\lambda]$. Then $A_n = F\lambda^n$, so each $d_n = 1$, and

$$H_A = \sum_{i \geq 0} \lambda^i = \frac{1}{1 - \lambda},$$

a rational function.

(ii) $A = F[\lambda_1, \ldots, \lambda_n] = \otimes_{i=1}^{n} F[\lambda_i]$. Then $H_A = \frac{1}{(1-\lambda)^n}$.

(iii) $A = F\{x_1, \ldots, x_\ell\}$, the free algebra. Then A_n has a base comprised of words of length n, so $d_n(A) = \ell^n$, which is exponential in n, and thus destroys any hope of the Hilbert series being rational.

Fortunately, the imposition of a polynomial identity makes the growth behavior more like (ii) than like (iii).

Definition 9.5. The *associated graded algebra* $\mathrm{Gr}(A)$ of a filtered algebra A is defined as the vector space $\oplus(\tilde{A}^n/\tilde{A}^{n-1})$ under multiplication

$$(r_n + \tilde{A}^{n-1})(s_m + \tilde{A}^{m-1}) = r_n s_m + \tilde{A}^{n+m-1},$$

for $r_n \in \tilde{A}^n$ and $s_m \in \tilde{A}^m$.

Unless otherwise indicated, $A = F\{a_1, \ldots, a_\ell\}$ will denote an affine algebra, which we shall study in terms of the vector space $V = \sum_{i=1}^{\ell} Fa_i$. V is called the *generating space* of A, and clearly depends on the choice of generators $a_1, \ldots, a_\ell$. Then A is endowed with the *natural filtration* given by

$$\tilde{A}_n = F + V + V^2 \cdots + V^n. \tag{9.2}$$

(Here V^n is spanned by all $a_{i_1} \cdots a_{i_n}$.) Clearly, V depends on the choice of generators; when V is ambiguous, we write $H_{A;V}$ for H_A. Note that

$$\tilde{A}_n/\tilde{A}_{n-1} \approx V^n/(V^n \cap \tilde{A}_{n-1}),$$

so its dimension d_n measures the growth of A in terms of its generators. This will be our point of view throughout, and we are interested in the Hilbert series as a way of determining the asymptotic behavior of the d_n.

Remark 9.6. [KrLe00, Proposition 7.1 and Lemma 6.7] $\mathrm{Gr}(A)$ is affine, and every a_i is homogeneous of degree 1, by definition. $H_A = H_{\mathrm{Gr}(A)}$, seen immediately from the definition.

In view of Remark 9.6, in studying general properties of growth, we may assume A is graded, and that A_1 is the generating set. (Later, however, we shall assume the generating set includes 1, which conflicts with this assumption since $1 \in A_0$ when A is graded.) We can push this observation further:

Definition 9.7. A *monomial algebra* is an algebra defined by generators and relations, i.e., $F\{X\}/I$, where the relation ideal I is generated by monomials.

Since monomials are homogeneous in the free algebra under the natural grade given by degree (Example 1.34), any monomial algebra inherits this grading. Let us strengthen Remark 9.6.

Proposition 9.8. *Every affine algebra* $A = F\{a_1, \ldots, a_\ell\}$ *has the same Hilbert series as an appropriate monomial algebra.*

Proof. We need a total order on words, so we order the words first by length and then by the lexicographic order of Chapter 2. (We saw there that any two words of the same length are comparable.) We say a word in A is *reducible* if it can be written as a linear combination of words of smaller order. Clearly, any word containing a reducible word is reducible, so the linear combinations of reducible words (for A) comprise an ideal I in the free algebra $F\{x_1, \ldots, x_\ell\}$, and $\tilde{A} = F\{x_1, \ldots, x_\ell\}/I$ is by definition a monomial algebra that clearly has the same growth as A. (In the above notation, V^n/V^{n-1} is spanned by elements corresponding to irreducible words; conversely, any dependence of elements of A can be expresssed as writing the largest word as a linear combination of the others, i.e., the largest word is reducible.) $\square$

Unfortunately, the monomial algebra $\tilde{A}$ formed in the proof need not be finitely presented, even when A is finitely presented. Ufnarovsky found such an example, cf. [BBL97]. The theory of monomial PI-algebras was studied carefully in [BBL97], and we shall see interesting examples of monomial PI-algebras below and in Exercise 12.

9.2   Gelfand-Kirillov Dimension

As above, we work with an affine algebra $A = F\{a_1, \ldots, a_\ell\}$ under the natural filtration of (9.2).

Definition 9.9. The *Gelfand-Kirillov* dimension

$$\mathrm{GKdim}(A) = \varlimsup_{n \to \infty} \log_n \tilde{d}_n, \tag{9.3}$$

where $\tilde{d}_n = \dim_F \tilde{A}_n$, and $\tilde{A}_n = \sum_{i=0}^n V^i$ is as in (9.2).

GK dimension also can be defined for non-affine algebras, but we stick to the affine case.

Throughout this chapter, we shall use $\tilde{d}_n$ for dim $\tilde{A}_n$, as in the definition. GKdim(A) can be computed at once from its Hilbert series with respect to a generating set.

The Gelfand-Kirillov dimension was originally introduced by Gelfand and Kirillov to study growth of finite dimensional Lie algebras, but quickly became an important invariant of any affine algebra A, being independent of the choice of its generating set (in contrast to the Hilbert series). In particular, we may assume that $1 \in V$, in which case $V^i \subseteq V^j$ for all $j \geq i$, and thus $\tilde{A}_n = V^n$. At times it is convenient to make this assumption; at other times we want to work with a specific generating set which may not include 1.

The standard reference on GK dimension is the book by Krause and Lenagan [KrLe00], which also contains the main facts about GK dimension for PI-algebras. Our goal here is to enrich that account with some recent observations, largely based on the theory of Chapters 2 and 4; we shall also study Hilbert series of PI-algebras. Let us recall some basic facts.

Example 9.10. The GK dimension of the commutative polynomial algebra $A = F[\lambda_1, \ldots, \lambda_\ell]$ is ℓ. To see this, note that the space $\tilde{A}_n$ of polynomials of degree $\leq n$ is spanned by the monomials $\lambda_1^{n_1} \ldots \lambda_\ell^{n_\ell}$, where $n_1 + \cdots + n_\ell \leq n$; hence, its dimension is at least $(n/\ell)^\ell$ (choosing only those exponents up to $\frac{n}{\ell}$) and at most n^ℓ; thus,

$$\ell(1 - \log_n \ell) \leq \log_n(\tilde{d}_n) \leq \ell,$$

and the limit as $n \to \infty$ is ℓ on both sides.

From this one obtains without difficulty, for A a (commutative) integral domain, that $\mathrm{GKdim}(A) = \mathrm{trdeg}_F A$, an integer. For A noncommutative the situation becomes much more complex—$\mathrm{GKdim}(F\{X\})$ is infinite, for example. (There are different levels of infinite growth, which we shall not discuss here, since we shall see that Shirshov's Height Theorem implies that all affine PI-algebras have finite GK dimension.)

GK dimension does not increase under central localization; hence, for A prime PI, one can localize and appeal to Theorem H to show $\mathrm{GKdim}(A) = \mathrm{trdeg}_F \mathrm{Cent}(A)$, an integer; it follows that $\mathrm{GKdim}(A)$ is an integer when A is semiprime PI.

Obviously $\mathrm{GKdim}(A) = 0$ if $[A : F] < \infty$. On the other hand, if $[A : F]$ is infinite, then $V^{n+1} \supset V^n$ for each n, so $\tilde{d}_n \geq n$ and $\mathrm{GKdim}(A) \geq \log_n n = 1$.

9.2.1 Bergman's gap

Having just seen that no affine algebra has GK dimension properly between 0 and 1, we are ready for an interesting combinatoric result due to Bergman.

Theorem 9.11 (Bergman's gap). If $\mathrm{GKdim}(A) > 1$, then $\mathrm{GKdim}(A) \geq 2$.

Proof. There are several proofs in the literature, including [KrLe00, Lemma 2.4], but here is a quick proof using the ideas from Chapter 2.

By Proposition 9.8, we may assume $A = F\{a_1, \ldots, a_\ell\}$ is a monomial algebra, so $\tilde{d}_k$ now is the number of nonzero words of length $\leq k$ in $a_1, \ldots, a_\ell$.

Assume $\mathrm{GKdim}(A) > 1$. We claim that there is some hyperword in the a_i that is not quasiperiodic. Otherwise, if all hyperwords were quasiperiodic, then by the König Graph Theorem (since we have a finite number of

a_i), there would also be a bound k on the preperiodicity. But then starting after the k position and applying the König Graph Theorem to the periods, we also would have a bound m on the periodicity. Hence, the number of periods would be at most the number of words of length $\leq m$, which is $(\ell+1)^m$, so any word of length $\geq k$ on the a_i could be extended in at most $(\ell+1)^m$ possible ways, implying

$$\tilde{d}_{k+n} = \dim \tilde{A}_{k+n} \leq n(\ell+1)^m \tilde{d}_k, \qquad (9.4)$$

so $\log_n \tilde{d}_{k+n} \leq \log_n n + m \log_n(\ell+1) + \log_n \tilde{d}_k$, whose limit as $n \to \infty$ is 1, contrary to GKdim$(A) > 1$.

Since there is some hyperword h that is not quasiperiodic, Lemma 2.15 implies $\nu_n(h) < \nu_{n+1}(h)$ for every n. But $\nu_1(h) > 1$, so this means that each V^n has dimension $> n$, and

$$\tilde{d}_n \geq 2 + \cdots + (n+1) = \frac{n^2 + 3n}{2};$$

thus GKdim$(A) \geq 2$. $\qquad\qquad\qquad\qquad\qquad\qquad\qquad\qquad\qquad\square$

Remark 9.12.

(i) We proved a technically stronger result, that when GKdim$(A) > 1$, the $\tilde{d}_n$ grow at least at the rate $\frac{n^2+3n}{2}$.

(ii) We can turn the argument around to obtain more information about an algebra A with GK dimension 1. Namely, A cannot have a hyperword that is not periodic, since that would imply GKdim$(A) \geq 2$. Hence, the inequality (9.4) holds, implying the $\tilde{d}_n$ grow at most linearly in n, i.e., there is a constant N such that

$$\tilde{d}_n - \tilde{d}_{n-1} \leq N, \quad \forall n.$$

(For example take $N = (\ell+1)^m \tilde{d}_k$.) The smallest such constant N, which we call the *linear growth rate*, plays an important role, cf. Exercise 2.

(iii) To avoid hyperwords and the König graph theorem, and present the proof of Bergman's gap as quickly as possible from start to finish, one could define a word w to be *unconfined* if it occurs infinitely many times as a nonterminal subword in the various $\tilde{A}_n$, and $\nu'(n)$ to be the number of unconfined subwords. As above, GKdim$(A) > 1$ implies $\nu'(1) > 1$ and $\nu'(n+1) > \nu'(n)$ for each n; hence,

$$\tilde{d}_n \geq \sum_{m=1}^{n} \nu'(m) \geq \frac{n^2 + 3n}{2}.$$

Unfortunately, Bergman's gap does not extend to higher GK dimension, and there is an example of a PI-algebra whose GK dimension is any given real number ≥ 2, cf. Example 9.18 and Exercise 12. Nevertheless, we shall see soon that GKdim(A) is an integer if A is representable.

Digression. Hyperwords of slow growth tie Bergman's gap in with dynamical systems and the theory of Sturmian sequences, both of which are the subject of extensive literature:

(i) Suppose V is a given arc of the unit circle U, and $T : U \to U$ is the rotation by some fixed angle θ. Starting with some point $\mathbf{p}$, we build the hyperword

$$i_0 i_1 i_2 \cdots$$

by $i_k = 1$ if $T^k(\mathbf{p}) \in V$, and $i_k = 0$ otherwise.

(ii) A *Sturmian sequence* is a hyperword in which any two subwords of equal length differ by at most one letter.

Small, Stafford, and Warfield proved that if GKdim(A) = 1, then A is a PI-algebra. Pappacena, Small, and Wald found a nice proof of this theorem for A semiprime, using the linear growth rate, which we shall sketch in the exercises. It follows for A prime of GK dimension 1 that A is f.g. over a central polynomial algebra.

There are well-known examples of algebras of GK dimension 2 which are not PI, for example, the Weyl algebra and the enveloping algebra of the nontrivial 2-dimensional Abelian Lie algebra. However, both of these examples are primitive rings. Recently a prime, non-PI example has been found [SmoVi03] that is not primitive.

Braun and Vonessen [BrVo92] extended the Small-Stafford-Warfield theorem to GKdim(A) = 2, in the sense that any affine prime PI-algebra of GK dimension 2 is "Schelter integral" over a commutative subalgebra K, by which we mean that any element satisfies an equation

$$a^n = f(a)$$

where f is in the free product $A\{x\}$ (i.e., the coefficients do not commute with x) and each monomial of f has total degree $< n$ in x. This form of integrality had been studied by Schelter [Sch76] in connection with Shirshov's theorem, and will be described in Section 11.3.6; Braun-Vonessen [BrVo92] shows that this concept is very useful in the study of extensions of arbitrary PI-algebras.

9.2.2   Examples of affine algebras and their GK dimensions

The structure theory of affine PI-algebras is so nice that one begins to expect all of the theory of commutative algebras to carry over to affine algebras. Lewin showed this is not the case, by producing nonrepresentable affine PI-algebras, although, as we have noted, Lewin's argument is based on countability, and does not display an explicit example. Our aim here is to present some general sources of examples: Algebras obtained via affinization, and monomial algebras.

Example 9.13. Suppose $R = F\{a_1, \ldots, a_\ell\}$ is as in (1.20) of Chapter 1, i.e.,

$$R \subseteq W = \begin{pmatrix} R/I_1 & M \\ 0 & R/I_2 \end{pmatrix}.$$

Write R_j for $R/I_j = e_{jj}We_{jj}$ for $j = 1, 2$, which are affine algebras, generated by some set V_j. Writing

$$a_i = \begin{pmatrix} a_{1i} & a_i' \\ 0 & a_{2i} \end{pmatrix},$$

i.e., $a_{ji} = e_{jj}a_ie_{jj}$ and $a_i' = e_{11}a_ie_{22}$, we see that the image of R is contained in the affine subalgebra of W generated by R_1, R_2, and $a_1', \ldots, a_\ell'$. Any word of length n has to start with, say, k elements from R_1, then (at most) one element from $\{a_1', \ldots, a_\ell'\}$, followed from elements from R_2, so the dimension is at most

$$\sum_{k=0}^{n} (\dim V_1^k)\ell(\dim V_2^{n-k-1});$$

taking logarithms and suprema gives

$$\text{GKdim}(R) \leq \text{GKdim}(R_1) + \text{GKdim}(R_2).$$

Clearly, equality holds in the case of a block triangular algebra, cf. Definition 1.71. Furthermore, the same kind of argument can be used for any PI-basic algebra, in the terminology of Chapter 4. This is an important example that plays a prominent role in the codimension theory of Giambruno and Zaicev.

Affinization. One of the ideas behind Lewin's example shows how to take a "bad" algebra and make it affine. Recall Example 1.82 from Chapter 1. Beidar and Small found a way of refining Examples 1.82 and 1.83, used to advantage by Bell [Bell03].

Example 9.14. Take $T = F\{x, y\}$, the free algebra in indeterminates x and y, and $L = Ty$, a principal left ideal. Then $F + L$ is a free algebra on the infinite set of "indeterminates" $\{x^i y : i \in \mathbb{N}\}$. Define R as in Example 1.82. Now suppose W is any prime, countably generated algebra, presumably not affine. We have a natural surjection

$$\phi : F + L \to W,$$

sending the $x^i y$ to the respective generators of W. Let $P = \ker \phi$, a prime ideal of $F + L$, and $Q_0 = RPe_{11}R$, an ideal of R. Clearly, $e_{11}Q_0 e_{11} = Pe_{11}$, so Zorn's Lemma yields an ideal $Q \supseteq Q_0$ of S maximal with respect to $e_{11}Qe_{11} = Pe_{11}$.

Let us consider some properties of Q.

First, Q is a prime ideal of R, by the maximality hypothesis. Furthermore, if $I \lhd R$ satisfies $e_{11}Ie_{11} \subseteq Pe_{11}$, then $e_{11}(Q + I)e_{11} \subseteq Pe_{11}$ implies $Q + I = Q$ by maximality of Q, implying $I \subseteq Q$. Also

$$e_{11}(R/Q)e_{11} \approx (F + L)/\ker \phi = W.$$

We call $A = R/Q$ the *affinization* of W with respect to ϕ.

Proposition 9.15. *[Bell03] The algebra A defined in Example 9.14 has GK dimension equal to* GKdim$(W) + 2$.

This proposition provides a host of interesting examples, but unfortunately (from our point of view) these need not be PI, and may even have properties which are impossible in affine PI-algebras.

Example 9.16. An affine prime F-algebra of GK dimension 3, having non-nil Jacobson radical. Notation as in Example 9.14, take W to be the localization of $F[\lambda]$ at the prime ideal $F[\lambda]\lambda$; since GKdim$(W) = 1$, we have GKdim$(A) = 3$. Also one sees easily $e_{11}J(A)e_{11} \subseteq J(W) = W\lambda$. Of course in view of Amitsur[Am57], this implies A is not PI.

Other examples are given in the exercises.

Monomial algebras. In view of Proposition 9.8, we should be able to display whatever GK dimensions that exist through a monomial algebra (although one could lose the PI in the process). Vishne [Vis99] contructed primitive monomial algebras over an arbitrary field that have arbitrary GK dimension ≥ 2; Belov and Ivanov [BelIv03] constructed large classes of monomial algebras with weird growth. Clearly, Vishne's examples cannot be PI, since otherwise they would be central simple (by Kaplansky's theorem) and thus would have the same GK dimension as their center, which must be integral (in fact 0 in the affine case).

Definition 9.17. A monomial algebra $F\{X\}/I$ has *Morse type* if the monomials generating I are defined by specifying a hyperword h, where a word $w \in I$ iff w is not a subword of h.

If v is not a subword of h and w is a word containing v, then certainly w is not a subword of h, so I is clearly an ideal. Monomial algebras of Morse type provide a rich assortment of examples, depending on the choice of hyperword; this was the class used by [Vis99].

There is another way of constructing monomial algebras that provides a rather easy example of a PI-algebra with non-integral GK dimension. The basic idea goes back to Warfield, cf. [KrLe00, pp. 19–20], and is carried out to fruition in [BBL97] and [BelIv03].

Example 9.18. (A PI-algebra with non-integral GK dimension.) Let A be the monomial algebra whose nonzero monomials are all subwords of

$$x_1^{k_1} y_1 x_2^{k_2} y_2 x_3^{k_3} y_3 x_4^{k_4}.$$

We claim that A satisfies the PI

$$[x_1, x_2][x_3, x_4][x_5, x_6][x_7, x_8].$$

To see this, suppose on the contrary there were a nonzero word substitution

$$[v_1, v_2][v_3, v_4][v_5, v_6][v_7, v_8]$$

in A. Rearranging the indices, we may assume $v_1 v_2 \ldots v_8 \neq 0$. Let $m_i = \deg(v_i)$. If $m_1 + m_2 \leq k_1$, then $v_1 = x_1^{m_1}$ and $v_2 = x_1^{m_2}$, so $[v_1, v_2] = 0$. Hence, $m_1 + m_2 > k_1$, and v_3 must start with x_2 (at least). The same argument applied to $[v_2, v_3]$ shows $m_1 + m_2 + m_3 + m_4 > k_1 + 1 + k_2$, and so v_5 must start with x_3 (at least), and likewise v_7 must start with x_4. But then v_7, v_8 both are powers of x_4, so $[v_7, v_8] = 0$. Thus, one of the factors must make the word substitution zero.

We can control GKdim(A) by adding stipulations on the k_i. For example, we shall now construct an affine PI-algebra A with GKdim$(A) = 3\frac{1}{2}$. Towards this end, we require $k_3^2 \geq k_2$. Any nonzero monomial thus is determined by the sequence (k_1, k_2, k_3, k_4), so arguing as in Example 9.10, the number of nonzero words of degree $\leq n$ is less than $n^{3 + \frac{1}{2}} = n^{7/2}$. On the other hand, we can bound the number of nonzero words of degree $\leq n$ from below, by letting k_1, k_4 run from 0 to $\frac{n}{4}$ and k_3 from 0 to $\sqrt{n}$, and thus get

$$(n/4)^2 \sum_{k_3=0}^{\sqrt{n}} \sum_{k_2=0}^{k_3^2} 1 = (n/4)^2 \sum_{k_3=0}^{\sqrt{n}} k_3^2 \geq \frac{\sqrt{n^7}}{96},$$

so as $n \to \infty$ we see $\frac{7}{2} \leq$ GKdim$(A) \leq \frac{7}{2}$, and thus GKdim$(A) = \frac{7}{2}$.

We shall see below (Corollary 9.29) that this example is nonrepresentable. It not difficult to generalize this to arbitrary GK dimension for affine PI-algebras, and furthermore permits us to produce affine PI-algebras having arbitrary upper and lower Gelfand-Kirillov dimensions, cf. Exercise 13. A finitely presented PI-algebra with non-integral GK dimension is given in [BelIv03].

9.2.3 The Shirshov height and Gelfand-Kirillov dimension

As indicated earlier, Shirshov's Height Theorem 2.3 is an important tool in studying GK dimension. Our first observation, due to Amitsur and Drensky, is that Shirshov's height yields a bound on the GK dimension of any PI-algebra.

Theorem 9.19. *If $\mu = \mu(\ell, d)$ is as in Shirshov's height theorem, then*

$$\mathrm{GKdim}(A) \le \mu$$

for any affine algebra $A = F\{a_1 \dots, a_\ell\}$ of PI-degree d.

Proof. By Shirshov's Height Theorem we could take our generating vector space V to be a Shirshov base. Furthermore, by Remark 2.9, the reductions used in its proof do not change the degree of a word. So for $n \ge \mu$, $\dim V^n$ is at most the number of products $w_1^{k_1} \dots w_\mu^{k_\mu}$ for which $\sum k_j = n$. This is $m^\mu t_{n,\mu}$ where $m = \ell^d + \ell^{d-1} + \dots 1$ is the number of words of length at most d in the a_i, and $t_{n,\mu}$ is the number of ways of picking $k_1, \dots, k_\mu$ such that $\sum k_j = n$. This is at most n^μ, so

$$\dim V^n \le m^\mu n^\mu = (mn)^\mu.$$

Hence, $\mathrm{GKdim}(A) \le \lim_{n \to \infty} \log_n((mn)^\mu) = \lim_{n \to \infty} (\mu(1 + \log_n m)) = \mu.$ $\square$

(Note: It is well-known from combinatorics that $t_{n,\mu} = \binom{n+\mu-1}{n}$, but our estimate is good enough, as seen in the discussion of Example 9.10.) In particular, this gives Berele's result that any affine PI algebra has finite GK dimension, and has immediate consequences:

Corollary 9.20. *If A is an affine PI-algebra over a field, then any chain of prime ideals has bounded length $\le \mu$. In other words, A satisfies the ascending and descending chain conditions on prime ideals.*

Proof. (This is a well-known fact about rings with finite GK dimension, [KrLe00, Corollary 3.16].) $\square$

The formula $\text{GKdim}(A) \leq \mu$ of Theorem 9.19 cannot be an equality in general, since we saw examples of affine PI-algebras whose GK dimension is not an integer. But equality holds in a certain sense for representable PI-algebras. To see this, we extend the notion of Shirshov base. (We also modify the notation a bit, writing W where in Chapter 2 we would have used $\bar{W}$.)

Definition 9.21. A subset $W \subseteq A$ is called an *essential Shirshov base* for the algebra A, if there exist a number ν and a finite set $W' \subset A$, such that A is spanned by words in A of the form

$$v_0 w_1^{k_1} v_1 \ldots w_\nu^{k_\nu} v_\nu, \tag{9.5}$$

where each $w_i \in W$, $v_i \in W'$, and $k_i \geq 0$.

The minimal number $\nu(A)$ which satisfies the above conditions is called the *essential height* of A over W.

Of course, if $\dim_F A < \infty$, then we could take W' to be a base of A, and we get $\nu = 0$.

Intuitively, bounded height means we can span a subspace by products of powers of given subwords (evaluated on the generators), together with an extra finite set. The essential height is the minimal such bound. In Shirshov's Theorem, for example, W is the set of words of length $\leq d$ in the generators, and $W' = \{1\}$. However, in Shirshov's original proof, cf. Section 2.6.1, it was convenient to take W to be the words in $a_1, \ldots, a_{\ell-1}$ of length $\leq d$, and W' to be various powers of a_ℓ.

If W and W' are given as in Definition 9.21, then the number of elements of (9.5) such that $\sum k_i \leq n$, grows at most at the order of $|W|^\nu n^\nu$, where $\nu = \nu(A)$. Thus, we have the following bound.

Proposition 9.22. $\text{GKdim}(A) = \varlimsup_{n \to \infty} \log_n(|W|n)^\nu \leq \nu$. *In particular,* $\text{GKdim}(A) \leq \nu(A)$.

We aim to prove $\text{GKdim}(A) = \nu(A)$, for A representable. Towards this end, we need some preliminaries.

Proposition 9.23. *Suppose* $A \subseteq M_t(C)$, *where* C *is a commutative F-algebra. Suppose that for suitable elements* $v_0, \ldots, v_j, w_1, \ldots, w_j \in A$,

$$\sum_{k_1+\cdots+k_j \leq t} \alpha_I v_0 w_1^{m_1+k_1} v_1 \ldots w_j^{m_j+k_j} v_j = 0 \tag{9.6}$$

holds for all $m_1, \ldots, m_j$ *satisfying* $0 \leq m_i \leq t-1$ *for all i (the coefficients* $\alpha_I \in F$ *not depending on the* m_i). *Then we also have*

$$\sum_{k_1+\cdots+k_j \le t} \alpha_I v_0 w_1^{q_1+k_1} v_1 \ldots w_j^{q_j+k_j} v_j = 0 \qquad (9.7)$$

for all $q_1, \ldots, q_j \in \mathbb{N}$.

Proof. Induction on each q_i in turn. Indeed, if each $q_i \le t - 1$, then (9.7) holds by (9.6). So assume some $q_i \ge t$. Then the Hamilton-Cayley Theorem lets us rewrite $w_i^{q_i+k_i}$ as a C-linear combination of $w_i^{q_i-n+k_i}, \ldots, w_i^{q_i-1+k_i}$. But by induction Equation (9.7) holds for each of these smaller values, so therefore Equation (9.7) holds with the original q_i. $\square$

Proposition 9.24. *Let U be the F-linear span of all elements of the form*

$$a = v_0 w_1^{q_1} v_1 \ldots w_j^{q_j} v_j, \quad q_1, \ldots, q_j \in \mathbb{N}.$$

If Equation (9.7) holds, then the F-linear span of those elements for which one of the q_i is smaller than t, is the same as U.

Proof. One might want to repeat the Hamilton-Cayley argument used above, but a difficulty now arises insofar as we want to take the F-linear span and not the C-linear span. (In the previous result, we only needed to take the linear span of 0, which remains 0.) Thus, we need a slightly more intricate argument, and turn to an easy version of pumping (discussed in earlier chapters).

Given a as above, we call $(q_1, \ldots, q_j)$ the *power vector* of a. We call a *good* if $q_i < t$ for some i. It is enough to show that any element $a \in U$ is a linear combination of good elements.

We order the power vectors $(q_1, \ldots, q_j)$ lexicographically, and proceed by induction on the power vector. Suppose a is "bad," i.e., each $q_i > t$. In (9.6) take that term with $(\hat{k}_1, \ldots, \hat{k}_t)$ maximal among the $(k_1, \ldots, k_t)$. Let $m_i = q_i - \hat{k}_i$. Since $q_i \ge t$, we see $m_i \ge 0$ and $y_i^{q_i} = y_i^{m_i} y_i^{\hat{k}_i}$. Now Equation (9.7) enables us to write $v_0 w_1^{q_1} v_1 \ldots w_j^{q_j} v_j$ as a linear combination of terms

$$v_0 w_1^{q_1+k_1-\hat{k}_1} v_1 \ldots w_j^{q_j+k_j-\hat{k}_j} v_j,$$

i.e., with smaller power vectors $(q_1 + k_1 - \hat{k}_1, \ldots, q_j + k_j - \hat{k}_j)$, which by induction yield linear combinations of good elements. Thus, a also is a linear combination of good elements, as desired. $\square$

Corollary 9.25. *Write $S(q)$ for the dimension of the linear span of all elements of the form $v_0 w_1^{q_1} v_1 \ldots w_j^{q_j} v_j$, such that each $q_i \le q$. If Equation (9.6) holds, then there exists a constant κ, such that $S(q) < \kappa q^{j-1}$.*

Proof. In view of Proposition 9.24, we may take a base of "good" elements; i.e., we have degree at most t in one of the positions, and we have q^{j-1} choices for the powers $w_i^{q_i}$, in the other $j - 1$ positions. Thus, we have at most $|W'|^j q^{j-1}$ possibilities for our base elements, and we may take $\kappa \leq |W'|^j$. $\square$

Theorem 9.26. *Suppose $A \subseteq M_n(C)$ is affine, and $\mathrm{GKdim}(A) > 0$. Then $\mathrm{GKdim}(A) = \nu(A)$.*

Proof. Let ν be the essential height of $A = F\{a_1, \ldots, a_\ell\}$, and we take the v_j, w_i as in Definition 9.21. We already know $\mathrm{GKdim}(A) \leq \nu$, so our aim is to show that if $\mathrm{GKdim}(A) < \nu$, then we can reduce ν, contrary to choice of ν.

Let $\mathcal{V}$ denote the (at most) t^j-dimensional vector space with base elements of the form

$$v_0 w_1^{m_1+k_1} v_1 \ldots w_j^{m_j+k_j} v_j, \quad 0 \leq m_i < t.$$

(This base is determined by the vectors $(m_1, \ldots, m_j)$ of nonnegative integers, $m_i < t$, since the v_i and w_i are given.) We use this as our generating space for A, since we may choose whichever one we like in calculating $\mathrm{GKdim}(A)$.

If $\mathrm{GKdim}(A) < j$, then given ε, $\dim_F \mathcal{V} < n^{j-\varepsilon}$ for large enough n, thereby yielding Equation (9.6). Define a new set

$$W'' = \{v_i w_i^{k_i} v_{i+1} : 1 \leq i \leq \nu, 1 \leq k_i \leq t\}.$$

This is finite, of course, having $\leq (t+1)|W'|^2 \nu$ elements. But each "good" word can be written in length $\nu - 1$, since the small power of one of the w_i has been absorbed in W''. Hence, we have reduced ν, as desired. $\square$

We can push this argument a bit further.

Remark 9.27. A slight modification of the proof shows

$$\nu \leq \liminf_{n \to \infty} \log_n(\dim(\mathcal{V}^n)).$$

This proves also $\mathrm{GKdim}(A) = \liminf_{n \to \infty} \log_n(\dim(\mathcal{V}^n))$.

Corollary 9.28. *The essential height of a representable algebra is independent of the choice of W, and its Gelfand-Kirillov dimension is an integer.*

Corollary 9.29. *Any affine PI-algebra with non-integral GK dimension is nonrepresentable. (Compare with Example 9.18.)*

V.T. Markov first asserted the GK dimension of a representable algebra is an integer; A proof of Belov was published in [BBL97]. It would be interesting to construct a (non-representable) PI-algebra in which the essential height depended on the choice of essential Shirshov base. Soon we shall obtain a polynomial bound on Gelfand-Kirillov dimension of a PI-algebra (and hence on the essential height), first noted by Berele.

9.3 Upper Bounds for the Gelfand-Kirillov Dimension

Suppose $A = F\{a_1, \ldots, a_\ell\}$ has PI-degree d. In Theorem 9.19, we saw that GKdim(A) is bounded by the essential height (i.e., the lowest possible Shirshov height) and we also saw in Theorem 9.26 that equality holds for representable algebras. On the other hand, every relatively free algebra is representable, by Theorem 9.44 (which was also proved by Belov in characteristic p). So, given any T-ideal $\mathcal{I}$ of the free algebra $F\{x_1, \ldots, x_\ell\}$, we have an effective bound for the GK dimension of any PI-algebra A for which id(A) = $\mathcal{I}$, namely the essential height of the relatively free algebra.

This motivates us even more to compute a formula for GKdim(A), which would also give a lower bound for the essential height. One special case has been known for a long time—Procesi showed in 1967 ([Pro73]) for A of PI-class n that GKdim(A) $\leq (\ell - 1)n^2 + 1$, equality holding iff A is isomorphic to the algebra of generic $n \times n$ matrices. Once Kemer's theory became available, Berele [Ber94] was able to generalize this formula for all PI-algebras, using a simple structure-theoretical argument. Let us recall:

Remark 9.30. [KrLe00, Corollary 3.3] If A is a subdirect product of $A_1, \ldots, A_q$, then

$$\text{GKdim}(A) = \max\{\text{GKdim}(A_j) : 1 \leq j \leq q\}.$$

This observation fits in very well with the theory of Chapter 4.

Proposition 9.31. If char(F) = 0 and $A = F\{a_1, \ldots, a_\ell\}$ is PI, then GKdim(A) $\leq s((\ell - 1)t + 1)$, where index(A) = (t, s) is the Kemer index, cf. Definitions 4.15, 4.18, and 4.19.

Proof. Since A is a homomorphic image of its relatively free algebra, we may assume A is relatively free; in view of Corollary 6.74, we may assume A is the relatively free algebra of a PI-basic algebra of some Kemer index $\kappa = (t, s)$. But then $J^s = 0$ and A/J is relatively free semiprime and thus prime PI, of PI-class n where $n^2 = t$. [KrLe00, Corollary 5.10] shows

$$\mathrm{GKdim}(A) \leq s\,\mathrm{GKdim}(A/J),$$

so we conclude with Procesi's result. $\qquad\square$

Berele's original proof did not use the Kemer index, but did require some more computation, cf. Exercise 25. One can tighten this result a bit, using Example 1.71.

9.4　Rationality of Certain Hilbert Series

By definition, the GK dimension is determined by the Hilbert series under the natural filtration; since the GK dimension of representable PI-algebras is so well behaved, this leads us to look more closely at the Hilbert series. Unfortunately, unlike the GK dimension, the Hilbert series depends on the choice of generating space V, and we must be careful in selecting V. To emphasize this point, we write $H_{A;V}$ (resp. $H_{M;V}$) instead of H_A (resp. H_M), when appropriate.

Furthermore, in this case, $V^n = V^{n+1}$ iff $V^n = A$. Hence, H_A is a polynomial iff $[A : F] < \infty$, in which case, $\mathrm{GKdim}(A) = 0$.

The next question is whether H_A is a rational function, i.e., the quotient $\frac{p(\lambda)}{q(\lambda)}$ of two polynomials.

We start with a classical result, given in [Jac80, p. 442]:

Lemma 9.32. *If C is an $\mathbb{N}$-graded affine commutative algebra, then any homogeneous generating set is rational, and also the corresponding Hilbert series of any f.g. graded C-module is rational.*

This yields an ungraded result, by means of Remark 9.6.

Proposition 9.33.

(i) *Any generating set of a commutative affine algebra R is rational.*

(ii) *Any finitely generated module M over a commutative affine algebra A has a rational Hilbert series.*

Proof. Reduce to the graded case by passing to $\mathrm{Gr}(A)$ and $\mathrm{Gr}(M)$. $\qquad\square$

Stafford has produced an example of an affine PI-algebra which has a rational Hilbert series with respect to one set of generators, but which is nonrational with respect to a different set of generators. So our goal is to find a suitable finite generating set for which the Hilbert series is rational. We shall call this a *rational generating set*.

Proposition 9.34. *Suppose R is an affine algebra that is a f.g. module over a commutative subalgebra A. Then any finite subset of R can be expanded to a rational generating set V of R (as an affine algebra).*

Proof. A is affine by the Artin-Tate lemma, [Row88]. Taking a generating space V_0 of A, we write $R = \sum_{j=1}^{t} Ar_j$ and write $r_j r_k = \sum_{m=1}^{t} a_{jkm} r_m$. Then letting $V = V_0 + \sum_{j,k,m} Fa_{jkm}$, we see that R now is a filtered module over A, whose Hilbert series as a module with respect to V is the same as the Hilbert series as an algebra with respect to $V + \sum Fr_j$. $\square$

From now on, we assume $1 \in V$ (since we can clearly expand V by the element 1), so that the natural filtration of A is now given by $\tilde{A}_n = V^n$.

Remark 9.35. Suppose $R = A[s^{-1}]$, the localization of A by a regular central element s. Given generators $a_1, \ldots, a_\ell$ of A, we have generators $a_1 s^{-1}, \ldots, a_\ell s^{-1}, s^{-1}, 1, s$ of R, so $W = Vs^{-1} + Fs^{-1} + F + Fs$ is a generating subspace of R. Then letting $V' = V + F + Fs + Fs^2$, we have $(V')^n = (sW)^n = s^n W^n$. In other words, left multiplication by s^n gives us an isomorphism from W^n to $(V')^n$, so $\dim_F (V')^n = \dim_F W^n$ and

$$H_{R;W} = H_{A;V'}.$$

We strengthen slightly a theorem of Bell [Bell04].

Theorem 9.36. *Suppose A is a prime affine PI-algebra. Then any generating set of A can be expanded to a rational generating set.*

Proof. Notation as in Theorem I(ii), we could invert any single element c in $h_n(A)$ to get an algebra that is f.d. free over its center, so we are done by Proposition 9.33 and Remark 9.35. $\square$

Having proved that, at times, the Hilbert series of a PI-algebra is rational, we want to know if it must have certain form. To this end we quote [KrLe00, p. 175]:

Theorem 9.37. *If A is an affine PI-algebra and $H_A = 1 + \sum \tilde{d}_n \lambda^n$ is rational, then there are numbers m, d such that*

$$H_A = \frac{p(\lambda)}{(1 - \lambda^m)^d},$$

with $p(1) \neq 0$. In this case, there are polynomials $p_1, \ldots, p_{m-1}$ such that $\tilde{d}_n = p_j(n)$ for all large enough $n \equiv j \pmod{m}$, and $\max \deg p_j = d - 1$; then $\mathrm{GKdim}(A) = d$.

Thus, a non-rational Hilbert series implies non-integral GK dimension. On the other hand, even if a PI-algebra has integer GK dimension, the growth of the subspaces might be sufficiently grotesque as to keep it from having a rational Hilbert series, cf. Exercise 12. Furthermore, although the GK dimension of a representable algebra is finite, there are representable algebras whose Hilbert series are non-rational. One way of seeing this is via monomial algebras. We quote a fact from [BBL97]:

Proposition 9.38. *A monomial algebra is representable iff the following two conditions hold:*

(i) *A is PI (which for monomial algebras equivalent to Shirshov's theorem holding);*

(ii) *the set of all defining relations of A is a finite union of words and relations of the form "$v_1^{k_1} v_2^{k_2} \ldots v_s^{k_s} = 0$ iff $p(k_1, \ldots, k_s) = 0$," where p is an exponential diophantine polynomial, e.g., $p = \lambda^{k_1} + k_2 + k_3^4 3^{k_2}$.*

The proof is not too difficult—if A is representable, then one can examine elements in Jordan form and, by computing their k power, obtain any given set of exponential diophantine polynomials with respect to k; the converse is an inductive construction. Using this result, we have:

Example 9.39. A representable affine PI-algebra A that does not have a rational Hilbert series. Consider the monomial algebra supported by monomials of the form

$$h = x_1 x_2^m x_3 x_4^n x_5$$

and also satisfying the extra relation $h = 0$ if $m^2 - 2n^2 = 1$. This is representable by Proposition 9.38 but the dimensions u_n grow at the approximate rate of polynomial minus $c \ln n$, contrary to the conclusion of Theorem 9.37.

The dimension argument shows this algebra cannot have any rational Hilbert series.

9.4.1   Hilbert series of relatively free algebras

Our other result along these lines has a slightly different flavor, since it deals with a generating set that comes naturally. Namely, any relatively free algebra is graded naturally by degree, and our generating set is the elements of degree 1, i.e., the images of the letters. Let us record an observation that enables us to apply induction of sorts.

Lemma 9.40. *Suppose $I \lhd A$ with the filtration $I_n = I \cap \tilde{A}_n$.*

(i) $H_A = H_{A/I} + H_I$.

(ii) If $I, I' \lhd A$ are graded, then
$$H_{A/I\cap I'} = H_{A/I} + H_{A/I'} - H_{A/(I+I')}.$$

Proof. (i) Immediate from the definition.

(ii) $A/(I + I') \approx (A/I')/((I + I')/I')$ and $(I + I')/I' \approx I/(I \cap I')$, so $H_{A/I'} = H_{A/(I+I')} + H_{(I+I')/I'} = H_{A/(I+I')} + H_{I/(I\cap I')}$, implying

$$H_{A/I\cap I'} = H_{A/I} + H_{I/(I\cap I')} = H_{A/I} + H_{A/I'} - H_{A/(I+I')}.$$

$\square$

Another general technique that is available is the trace ring of Proposition 2.38; since we assumed characteristic 0 there, we shall continue this assumption here. Let I be as in Remark 2.34. Since $\hat{A}$ is Noetherian and $I \subseteq \hat{A}$, we see I is a f.g. $\hat{A}$-module. On the other hand, the natural filtration of $\hat{A}$ might not produce the same Hilbert series as the natural filtration on A.

Definition 9.41. Suppose $A = F\{a_1, \ldots, a_\ell\}$ is representable, and let I be as in Remark 2.34. We say the trace ring $\hat{A}$ has a *compatible* filtration to A if
$$A \cap \hat{A}_n = \tilde{A}_n.$$

Proposition 9.42. *Notation as in Definition 9.41, suppose A has a compatible filtration with $\hat{A}$, with respect to which A/I has a rational Hilbert series. Then A has a rational Hilbert series.*

Proof. Take the filtration $I_n = I \cap \hat{A}_n$ of Lemma 9.40(i). Since $\hat{A}_n$ is f.g. as a module over its affine center, H_I is rational, so we can apply Lemma 9.40. $\square$

Corollary 9.43. *If A is representable and graded, having a set of homogeneous generators with respect to which A/I has a rational Hilbert series (I as in Remark 2.34), then A has a rational Hilbert series.*

Proof. We can grade $\hat{A}$ by giving $\mathrm{tr}(w)$ the same grade as w. In view of Theorem I, this grading is compatible with that of A, so we can apply the proposition. (Note that I is a graded ideal, yielding a grade on A/I.) $\square$

Using these facts, we have a nice result for relatively free PI-algebras:

Theorem 9.44 (Belov). *In characteristic 0, every relatively free, affine PI-algebra A has a rational Hilbert series.*

Proof. In view of the ACC on T-ideals, we may assume by Noetherian induction (cf. Remark 1.84) that A/I has a rational Hilbert series for every T-ideal $I \neq 0$ of A. In view of (iii) and Proposition 6.73, we may assume A is varietally irreducible; we take the generators to be the generic elements $\bar{x}_1, \ldots, \bar{x}_\ell$, i.e., $V = \sum F\bar{x}_i$. Then A has a Kemer polynomial f, and let I be the T-ideal of A generated by $\langle f(A) \rangle$. I is also an ideal of the trace algebra $\hat{A}$, defined in the proof of Proposition 4.59.

We grade $\hat{A}$ by degree, where $\deg \operatorname{tr}(f) = \deg(f)$. For our generating set of $\hat{A}$, we take the generating set $\hat{V}$ of $\hat{A}$ consisting of V together with the homogeneous parts of the traces we added to A (which are finite in number). Since our algebras are relatively free, they are all graded, so clearly the compatibility hypothesis in Proposition 9.42 is satisfied. We thus can apply induction to Lemma 9.40(i) in conjunction with Proposition 9.33(ii). $\quad\square$

The same general idea holds in nonzero characteristic.

Proposition 9.45. *Any relatively free, representable affine algebra has a rational Hilbert series.*

Proof. In characteristic 0, this has already been proven, so we assume $\operatorname{char}(F) = p > 0$. We define the characteristic closure $\hat{A}$ as in Definition 2.37, i.e., using suitably high powers $\bar{w}^q$ of the words $\bar{w}$. Now Proposition 2.39 is available, so we just take I to be the ideal generated by evaluations in A of the appropriate Capelli polynomial, and conclude with the last paragraph of the previous proof. $\quad\square$

Note that we bypassed difficulties in the last assertion by assuming the algebra is representable. Belov has shown that any relatively free affine algebra of arbitrary characteristic is representable, but the proof is too long for this monograph.

Here is another result connected to Noetherian algebras.

Theorem 9.46. *If A is affine and its graded algebra is Noetherian PI, then A has a rational generating set.*

Proof. Since A has the same Hilbert series as its graded algebra, we may assume A is graded Noetherian. But now A is representable, so we take I as above. By Noetherian induction, A/I has a generating set for which the Hilbert series with respect to any larger set is rational. We lift this to A and apply Corollary 9.43. $\quad\square$

Remark 9.47. It is still open whether the Hilbert series of a left Noetherian PI-algebra A is rational. In view of Anan'in's theorem, A is representable, so Proposition 2.38 is available, and a Noetherian induction argument would enable one to assume $J^{s-1}I \neq 0$, and thus is a common ideal with A and $\hat{A}$. In order to conclude using Corollary 9.43, it would suffice to show $\hat{A}$ has a filtration compatible with the natural filtration on A. The natural candidate would be for $\mathrm{tr}(a_n) \in \hat{A}_n$ for each $a \in A_n$, but it is not clear without some extra hypothesis that this filtration is compatible with that of A.

9.5 The Multivariate Poincare-Hilbert Series

The concept of Hilbert series can be refined even further, cf. [KrLe00, pp. 180–182]. We consider such a situation. Suppose A is filtered, generated by F and $V = A_1$, and we assume V has a base $a_1, \ldots, a_\ell$. For example, if $A = F\{x_1, \ldots, x_\ell\}$, we could take $V = \sum_{i=1}^\ell Fx_i$. (Compare with Exercise 15.)

Given an ℓ-tuple $\mathbf{k} = (k_1, \ldots, k_\ell)$, define the *length* $|\mathbf{k}| = k_1 + \cdots + k_\ell$; if $|\mathbf{k}| = n$ define

$$B_{\mathbf{k}} = \{b = a_{i_1} \ldots a_{i_n} : \deg_i(x_{i_1} \ldots x_{i_n}) = k_i, \forall i\}.$$

In other words, $b = a_{i_1} \ldots a_{i_n} \in B_{\mathbf{k}}$ iff the letter a_i appears exactly k_i times for each i. We write $\deg_i b$ for the number of times a_i occurs in this product. Thus, $(\deg_1 b, \ldots, \deg_\ell b)$ is the multi-degree. $A_n = V^n$ is spanned by the $B_{\mathbf{k}}$. (Note, however, that a certain b could occur in $B_{\mathbf{k}}$ for several different $\mathbf{k}$ of the same length, so the multi-degree is not well-defined without a further assumption, such as A being graded.)

Definition 9.48. Ordering the $\mathbf{k}$ first by length and then by lexicographic order, define $\tilde{V}_{\mathbf{k}}$ to be the vector space spanned by all $B_{\mathbf{k}'}$, $\mathbf{k}' \leq \mathbf{k}$, and $\tilde{V}'_{\mathbf{k}}$ to be the vector space spanned by all $B_{\mathbf{k}'}$, $\mathbf{k}' < \mathbf{k}$. Define $d_{\mathbf{k}} = \dim_F(\tilde{V}_{\mathbf{k}}/\tilde{V}'_{\mathbf{k}})$, intuitively the number of base elements not occurring in "lower" components, and the *multivariate* Poincaré-Hilbert series

$$H(A) = \sum_{\mathbf{k} \in \mathbb{N}^{(\ell)}} d_{\mathbf{k}} \lambda_1^{k_1} \ldots \lambda_\ell^{k_\ell}.$$

Remark 9.49. Specializing all $\lambda_i \mapsto \lambda$ in the multivariate $H(A)$ gives the original Hilbert polynomial.

As in Example 9.4, it is easy to see that

$$H(F[\lambda_1, \ldots, \lambda_\ell]) = \prod_{i=1}^\ell \frac{1}{1 - \lambda_i},$$

and likewise Remark 9.3 carries over. Accordingly, one can also obtain the results of Section 9.4.1, and thus generalize Belov's theorem to prove the multivariate Poincaré-Hilbert series of any relatively free PI-algebra is rational, cf. Exercise 26.

This generalization is not merely a formal exercise. There is another, completely different, approach, to the Hilbert series of relatively free algebras, namely, by using the representation theory described briefly in Chapter 5 (and to be elaborated in Chapter 10.) Formanek describes this approach in detail in [For84], especially cf. Theorem 7, which identifies the multivariate Hilbert series with the series of codimensions. This enables him to compute the Hilbert series for the trace algebra of generic $n \times n$ matrices (cf. Chapter 12 below), and in the case of $n = 2$, he derives the Hilbert series for generic 2×2 matrices (due to Formanek-Halpin-Li for two matrices, and due to Drensky for an arbitrary number of generic 2×2 matrices). The formula for generic $n \times n$ matrices remains unknown, for $n > 2$, but the technique is powerful, as we shall mention at the end of Chapter 12.

Chapter 10

More Representation Theory

In this chapter we introduce finer tools from representation theory (in characteristic 0), in order to gain even more precise information about the nature of polynomial identities. In particular, we refine some of the results of Chapter 5 on $F[S_n]$-modules (also called S_n-*modules*). We appeal again to the theory of Young diagrams, which describes the irreducible S_n-representations in terms of partitions λ of $\{1, \ldots, n\}$. Recall $V_n = V_n(x_1, \ldots, x_n)$ is the F-vector space of the multilinear polynomials in $x_1, \ldots, x_n$, viewed naturally as an S_n-module, and, fixing a PI-algebra A, we define

$$\Gamma_n = \Gamma_n(A) = \mathrm{id}(A) \cap V_n. \tag{10.1}$$

The main innovations here are the more systematic use of the character of the module V_n/Γ_n itself, rather than just its dimension (the n-*codimension* c_n of A), and the use of GL-representations to study homogeneous identities. Much of this theory (in its PI-formulation) is due to Regev. Perhaps surprisingly, the same theory can be applied to trace identities, to be discussed in Chapter 12.

10.1 Cocharacters

Definition 10.1. The S_n-character afforded by the quotient S_n-module V_n/Γ_n is called the n-th *cocharacter* of the algebra A, and is denoted by

$$\chi_n(A) = \chi_{S_n}(V_n/\Gamma_n).$$

Obviously, $c_n(A) = \deg \chi_n(A)$.

Remark 10.2. As usual, we assume that $\mathrm{char}(F) = 0$; it follows that all S_n-characters are completely reducible. This implies that

$$\chi_n(A) = \sum_{\lambda \in \mathrm{Par}(n)} m_\lambda(A)\chi^\lambda$$

for suitable *multiplicities* $m_\lambda(A)$. Determining the multiplicities would essentially determine the structure of $\mathrm{id}(A)$. Accordingly, the problem of computing the $m_\lambda(A)$ for a given PI-algebra A is perhaps the major open question in PI-theory. Recently, using the theory of Formanek [For84], Berele [Ber04] has shown for $M_k(\mathbb{Q})$ that $\sum_{\lambda \in \mathrm{Par}(n)} m_\lambda(A)$ asymptotically is $\alpha n^{\binom{k^2}{2}}$ for some constant α, and has computed them for $k = 2$.

Remark 10.3. We can view the standard identity

$$s_n = \sum_{\sigma \in S_n} \mathrm{sgn}(\sigma)\sigma$$

as the (central) unit in the two-sided ideal J_λ, where $\lambda = (1^n)$.

As an example of the usefulness of cocharacters, let us show how the Capelli identity c_m is characterized in terms of strips of Young diagrams, following Regev [Reg78].

Proposition 10.4. *The algebra A satisfies the Capelli identity c_m if and only if its cocharacters are supported on a strip of Young diagrams of height $< m$:*

$$\chi_n(A) = \sum_{\lambda \in H(m,0;n)} m_\lambda(A)\chi^\lambda$$

(cf. Definition 5.33); namely, $m_\lambda(A) = 0$ if the length λ'_1 of the first column of λ is $\geq m$.

Proof. ($\Rightarrow$) Assume A satisfies c_m. Let T_λ be a tableau of shape λ with corresponding semi-idempotent $e'_{T_\lambda} = R^+_{T_\lambda} C^-_{T_\lambda} \in F[S_n] \equiv V_n$. By construction, $C^-_{T_\lambda}$ is alternating in λ'_1 variables, so e'_{T_λ} is a linear combination of such polynomials.

In general, suppose $f = f(x_1, \ldots, x_n) \in V_n$ is t-alternating. If f contains the monomial $\alpha \cdot w_0 x_1 w_1 x_2 \cdots w_{t-1} x_t w_t$ with coefficient $\alpha \neq 0$, where $w_0, \ldots, w_t$ are words in $x_{t+1}, \ldots, x_n$, some possibly equal to 1, then f contains

$$\alpha \cdot \sum_{\sigma \in S_t} \mathrm{sgn}(\sigma) w_0 x_{\sigma(1)} w_1 x_{\sigma(2)} \cdots w_{t-1} x_{\sigma(t)} w_t.$$

This is a consequence of c_m, provided $t \geq m$. Thus, if $\lambda_1' \geq m$, all the elements of the two-sided ideal $I_\lambda \subseteq F[S_n]$ are identities of A, which implies that $m_\lambda(A) = 0$.

($\Leftarrow$) Assume $\chi_n(A) = \sum_{\lambda \in H(m,0;n)} m_\lambda(A)\chi^\lambda$. We claim that A satisfies c_m. Indeed, given any n, let λ be a partition of n with the corresponding two-sided ideal I_λ in $F[S_n]$. By assumption, the elements of I_λ are identities of A if $\lambda_1' > m$. Let $\mu = (1^m)$, and let J denote the right ideal $J = s_m F[S_{2m}]$ in $F[S_{2m}]$. Then Corollary 5.22 implies

$$J \subseteq \bigoplus_{\mu \leq \lambda} I_\lambda.$$

But $\mu \subseteq \lambda$ implies that $\lambda_1' \geq m$, and it follows that the elements of J are identities of A. In particular, $e_m \pi$ is an identity of A for any $\pi \in S_{2m}$.

Choose $\pi \in S_{2m}$ such that

$$x_1 \cdots x_{2m} \pi = x_1 x_{m+1} x_2 x_{m+2} \cdots x_m x_{2m},$$

cf. Lemma 5.9. Then $c_m = e_m \pi \in \mathrm{id}(A)$. $\qquad\square$

See Exercise 1 for an application of this result. Also, in Exercise 3, we see how Amitsur and Regev [AmRe82]) used the polynomials p^* from Definition 5.8 to create an explicit sparse identity with very nice properties.

10.1.1   A hook theorem for the cocharacters

In this subsection we elaborate Proposition 5.54. Here e always denotes the number $2.718 \cdots$.

Theorem 10.5. *Suppose A is a PI-algebra over a field F of characteristic zero. Then there exist positive integers k and ℓ such that the cocharacters $\chi_n(A)$ all are supported on the (k, ℓ) hook of Young diagrams; that is,*

$$\chi_n(A) = \sum_{\lambda \in H(k,\ell;n)} m_\lambda(A)\chi^\lambda.$$

Equivalently, $E_\mu^* \in \mathrm{id}(A)$ where μ is the $(k + 1) \times (\ell + 1)$ rectangle. Such k and ℓ can be found explicitly: If A satisfies an identity of degree d, then we can choose any k and ℓ satisfying

$$\ell > \frac{e}{2} \cdot (d - 1)^4 = a \qquad and \qquad k > \frac{a\ell}{\ell - a};$$

or in particular, we can take $k = \ell > e(d - 1)^4$.

We need some preliminaries for the proof.

Remark 10.6. Although in general Γ_n is not a two-sided ideal, Theorem 5.38 and Lemma 5.45 indicate that when n is large, Γ_n contains a large two-sided ideal of V_n, since for most $\lambda \vdash n$, $f^\lambda \geq (d-1)^2 > c_n(A)$.

Proof of Theorem 10.5. Assume A satisfies an identity of degree d. By Theorem 5.38, $c_n(A) \leq (d-1)^{2n}$. Let $\alpha = (d-1)^4$ and choose integers u, v such that $uv/(u+v) > e\alpha/2 = e(d-1)^4/2$. For example, choose (any) $u = v > e(d-1)^4$. Let $\lambda = (v^u) \vdash n$, where $n = uv$. Then, by Lemma 5.46,

$$f^\lambda > \left(\frac{2}{e} \cdot \frac{uv}{v+v}\right)^n > \alpha^{4n} = (d-1)^{4n} \geq (d-1)^{2n} > c_n(A).$$

Hence by Lemma 5.45, $I_\lambda \subseteq \mathrm{id}(A)$, and in particular, $E_\lambda \in \mathrm{id}(A)$.

Moreover, if $\mu \vdash m$, $n \leq m \leq 2n - 1$ and $\mu \geq \lambda$ then

$$f^\mu \geq f^\lambda \geq (d-1)^{4n} \geq (d-1)^{2m};$$

hence, also $I_\mu \subseteq \mathrm{id}(A)$. By Corollary 5.22(i) $F[S_m]I_\lambda F[S_m] \subseteq \mathrm{id}(A)$, and by Theorem 5.50 it follows that $E_\lambda^* \in \mathrm{id}(A)$. The proof now follows by Proposition 5.54, with $k = u - 1$ and $\ell = v - 1$. $\square$

10.2 $GL(V)$-Representation Theory

Let $V = \sum_{i=1}^k Fx_i$. Recall $T^n(V)$ from Definition 5.30, and the identification

$$F\{x_1, \ldots, x_k\} = T(V) = \oplus T^n(V).$$

Also $E_{n,k}$ denotes $\mathrm{End}_F(T^n(V))$. Let $GL(V)$ be the corresponding general linear group. The *diagonal* action of $GL(V)$ on $T^n(V)$ is given by:

$$g \mapsto \bar{g}, \quad \bar{g}(v_1 \otimes \cdots \otimes v_n) := g(v_1) \otimes \cdots \otimes g(v_n),$$

which, extended by linearity, clearly makes $T^n(V)$ into a left $GL(V)$-module. The $GL(V)$-representations obtained in this way are called *polynomial representations*.

Analogously, the Lie algebra $gl(V)$ acts on $T^n(V)$ by derivations, as follows:

$$\tilde{g}(v_1 \otimes \cdots \otimes v_n) = \sum_{i=1}^n (v_1 \otimes \cdots \otimes v_{i-1} \otimes gv_i \otimes v_{i+1} \otimes \cdots \otimes v_n).$$

We saw in Definition 5.31 that the group S_n also acts naturally on $T^n(V)$ via

$$\pi \mapsto \hat{\pi}, \quad \hat{\pi}(v_1 \otimes \cdots \otimes v_n) := v_{\pi(1)} \otimes \cdots \otimes v_{\pi(n)},$$

reversing the order of multiplication. (This action does not preserve identities.) Thus, $T^n(V)$ is a representation for $GL(V)$ as well as for S_n, and hence, for $GL(V) \times S_n$. We use the notation $\bar{g}$ and $\hat{\pi}$ to avoid ambiguity concerning the actions.

Remark 10.7.

(i) Suppose $p(x) = p(x_1, \ldots, x_k) \in F\{X\}$, and $g \in GL(V)$. Then

$$\bar{g}p(x_1, \ldots, x_k) = p(g(x_1), \ldots, g(x_k))$$

Thus, if $p(x) \in \mathrm{id}(A)$, then also $\bar{g}p(x) \in \mathrm{id}(A)$.

(ii) Clearly, $\bar{g} \in E_{n,k}$ and the map $\psi_{n,k} : GL(V) \to E_{n,k}$, given by $g \mapsto \bar{g}$, is multiplicative: $\overline{g_1 g_2} = \bar{g}_1 \bar{g}_2$. Similarly, the map $\Psi_{n,k} : g \mapsto \tilde{g}$ is a Lie algebra homomorphism.

10.2.1 Applying the Double Centralizer Theorem

Definition 10.8. Denote by

$$B(n, k) = \langle \psi_{n,k}(GL(V)) \rangle \subseteq E_{n,k},$$

the subalgebra generated by the image $\psi_{n,k}(GL(V))$ in $E_{n,k}$.

Note that for $\sigma \in S_n$,

$$(\bar{g}\hat{\sigma})(v_1 \otimes \cdots \otimes v_n) = (\hat{\sigma}\bar{g})((v_1 \otimes \cdots \otimes v_n)).$$

The celebrated "Double Centralizer Theorem" of I. Schur connects the S_n and GL actions:

Theorem 10.9 (Schur). *Let S_n and $GL(V)$ act on $T^n(V)$ as above, with corresponding subalgebras $A(n,k)$ and $\bar{B}(n,k)$ in $E_{n,k}$. Then $B(n,k) = \bar{B}(n,k)$; namely, $\bar{B}(n,k)$ is the centralizer of $A(n,k)$ in $E_{n,k}$, and $A(n,k)$ is the centralizer of $\bar{B}(n,k)$ in $E_{n,k}$.*

Also $B(n,k) = \langle \Psi_{n,k}(gl(V)) \rangle$, [BeRe87]. This algebra is called the *enveloping algebra* of $GL(V)$—or $gl(V)$—in $\mathrm{End}_F(T^n(V))$.

Part of Schur's Double Centralizer Theorem can be obtained through easy considerations about modules, so we pause for a moment to collect these results.

Proposition 10.10. *Assume F is algebraically closed. Let M be a finite-dimensional F vector-space and let $A \subseteq \mathrm{End}_F(M)$ be a semisimple subalgebra. View M naturally as a left A-module. Write $A = \oplus_{i=1}^r A_i$ where $A_i \approx M_{d_i}(F)$, and let $J_i \subseteq A_i$ be minimal left ideals. Let $B = \mathrm{End}_A(M)$. Then*

(i) B is semisimple and has an isotypic decomposition similar to that of A (in Remark 5.11):

$$B = \bigoplus_{i=1}^r B_i \quad where \quad B_i \approx \mathrm{End}_{A_i}(J_i^{\oplus m_i}) \approx M_{m_i}(F).$$

(ii) Notation as in Remark 5.11, $M_i = B_i M$. If $L_i \subseteq B_i$ are minimal left ideals, then $\dim L_i = m_i$, and as a left B-module,

$$M \approx \bigoplus_{i=1}^r L_i^{\oplus d_i}.$$

Thus, the decomposition in Remark 5.11) is also the isotypic decomposition of M as a B-module.

(iii) $A = \mathrm{End}_B(M)$ and $A_i \approx \mathrm{End}_{B_i}(M_i)$. Also $\mathrm{End}_F(M_i) \approx A_i \otimes_F B_i$.

Remark 10.11. If M is a left A-module and $B = \mathrm{End}_A(M)$, then aM is a B-submodule of M, $\forall a \in A$. Similarly, bM is an A-submodule of M, $\forall b \in B$.

The following lemma will be useful in the sequel.

Lemma 10.12. *As in Proposition 10.10, let $A, B \subseteq \mathrm{End}_F(M)$ be two semisimple algebras satisfying $B = \mathrm{End}_A(M)$ and $A = \mathrm{End}_B(M)$.*

(i) Let $N \subseteq M$ be a left B-submodule of M. Then $N = eM$ for some idempotent $e \in A$.

(ii) Let $J_1, J_2 \subseteq A$ be left ideals with $J_1 \cap J_2 = 0$. Then $J_1 M \cap J_2 M = 0$. Similarly for any direct sum of left ideals.

(iii) If $J \subseteq A$ be a minimal right ideal in A, then JM is a simple left B-submodule of M. Thus, if $e \in A$ is a primitive idempotent, then eM is B-simple.

Proof. (i) By Proposition 10.10 B is semisimple, so there exists a B-submodule $M_2 \subseteq M$ with $M = N \oplus N'$. Let $e : M \to N$ be the projection map, so $e \in \mathrm{End}_B(M) = A$. Clearly, $N = eM$ and also, since e is a projection, $e = e^2$.

(ii) By Proposition 5.13, there are idempotents $e_1, e_2 \in A$ such that $J_i = e_i A$ and therefore, $J_i M = e_i M$, $i = 1, 2$. Suppose $a = e_1 w_1 = e_2 w_2 \in J_1 M \cap J_2 M$. Then $0 = e_1 e_2 w_2 = e_1^2 w_1 = e_1 w_1 = a$.

(iii) J is in some simple component A_i of A. Replacing A by A_i and B by B_i, we may assume A, B are simple, say $A \approx M_s(F)$ and $B \approx M_t(F)$. Then $\dim M = st$. Decompose $A = \bigoplus_{j=1}^{i} J_i$ with $J_1 = J$ and all $J_i \approx J$. By (ii), $M = \oplus_i J_i M$ and each $J_i M \neq 0$. Since $J_i M$ is a left B-module, $\dim J_i M \geq t$. Hence $st = \dim M = \sum \dim J_i M \geq st$, so equality holds and $\dim J_i M = t$ for each $1 \leq i \leq s$, proving JM is a simple module. □

10.2.2 Weyl modules

Remark 10.11 allows the construction of irreducible representations of $GL(V)$ and $gl(V)$ from those of S_n. This follows from the obvious fact that a subspace $M \subseteq T^n(V)$ is a $GL(V)$-module ($gl(V)$-module) exactly when it is a $\bar{B}(n, k)$-module. If $a \in F[S_n]$ (or $a \in A(n, k)$), then $aT^n(V)$ is a $B(n, k)$-module, hence a $GL(V)$-module ($gl(V)$-module). This leads us to another notion.

Definition 10.13. Given a tableau T_λ, recall from Section 5.2 that e_{T_λ} is its corresponding semi-idempotent. Define the *Weyl module*

$$V_k^\lambda = e_{T_\lambda} T^n(V).$$

These are parametrized by the strips $H(k, 0; n)$, cf. Definition 5.33, in the next result.

Theorem 10.14. *Let* $\dim V = k$ *and let* $\lambda \in H(k, 0; n)$ *with* T_λ *a tableau of shape* λ, *then*

(i) *The Weyl module* V_k^λ *is a simple* $GL(V)$-(or $gl(V)$-)module. *Moreover, if* $\mu \vdash n$ *and* $\mu \notin H(k, 0; n)$, *then* $I_\mu T^n(V) = 0$.

(ii) *All simple* $GL(V)$-(or $gl(V)$-)submodules of $W_\lambda = I_\lambda T^n(V)$ *are isomorphic to* V_k^λ.

(iii) *As a* $GL(V)$-(or $gl(V)$-)module, $W_\lambda \approx (V_k^\lambda)^{\oplus f^\lambda}$, *where* f^λ *equals the number of standard tableaux of shape* λ. *Moreover,*

$$W_\lambda = \bigoplus_{T_\lambda} e_{T_\lambda} T^n(V),$$

where the sum is over all standard tableaux of shape λ.

(iv) Let S^λ denote the corresponding Specht module. Then, as an S_n-module, $W_\lambda \approx (S^\lambda)^{\oplus f_k^\lambda}$, where $f_k^\lambda = \dim V_k^\lambda$. (This clearly follows from Proposition 10.10.)

(v) If the λ are tableaux of different shapes λ, then

$$\sum_\lambda V_k^\lambda = \bigoplus_\lambda V_k^\lambda.$$

A straightforward application of Proposition 10.10 yields the following analog to Theorem 5.34:

Proposition 10.15.

$$B(n,k) = \bigoplus_{\lambda \in H(k,0;n)} B_\lambda.$$

(i) Also $W_\lambda = B_\lambda T^n(V)$, and the decomposition in Theorem 5.34(iii) is also the isotypic decomposition of $T^n(V)$ as a $B(n,k)$-module.

(ii) B_λ is the centralizer of A_λ in $\operatorname{End}_F(W_\lambda)$ and also is a simple algebra.

(iii) $\operatorname{End}_F(W_\lambda) \approx A_\lambda \otimes_F B_\lambda$.

Let us adjust Lemma 10.12 to the anti-homomorphism $\varphi_{n,k}$.

Lemma 10.16. Let $\varphi_{n,k}(F[S_n]) = A(n,k) = A \subseteq E_{n,k} = \operatorname{End}_F(T^n(V))$, and let $B = B(n,k)$ be its centralizer in $E_{n,k}$.

(i) Let $J_1, J_2 \subseteq A(n,k)$ be two left ideals. If $J_1 \cap J_2 = 0$, then also $J_1 T^n(V) \cap J_2 T^n(V) = 0$. Similarly for any direct sum of left ideals in $A(n,k)$.

(ii) If $L \subseteq A$ is a minimal right ideal in $A(n,k)$, then $LT^n(V)$ is a simple left $B(n,k)$-submodule of $T^n(V)$. Thus, if $e \in A(n,k)$ is a primitive idempotent, then $eT^n(V)$ is $B(n,k)$-simple. In particular, the Weyl modules are simple.

Putting everything together, we have the following result.

Theorem 10.17.

(i) The $GL(V) \times S_n$ decomposition of $T^n(V)$ is given by

$$T^n(V) \approx \bigoplus_{\lambda \in H(k,0;n)} V_k^\lambda \otimes S^\lambda,$$

where S^λ is the corresponding Specht module.

(ii) Thus, as a GL(V)-module,

$$T^n(V) \approx \bigoplus_{\lambda \in H(k,0;n)} (V_k^\lambda)^{\oplus f^\lambda},$$

and as S_n-modules,

$$T^n(V) \approx \bigoplus_{\lambda \in H(k,0;n)} (S^\lambda)^{\oplus f_k^\lambda},$$

where $f_k^\lambda = \dim V_k^\lambda$.

The dimension of the Weyl module V_k^λ is given by tableaux as follows.

Definition 10.18. Let T_λ be a tableau λ filled with integers ≥ 0 (repetitions allowed!). T_λ is called *semi-standard* if its rows are weakly increasing from left to right, and its columns are strictly increasing from top to bottom. Such a T_λ is a *k-tableau* if it is filled with the integers from the set $\{1, \ldots, k\}$.

For example, $\begin{array}{cc} 1 & 1 \\ 3 & \end{array}$ is a 3-(but not 2-)semi-standard tableau.

Theorem 10.19. $\dim(V_k^\lambda) = \dim(e_{T_\lambda} W_{n,k}) = f_k^\lambda$, *where f_k^λ is the number of k-semi-standard tableaux of shape λ.*

The dimension f_k^λ equals the number of k-semi-standard tableaux of shape λ. A k-tableau is a tableau filled, possibly with repetitions, with some of the integers from $\{1, \ldots, k\}$.

There is also a hook-formula for $f_k^\lambda = \dim V_k^\lambda$. (For reference, see [MacD95].)

Theorem 10.20 (The GL hook formula). *For any partition λ,*

$$f_k^\lambda = \frac{\prod_{(i,j) \in \lambda} (k + j - i)}{\prod_{(i,j) \in \lambda} h_{i,j}(\lambda)}.$$

It easily follows that if $\lambda_{k+1} > 0$, then $f_k^\lambda = 0$. Thus, $f_k^\lambda \neq 0$ if and only if $\lambda \in H(k, 0; n)$.

10.2.3 Homogeneous identities

Let A be a PI-algebra. Let $\widehat{\Gamma}_n$ denote $T^n(V) \cap \mathrm{id}(A)$, the homogenous identities of degree n. Clearly, in view of Remark 5.60, $\widehat{\Gamma}_n$ is a $GL(V)$-submodule. As in the case of multilinear identities, we are led to consider the quotient space $T^n(V)/\widehat{\Gamma}_n$.

Definition 10.21. Let $\dim V = k$. The $GL(V)$-character of

$$T^n(V)/\widehat{\Gamma}_n,$$

which designates its isomorphism type as a $GL(V)$-module, is called the *homogeneous cocharacter* of the PI-algebra A, and is denoted by

$$\chi_{n,k}(A) = \chi_{GL(V)}(T^n(V)/\widehat{\Gamma}_n).$$

By Theorem 10.14(3), for suitable multiplicities $m'_\lambda(A) \leq f^\lambda$,

$$T^n(V)/(T^n(V) \cap \mathrm{id}(A)) \approx \bigoplus_{\lambda \in H(k,0;n)} (V_k^\lambda)^{\oplus\, m'_\lambda(A)} = \bigoplus_{\lambda \in H(k,0;n)} m'_\lambda(A) \cdot V_k^\lambda$$

and therefore

$$\chi_{n,k}(A) = \bigoplus_{\lambda \in H(k,0;n)} m'_\lambda(A) \cdot V_k^\lambda.$$

A theorem of Berele and Drensky shows that these homogeneous multiplicities $m'_\lambda(A)$ are the same as the multilinear multiplicities $m_\lambda(A)$ of Remark 10.2.

10.2.4 Multilinear and homogeneous multiplicities and cocharacters

As we have seen, the PI-theory studies multilinear identities S_n-representations, and homogeneous identities via GL-representations. It behooves us to indicate how the two methods coalesce.

Theorem 10.22 (Berele [Ber82], Drensky [Dr81b]). *The multiplicities m_λ in the homogeneous cocharacters are the same as in the multilinear cocharacters. More precisely:*

Let A be a PI-algebra and let

$$\chi_n(A) = \sum_{\lambda \vdash n} m_\lambda(A) \cdot \chi^\lambda$$

be its multilinear cocharacter. Let V be a k-dimensional vector space and let

$$\chi_{n,k}(A) = \sum_{\lambda \in H(k,0;n)} m'_\lambda(A) \cdot V_k^\lambda$$

denote the corresponding homogeneous cocharacter of A.
Then, for every $\lambda \in H(k, 0; n)$, $m'_\lambda(A) = m_\lambda(A)$.

Note that the homogeneous cocharacters "capture" only the multiplicities m_λ where $\ell(\lambda) \leq \dim V$.

Proof. The argument here is due to Berele [Ber82]. See [Dr81b] for a proof based on the Hilbert series.

Since

$$\chi_n(A) = \sum_{\lambda \vdash n} m_\lambda(A) \cdot \chi^\lambda$$

therefore, up to isomorphism,

$$V_n \approx \Gamma_n \bigoplus \left(\bigoplus_{\lambda \vdash n} J_\lambda^{\oplus m_\lambda(A)} \right),$$

where J_λ is a minimal left ideal in I_λ, cf. Section 5.2. By Lemma 10.16, we have

$$T^n(V) = V_n T^n(V) = \Gamma_n T^n(V) \bigoplus \left(\bigoplus_{\lambda \in H(k,0;n)} (J_\lambda T^n(V))^{\oplus m_\lambda(A)} \right).$$

The proof now follows from the definition of $\chi_{n,k}(A)$ since, by Proposition 5.65, $\Gamma_n \cap T^n(V) = \widehat{\Gamma}_n$, and by Lemma 10.16(ii), $J_\lambda T^n(V)$ is isomorphic to the simple $GL(V)$-module V_k^λ, provided $\lambda \in H(k,0;n)$. $\square$

Furthermore, applying Shirshov's Height Theorem yields the following information.

Corollary 10.23. *The $m_\lambda(A)$ in the strip are polynomially bounded. Explicitly, let A be a PI-algebra with $\chi_n(A) = \sum_{\lambda \vdash n} m_\lambda(A)\chi^\lambda$ its cocharacters. Given k, there exists a polynomial $p(x)$ such that $m_\lambda(A) \leq p(n)$, for all n and all $\lambda \in H(k,0;n)$.*

Chapter 11

Unified Theory of Identities

Our study of polynomial identities has led us to introduce some variants of PI, namely superidentities, trace identities, and LGIs, which we have handled in an ad hoc fashion in order to push forward to our main objective, Kemer's solution of Specht's problem in characteristic 0, and the surrounding theory. Now that we can pause for breath, let us reformulate identities from the point of view of universal algebra (using the PI-theory for associative algebras as our point of departure). At the same time, we shall introduce varieties and note some of the parallels with algebraic geometry.

11.1 Identities in Universal Algebra

A good exposition of universal algebra can be found in [Jac80, Chapter 2]; we only review some basic properties. An *(abstract) algebraic system* is a collection of sets $A_1, A_2, \ldots$ endowed with a set Ω of various m-ary operations for $m \in \mathbb{N}$, each typically notated as

$$\omega : A_{u_1} \times \cdots \times A_{u_m} \to A_{u_{m+1}}. \tag{11.1}$$

The set Ω of such operations is called the *signature* of the system.

Usually we deal with a single set A, i.e., all $A_{u_i} = A$, although occasionally we need this wider context. The 0-ary operations are the distinguished elements. For example, a group $(G, e, \mathrm{inv}, \cdot)$ has distinguished element e, the unary operation "inv" of inverse, written $a \mapsto a^{-1}$, and the binary operation $\cdot$ of product; we customarily write ab for $a \cdot b$.

The main signature of interest for us is that of an algebra A over a commutative ring C. This can be described in terms of the two sets A and C, each of which has its distinguished elements 0,1, unary operation "negation," and binary operations $+, \cdot$; furthermore, there is a "scalar multiplication" $C \times A \to A$ that describes A as a C-module. (When convenient, we add a few more axioms to make C a field.)

An alternate way to describe the algebra A is to fix C, and describe scalar multiplication in terms of unary operations on A; namely, for each $c \in C$, we have the left multiplication map $\ell_c : A \to A$, which can be viewed as a unary operation. Although notationally simpler, this approach is easier to fit into a general conceptual framework.

Remark 11.1 (Aside). There is a more sophisticated way of formulating (11.1); if the A_i all are modules over a given commutative ring C, then we could require (11.1) to read

$$\omega : A_{u_1} \otimes_C \cdots \otimes_C A_{u_m} \to A_{u_{m+1}}. \tag{11.2}$$

This in turn leads to a generalization

$$\omega : A_{u_1} \otimes_C \cdots \otimes_C A_{u_m} \to A_{u_{m+1}} \otimes_C \cdots \otimes_C A_{u_{m+n}},$$

which has gained prominence in the study of Hopf algebras and monads. However, we shall not pursue this avenue here.

We need one more major ingredient, in order to define identities. A *formula* $f(x_1, \ldots, x_m)$ is a formal expression in variables $x_1, \ldots, x_m$, utilizing various operations from Ω (not specified in our notation).

Definition 11.2. An *identity* of an abstract algebra is an expression

$$\forall x_1 \ldots \forall x_m \, f(x_1, \ldots, x_m) = g(x_1, \ldots, x_m),$$

for suitable formulas f, g. (No conjunctions, disjunctions, or negations are used.)

(We can use the same subscript m on both sides, simply by not requiring each x_i to appear in both f and g.) Since $\forall$ is understood, we drop it from the notation; the identity is written

$$f(x_1, \ldots, x_m) = g(x_1, \ldots, x_m),$$

meaning that equality holds under all substitutions for $x_1, \ldots, x_m$.

Thus groups all satisfy the identities

$$ex = xe = x, \quad x^{-1}x = xx^{-1} = e, \quad (x_1 x_2)x_3 = x_1(x_2 x_3) \text{ (associativity)}. \tag{11.3}$$

Likewise, any associative monoid satisfies the associativity identity, as well as other consequences such as $(x_1 x_2^2)x_3 = (x_1 x_2)(x_2 x_3)$, but we want such identities to be "trivial," and to require an associative PI-algebra to satisfy some "nontrivial" identity. So we have to find some way to incorporate the "trivial" identities automatically into our theory. This is the point of our next definition.

11.2 Varieties

Definition 11.3. Let S be a class of abstract algebras of a given signature. The *variety* of S is the set of algebras satisfying all the identities of S.

Identities that hold for all algebras of a given structure type are called *trivial* (in that type). Intuitively, they are the formal consequences of the given identities. A *nontrivial* identity for an abstract algebra A is an identity that is not trivial (in the given signature).

For example, the group identities $(x^{-1})^{-1} = x$ and $(xy)^{-1} = y^{-1}x^{-1}$ are trivial group identities that are proven during the first day of a course in abstract algebra. An example of a nontrivial group identity is commutativity ($x_1 x_2 = x_2 x_1$), since not all groups are commutative (i.e., Abelian). Another example is for a group to have a given exponent p, which translates to the identity $x^p = e$. Of course, "triviality" depends heavily on the variety. For example, commutativity is well-known to be a trivial identity in the variety of groups of exponent 2.

Definition 11.4. id(A) denotes the set of identities of an algebra A. Given a set S of algebras of a given variety, we let id(S) denote the set of identities holding in each algebra in S i.e., $\bigcap_{A \in S}$ id(A).

We can compare two abstract algebras of some variety via *homomorphisms*, defined as functions that "preserve" the operations in the obvious way. An *isomorphism* is a bijective homomorphism whose inverse is also a homomorphism. By definition, any identity of an abstract algebra holds in every homomorphic image.

A *subalgebra* is a subset on which all the operations are defined. Since any universal sentence obviously holds for subsets, and also in direct products (with the operations defined componentwise), we have shown:

Remark 11.5. Any identity holding for a set of algebras of a given signature holds for every subalgebra, for every homomorphic image, and for every direct product. In particular, every variety is closed under subalgebras, homomorphic images, and direct products.

Let us formalize Remark 11.5.

Definition 11.6. A collection of abstract algebras (of a given signature) that is closed under subalgebras, homomorphic images, and direct products is said to be *varietally closed*.

(Strictly speaking, we must bound the cardinality of the sets under consideration, and deal with isomorphism classes of algebras rather than individual algebras, in order not to violate the axioms of set theory.)

We have proven that the class of algebras of a given variety is varietally closed. We would like to go in the opposite direction, and show that any varietally closed class $\tilde{V}$ is the class of algebras of a suitable variety V, i.e., determined by $\mathrm{id}(\tilde{V})$. We shall do this by showing that there is a single algebra in $\tilde{V}$ for which every element of $\tilde{V}$ is a homomorphic image!

To do this, we want to show that any given signature Ω has a certain "free" abstract algebra F_Ω that has no nontrivial relations.

First, we do this for a single set. We start with an alphabet $\{x_i : i \in I\}$, where usually I is countably infinite, and define *words* w inductively on their length, denoted $|w|$:

Each 0-ary operation is a word of length 0, and each x_i is a word of length 1; for each m-ary operation $\omega : A \times \cdots \times A \to A$, and all previously defined words $w_{i_1}, \ldots, w_{i_m}$, we introduce a new word $w = \omega(w_{i_1}, \ldots, w_{i_m})$, and define the length

$$|w| = |w_{i_1}| + \cdots + |w_{i_m}|.$$

Continuing inductively (stopping at the first limit ordinal), we arrive at a set of words, which is closed under all operations, and comprises an abstract algebra F_Ω, called the *free* of the signature Ω.

In one sense, F_Ω is too free, lacking any identities whatsoever (such as distributivity or associativity). Suppose now the variety V is defined by the identities $\mathcal{I}$ defined in the signature Ω. We define an equivalence relation on F_Ω, by stipulating $f(w_1, \ldots, w_m) \equiv g(w_1, \ldots, w_m)$ iff the sentence "$f(x_1, \ldots, x_m) = g(x_1, \ldots, x_m)$" is in $\mathcal{I}$. Using the theory of congruences, one sees easily that the set of equivalence classes constitutes an algebra U_V, clearly of signature Ω. We write $\bar{x}_i$ for the equivalence class of x_i.

Lemma 11.7. *Suppose $A \in V$. Then, for any a_i in A, there is a (unique) homomorphism $U_V \to A$ obtained by sending $\bar{x}_i \mapsto a_i$.*

The proof, which we delete since it is obvious in all cases of interest to us, is an exercise in formal induction, because the relations used in defining U_V must hold in A. (This is a special case of [Jac80, p. 65].) This justifies the following definition:

Definition 11.8. U_V as defined above is the *relatively free* algebra of the variety V.

Now suppose $\widetilde{V}$ is a varietally closed class of algebraic structures, and $\mathcal{I} = \mathrm{id}(\widetilde{V})$. Define V to be the variety of all algebras satisfying all identities from $\mathcal{I}$. Clearly $\widetilde{V} \subseteq V$, and thus also

$$\mathcal{I} = \mathrm{id}(V).$$

Proposition 11.9. *Notation as above.*

(i) *A sentence $f(x_1, \ldots, x_m) = g(x_1, \ldots, x_m)$ is an identity of V iff*

$$f(\bar{x}_1, \ldots, \bar{x}_m) = g(\bar{x}_1, \ldots, \bar{x}_m)$$

in U_V.

(ii) *U_V is in the variety V, and has no nontrivial identities (with respect to this variety).*

(iii) *$\widetilde{V}$ is precisely the variety V.*

Proof. (i) Direction ($\Rightarrow$) is obvious. Conversely, if $f(\bar{x}_1, \ldots, \bar{x}_m) = g(\bar{x}_1, \ldots, \bar{x}_m)$ and $\bar{w}_1, \ldots, \bar{w}_m$ are the images in U_V of arbitrary words, then Lemma 11.7 gives a homomorphism $\bar{x}_i \mapsto \bar{w}_i$, implying $f(\bar{w}_1, \ldots, \bar{w}_m) = g(\bar{w}_1, \ldots, \bar{w}_m)$, i.e., $f(x_1, \ldots, x_m) = g(x_1, \ldots, x_m)$ is an identity of U_V.

(ii) By definition, for any identity $f(x_1, \ldots, x_m) = g(x_1, \ldots, x_m)$ from $\mathcal{I}$, we have $f(\bar{x}_1, \ldots, \bar{x}_m) = g(\bar{x}_1, \ldots, \bar{x}_m)$ in U_V; hence U_V satisfies this identity, by (i). Conversely any identity of U_V must be satisfied by every A in V so lies in $\mathcal{I}$.

(iii) By Lemma 11.7, every algebra of the variety V is a homomorphic image of U_V, so it remains to show $U_V \in \widetilde{V}$. For each A in $\widetilde{V}$, we write $H(A)$ for the set of homomorphisms $\psi : U_V \to A$, and for each $\psi \in H(A)$, we write A_ψ for a copy of A. Define

$$\hat{V} = \prod_{A \in \widetilde{V}} \prod_{\psi \in H(A)} A_\psi,$$

and define a homomorphism $\Phi : U_V \to \hat{V}$ by sending $\bar{x}_i \mapsto (\psi(x_i))$, the tuple of all values of x_i under all homomorphisms. If

$$\Phi(f(\bar{x}_1, \ldots, \bar{x}_m)) = \Phi(g(\bar{x}_1, \ldots, \bar{x}_m)),$$

then $\psi(f(x_1, \ldots, x_m)) = \psi(g(x_1, \ldots, x_m))$ for all possible homomorphisms $U_V \to A$, which means $f(a_1, \ldots, a_m) = g(a_1, \ldots, a_m)$ for all a_i in A, i.e., $f = g$ is an identity, in $\mathcal{I}$. This means Φ is 1:1, i.e., U_V is a subalgebra of a direct product of algebras in $\widetilde{V}$, so is also in $\widetilde{V}$. $\square$

Thus we have identified the variety $\mathcal{V}$ with $\widetilde{V}$. So a variety can be defined either in terms of the class of all algebras satisfying a given set of identities, or as a class of algebras that is varietally closed.

Note that $\mathrm{id}(U_\mathcal{V}) = \mathrm{id}(\mathcal{V})$, a property reminiscent of varieties (and their zeroes) defined in algebraic geometry. Thus the theory of PIs, or equivalently, of varieties, has led to several formulations of noncommutative algebraic geometry, studied by Amitsur, Plotkin, Artin-Schelter, and others, but these lie outside the scope of this book.

Now that we have the framework of varieties in which to work, we would like to erect a uniform PI-theory. Even in the associative case, we have to contend with the identity px_1, which says merely that an algebra has characteristic p. We shall see below how to exclude such "improper" identities in the cases of algebras with multiplication and addition; for the time being we assume we have the correct notion of "proper" identity.

Definition 11.10. A *strong* identity of an algebra A is an identity that remains proper in every homomorphic image of A.

This matches what we did in the associative PI-theory, where in Definition 1.5, we required one of the coefficients to be 1.

11.3 Examples of Varieties of Algebras and their Theories

In case the reader feels uncomfortable with the generality in which varieties have been presented, let us pause to see how easily they are recognizable in the various contexts which we have used.

11.3.1 Varieties of Groups

Here a word in $\mathcal{F}_\Omega$ is a formal product of letters and their inverses taken any number of times; a typical element might be $x_1(((x_2)^{-1})^{-1}x_3)^{-1}$. The trivial group identities $(x^{-1})^{-1} = x$ and $(xy)^{-1} = y^{-1}x^{-1}$, followed by associativity, enable us to identify this word with $x_1x_3^{-1}x_2^{-1}$; in fact, any word can be reduced to a product of letters and their formal inverses, without parentheses. This yields the familiar description of the free group. A given variety of groups might contain additional identities, in which case it could become impossible to find a good reduction procedure to describe the relatively free algebra. Of course, a group identity $f = g$ can be rewritten $fg^{-1} = 1$, so the general form of a group identity is $f(x_1, \ldots, x_m) = 1$, where f is a word in the letters x_i and their inverses. There is an extensive literature of varieties of groups with additional identities; here are some basic examples.

1. Writing (a, b) for the group commutator $aba^{-1}b^{-1}$, we see that (x_1, x_2) is an identity of a group G iff G is Abelian. More generally, any solvable group satisfies an identity

$$(((x_1, x_2), (x_3, x_4)), ...);$$

   the converse is a deep theorem of Thompson, cf. [Fl95].

2. A nilpotent group of index n satisfies the identity

$$(x_1, (x_2, \ldots, (x_n, x_{n+1}))).$$

   The converse is true, but nontrivial, cf. [Fl95]. Recent work has been done in describing solvability and nilpotence of groups in terms of 2-variable identities, especially in terms of *Engel identities*.

$$E_1(x, y) = [x, y], \quad E_2(x, y) = (E_1(x, y), y), \quad \ldots,$$

$$E_n(x, y) = (E_{n-1}(x, y), y),$$

   cf. [BGGKPP].

   G is called an *n-Engel group* if it satisfies an identity $E_n(x, y) \equiv 1$ for some n. Wilson [Wil91] proved that any residually finite n-Engel group is nilpotent.

3. The theory becomes more manageable when one adds a hypothesis about embeddibility. A group G is called a *PI-group* if G can be embedded into the group of invertible elements of some PI-algebra A over a field F. For example, any linear group is a PI-group. Exercise 6 gives Procesi's solution of Burnside's problem for PI-groups.

   Engel conditions in PI-groups are studied in detail in [Plo04], which obtains several theorems analogous to structure theorems of PI-algebra. Let us describe some of the results. An element $g \in G$ is called *Engel* if there exists $n = n(g)$ such that $E_n(x, g) \equiv 1$ holds identically in G. An element $g \in G$ is called a *nil-element* if for every $a \in G$, there is $n = n(a, g)$ such that $E_n(a, g) = 1$. G is called a *nil-group* if every element is a nil-element. It follows quickly from Wilson's theorem [Wil91] that any nil PI-group is locally nilpotent. More generally, [Plo04] show for any PI-group G, there is a series of normal subgroups in G

$$1 = H_0 \subset H_1 \subset H_2 \subset G,$$

where H_1 is locally nilpotent, H_2/H_1 is nilpotent, group, and G/H_2 is embedded in $\mathrm{GL}_n(K)$, where K is a commutative ring with 1. The proof is obtained by exploiting a corresponding chain of ideals in associative PI-algebras.

Note. All of our other examples are signatures with (at least) two operations: Addition $(+)$ and multiplication $(\cdot)$. In this case, any identity $f = g$ can be rewritten in the form $f - g = 0$, so takes the form of the identity $f(x_1, \ldots, x_m) = 0$ as defined in Chapter 1.

11.3.2   Varieties of associative algebras (with unit element 1)

Associativity comes to our aid, and we can write any word without parentheses. Hence the free associative algebra is $C\{X\}$; we get the free associative ring by taking $C = \mathbb{Z}$. We described the elements of the free algebra as linear combinations of words, which we call *monomials*. Any identity which is a linear monomial is called *improper*; thus the identity px_1 is improper. we say an identity is *proper* if it is not the consequence of improper identities. (For multilinear identities, this is equivalent to one of its monomials not being an identity.) Kaplansky's theorem works for proper identities, enabling one can build up the PI-theory. Amitsur [Am71] proved that any algebra satisfying a strong identity (as defined above) satisfies a power of s_n for some n, and thus is a PI-algebra in the sense that we defined earlier.

Improper identities often are taken into consideration as some of the identities defining the variety. For example, we treated varieties of algebras of characteristic > 0 in Chapter 7.

11.3.3   Varieties of associative algebras without 1

The free associative algebra without 1 is the subalgebra without 1 of $C\{X\}$ consisting of polynomials with constant term 0; we saw in Remark 3.21 that this theory can be embedded into the PI-theory of associative algebras by means of S-ideals. Although we have focused on algebras with 1, many expositions of the PI-theory (including all the previously published proofs of Kemer's theorem) deal with algebras without 1, to facilitate the treatment of nil PI-algebras (and Nagata's theorem).

11.3.4   Associative algebras with extra linear unary operations

Suppose we are given a single extra unary operation $T : A \to A$ in the structure. Thus a homomorphism $f : A_1 \to A_2$ must satisfy $f(T(a)) = T(f(a))$, $\forall a \in A$. If we stipulate that T is a linear transformation, then the notation becomes quite manageable; we expand our alphabet X to include formal symbols $T^j(x_i)$, for all $j \in \mathbb{N}$, and the free object is just the usual free algebra in all the $T^j(x_i)$, where the action of T is induced by the relation

$$T(T^j(x_i)) = T^{j+1}(x_i).$$

The set of letters used in monomials then has to be expanded to include all these $T^j(x_i)$. One could just as easily deal with a set of linear operators (and their formal compositions). The definition of improper identity must now be expanded to include identities of degree 1; for example, $T^2(x_1) - x_1$ says that $T^2 = I$, but does not yield structural information about the algebra. This is a touchy point, as we shall see. There are many important instances:

1. **Trace identities.** These already have been used in the proof of the Amitsur-Levitzki theorem, and, as we have seen, arise naturally by applying Newton's formulas to the Hamilton-Cayley characteristic polynomial of a matrix. We shall treat trace identities in depth in Chapter 12, so skip the details here.

2. **Identities with involution.** An *involution* of an algebra is an anti-automorphism of degree 2; algebras with involution arise naturally in matrix theory, the theory of quadratic forms, and the structure theory of finite dimensional algebras. The free algebra with involution has two sets of indeterminates, the x_i and the x_i^*, and using the identities $(x_1 x_2)^* = x_2^* x_1^*$ and $(x^*)^* = x$, we can remove all parentheses from the discussion.

   Amitsur proved that any algebra satisfying a PI with involution is a PI-algebra (without the involution). The varieties of algebras with involution are of considerable interest. $M_n(\mathbb{C})$ has two kinds of involution fixing $\mathbb{C}$, each of which yields a corresponding variety; its relatively free algebra can be constructed as an algebra of generic matrices with involution, whose ring of fractions is another important example of a central simple algebra (being in fact the generic central simple algebra of exponent 2).

   Any algebra with involution satisfying the identity $x_1^* = x_1$ must be commutative, since

   $$ab = (ab)^* = b^* a^*.$$

3. **Identities with derivations and/or automorphisms.** These all are obtained by enriching the language by adjoining the appropriate 1-ary operations. Kharcenko proved that any algebra over a field satisfying a strong PI with derivations and/or automorphisms is a PI-algebra.

4. **Identities of an algebra with a Hopf algebra action.** This generalizes the previous paragraph, cf. Berele and Bergen [BerBer], and ties in with graded identities, which we are about to discuss.

11.3.5   Graded identities

In Section 1.3 we considered an algebra A that is *graded* by an Abelian monoid M, i.e.,

$$A = \bigoplus_{u \in M} A_u$$

with $A_u A_v \subseteq A_{u+v}$ for all $u, v \in M$.

Taking this into account, we could build the algebra structure in terms of the homogeneous components, $\{A_u : u \in M\}$. Thus a *graded identity* is a polynomial

$$f_{u_1,\dots,u_n}(x_{u_1,1},\dots,x_{u_1,m};x_{u_2,1},\dots,x_{u_2,m};\dots;x_{u_n,1},\dots,x_{u_n,m})$$

which vanishes for all substitutions of the $x_{u_i,j}$ to A_{u_i}.

The relatively free algebra requires a set of indeterminates $\{x_{uj} : j \in \mathbb{N}\}$ for each $u \in M$, but otherwise is constructed analogously to the free algebra, keeping track of components.

In terms of the theory presented in this book, the most important special case is for $M = \mathbb{Z}/2$; the ensuing theory was developed in Chapter 6, and is critical for Kemer's theorem.

When $M = \mathbb{Z}/m$ and the base field F contains an m-th root ζ of 1, we can define an automorphism σ of A by

$$\sigma\left(\sum a_u\right) = \sum \zeta^u a_u;$$

conversely, if A has an automorphism σ and $m^{-1} \in F$, then we define a grading on A by

$$a_u = \frac{1}{m}\sum \zeta^{-u}\sigma^u(a).$$

We utilized a special case of this in Remark 6.24 in studying the structure of superalgebras, and this remark likewise facilitates the study of $\mathbb{Z}/m$-graded identities in terms of the structure theory.

The theory of graded PIs received a boost when Vasilovsky [Vas99] proved that the graded identities of $M_n(F)$, under the grading of Example 1.34(ii), are all consequences of the following remarkably simple graded identities:

$$x_0 y_0 - y_0 x_0; \quad x_i y_{-i} z_i - z_i y_{-i} x_i \ (i \neq 0),$$

where the subscript denotes the grading component.

Berele and Bergen [BerBer] verified that the Kemer structure theory goes through for graded identities, and graded identities have been studied closely in the recent literature.

Another case of considerable interest is $M = \mathbb{Z}$, which has ties with projective geometry.

11.3.6   Generalized identities (GIs)

In the proof of central polynomials, we used "linear generalized identities." More generally, we could consider the algebraic structure of a pair of rings R, A with a multiplication $R \times A \to A$ which satisfies associativity and distributivity wherever possible, and $1x = x$, but not necessarily the usual algebra identity $(rx)y = x(ry)$ for r in R. The free structure F_Ω is now the free product of $W * \mathbb{Z}\{X\}$, i.e., the coefficients can occur anywhere in the monomials (not necessarily at the beginning), and the identities are now called *generalized identities* (GIs). Thus an LGI is a linear GI.

GIs are convenient in the usual PI-theory, because they permit us to deal with partial substitutions. For example, the proof of Theorem 2.55 could be framed in terms of GIs instead of requiring the different ψ.

Suppose $R = A$. Since the elements of the algebra can be thought of as coefficients, we can specialize $x_2, x_3, \ldots$ to arbitary elements and rewrite any GI in terms of an (infinite) number of linear LGIs. On the other hand, LGIs are improper, according to the definition given above, so we need a way of focusing on the proper GIs.

Every monomial has a "signature" obtained by erasing the coefficients, and a *generalized monomial* of a generalized polynomial is the sum of all its monomials having the same signature. A GI is *proper* if one of its generalized monomials is not a GI. Generalizing Kaplansky's Theorem, Amitsur proved that any primitive algebra with a proper GI has nonzero socle; Martindale proved a good generalization of Posner's theorem. Rowen proved any algebra with a strong GI is a PI-algebra. The theory is developed in depth in [Row80, Chapter 7] and [BeMM96].

Generalized polynomials also provide a very good generalization of integrality:

Definition 11.11. An element $a \in A$ is *Schelter integral* over R of *degree* m if there is a generalized polynomial $f(x_1) = x_1^m + g(x_1)$ where $\deg g < m$, such that $f(a) = 0$. A is *Schelter integral* over R if each element is Schelter integral over R.

This definition has turned out to be very effective when A is a PI-ring (in the usual sense) which is Schelter integral over R, as shown in [BrVo92].

(The special case where A is a central extension of R is treated in [Row80, Chapters 4 and 7].)

11.3.7   Varieties of nonassociative algebras

Lack of associativity requires us to write parentheses in our polynomials, so computations become more intricate.

1. *Lie algebras* are defined by anticommutativity ($[xx] = 0$) and the Jacobi identity ($[[xy]z] + [[yz]x] + [[zx]y] = 0$). The free Lie algebra has been well studied; the theory of Lie identities, cf. [Ba91], has very important applications in algebra, including Zelmanov's solution of the restricted Burnside problem. In fact, Zelmanov's original proof involved a translation to the theory of identities of Jordan algebras.

    In characteristic 0, Specht's problem has been solved for Lie varieties satisfying a Capelli identity by Iltyakov [Ilt03], but the general case remains open. In characteristic > 0, there are counterexamples by Vaughan-Lee [Vau70] (in characteristic 2) and Drensky [Dr74], [Dr00] (for all characteristics > 0).

2. Other well-known varieties of nonassociative algebras include alternative algebras and Jordan algebras. From the outset, alternative algebras have been studied in terms of their identities (such as the Moufang identities). The free alternative and Jordan algebras are familiar to specialists, and one early theorem (due to Glennie) for Jordan algebras was the existence of nontrivial Jordan identities for the family of all "special" Jordan algebras.

These theories can be studied in terms of Capelli identities, cf. [Zub95b], [Zub97], and thus Kemer's theory also is applicable in attempting to prove the finite basis property in these situations. Iltaykov [Ilt91] obtained an affirmative answer for affine alternative algebras, and Vais and Zelmanov [VaZel89] proved the result for affine Jordan algebras. However, the analog of Kemer's PI-Representability Theorem remains open in each case. The only obstacle to such a theorem is Step 1 of the program discussed in Chapter 4; so far there is no analogue of Lewin's embedding theorem.

11.3.8   Rational identities

These are obtained by adjoining formally a multiplicative inverse to the language used for associative varieties (II). Since the inverse is not distributive over addition, everything becomes much more complicated. Furthermore, the operator $^{-1}$ must be undefined on certain values of A, since 0^{-1} does not exist. Also, the notion of homomorphism is too restricted here, since division algebras are simple as rings. The theory is facilitated by adding a new, formally undefined, element ? to A. Then one replaces homomorphisms by "places" which are partial homomorphisms.

Amitsur [Am66] showed that rational identities correspond to intersection theorems for Desarguian projective geometries, thereby completing Dehn's longstanding program of characterizing such theorems in terms of conditions on the underlying division algebra. Although it is hard to detect when a rational identity is trivial, cf. Exercise 4, Amitsur also proved that over an infinite field, any simple algebra satisfying a rational identity must be PI and thus f.d. over its center. Bergman's proof of Amitsur's theorem, which involves a lovely use of *generalized* rational identities, is given in [Row80, Chapter 8]. Amitsur's theory also provides insights to multiplicative subgroups of algebras. For example, if R is a simple algebra whose multiplicative group of invertible elements is solvable, then by Thompson's theorem quoted above, R satisfies a rational identity, and thus is finite dimensional.

11.3.9   Related notions

Some sorts of identities do not fall into this general framework.

Identities of Hopf algebras. Recently there has been interest in Hopf algebras that satisfy a polynomial identity. However, this does not fit well into the framework of what we have described. The problem is that any Hopf algebra A has *comultiplication* $A \to A \otimes A$, which cannot be described generically in a naive way since elements of $A \otimes A$ can have an arbitrarily large number of summands.

Profinite identities. Our framework also misses a very recent area of study, that of infinite series whose specializations always converge in a certain topology. (For example, the series might eventually reach 0.)

Chapter 12

Trace Identities

Several times in this book we used results about traces, which we presented ad hoc as we needed them. Now we shall develop these *trace identities* formally, showing in one crucial respect that they behave better than polynomial identities. Razmyslov, Procesi, and Helling proved that all trace identities of $M_n(\mathbb{Q})$ formally are consequences of a single trace identity, which Razmyslov identified with the Hamilton-Cayley equation (rewritten as a trace identity via Newton's formulas.) We prove this basic result in this chapter, and also provide tools for further investigation. In the process, following Regev [Reg84], we see that the trace identities have a cocharacter theory analogous to that of the PI-theory, and the two can be tied together quite neatly.

12.1 Trace Polynomials and Identities

Definition 12.1. For any C-algebra A, a *trace function* is a C-linear map $\mathrm{tr} : A \to \mathrm{Cent}(A)$ satisfying

$$\mathrm{tr}(ab) = \mathrm{tr}(ba), \qquad \mathrm{tr}(a\,\mathrm{tr}(b)) = \mathrm{tr}(a)\,\mathrm{tr}(b), \qquad \forall a, b \in A.$$

It follows readily that

$$\mathrm{tr}(a_1 \ldots a_n) = \mathrm{tr}((a_1 \ldots a_{n-1})a_n) = \mathrm{tr}(a_n a_1 \ldots a_{n-1})$$

for any n.

According to the philosophy of Chapter 11, cf. Section 11.3.4, we view the trace function as a 1-ary symbol in the framework of universal algebra. In other words, we shall define a formal *trace symbol* Tr, linear over C, which will be stipulated to satisfy

(i) $\mathrm{Tr}(x_{i_1} x_{i_2} \ldots x_{i_n}) = \mathrm{Tr}(x_{i_n} x_{i_1} \ldots x_{i_{n-1}})$;

(ii) $\mathrm{Tr}(f) x_i = x_i \, \mathrm{Tr}(f)$;

(iii) $\mathrm{Tr}(f \, \mathrm{Tr}(g)) = \mathrm{Tr}(f) \, \mathrm{Tr}(g)$ for all $f, g \in C\{X\}$.

Note that (iii) enables us to eliminate Tr^2, since

$$\mathrm{Tr}^2(f) = \mathrm{Tr}(1 \, \mathrm{Tr}(f)) = \mathrm{Tr}(1) \, \mathrm{Tr}(f).$$

Definition 12.2. A *simple trace monomial* is an expression of the form

$$\mathrm{Tr}(x_{i_1} \ldots x_{i_t}).$$

A *pure trace monomial* is a product of simple trace monomials, i.e., of the form

$$\mathrm{Tr}(x_{i_{t_1+1}} \ldots x_{i_{t_1+2}} \ldots x_{i_{t_2}}) \, \mathrm{Tr}(x_{i_{t_2+1}} \ldots x_{i_{t_3}}) \cdots ,$$

In general, a *(mixed) trace monomial* is the product of a regular monomial with an arbitrary number of pure trace monomials, i.e., of the form

$$h = x_{i_1} \ldots x_{i_{t_1}} \, \mathrm{Tr}(x_{i_{t_1+1}} \ldots x_{i_{t_1+2}} \ldots x_{i_{t_2}}) \, \mathrm{Tr}(x_{i_{t_2+1}} \ldots x_{i_{t_3}}) \cdots .$$

Thus, the trace monomial is a pure trace monomial iff $t_1 = 0$.

The *degree* $\deg_i h$ of a trace monomial is computed by counting each x_i, regardless of whether or not it appears inside a Tr symbol.

Definition 12.3. The *free algebra with traces* $T'\{X\}$ is spanned by the the trace monomials, with the obvious multiplication (juxtaposition) satisfying (i), (ii), and (iii) stated before Definition 12.2; its elements are called *trace polynomials*. The *algebra of pure trace polynomials* $T = T\{X\}$ is the subalgebra of $T'\{X\}$ generated by the simple trace monomials.

Thus, any element of $T'\{X\}$ is written as a linear combination of

$$w_1 \, \mathrm{Tr}(w_2) \cdots \mathrm{Tr}(w_k) \quad \text{or} \quad \mathrm{Tr}(w_2) \cdots \mathrm{Tr}(w_k),$$

where $w_1, \ldots, w_k$ are words in the x_i, permitted to be 1. It is easy to see that $T'\{X\}$ is an associative algebra, on which Tr acts linearly via

$$\mathrm{Tr}(w_1 \, \mathrm{Tr}(w_2) \cdots \mathrm{Tr}(w_k)) = \mathrm{Tr}(w_1) \, \mathrm{Tr}(w_2) \cdots \mathrm{Tr}(w_k);$$

in particular. $\mathrm{Tr}(\mathrm{Tr}(w_2) \cdots \mathrm{Tr}(w_k)) = \mathrm{Tr}(1) \, \mathrm{Tr}(w_2) \cdots \mathrm{Tr}(w_k)$. Note that T is a commutative subalgebra of $T'\{X\}$.

We define *homogeneous* and *multilinear* trace polynomials, according to the degrees of their monomials obtained by erasing the Tr. For example,

$$\text{Tr}(x_1 x_2)\,\text{Tr}(x_1 x_3^2) - \text{Tr}(x_1^2 x_2 x_3^2)$$

is a homogeneous pure trace polynomial. Also

$$\text{Tr}(x_1 x_2)\,\text{Tr}(x_3) + \text{Tr}(x_1 x_2 x_3)$$

is a multilinear pure trace polynomial.

Definition 12.4. A *trace identity* of an algebra A with trace function tr is a trace polynomial that vanishes identically on A, whenever we substitute elements of A for the x_i, and tr for Tr.

Remark 12.5. As with ordinary identities, one can multilinearize trace identities by applying the operators Δ_i of Definition 1.11; when the field F has characteristic zero, the trace identities can be recovered from the multilinear trace identities.

Our interest in this chapter is in $A = M_n(F)$, for a field F, where tr is the usual trace function; hence $\text{tr}(1) = n$. A trace identity is *trivial* if it holds for $M_n(F)$ for all n; examples were given in Remark 1.48.

Example 12.6. Assume $\text{char}(F) \neq 2$. In $M_2(F)$ we know

$$\det(a) = \frac{\text{tr}(a)^2 - \text{tr}(a^2)}{2},$$

so the Hamilton-Cayley theorem yields the trace identity

$$x^2 - \text{tr}(x)x + \frac{\text{tr}(x)^2 - \text{tr}(x^2)}{2},$$

which multilinearizes (via the usual technique of substituting $x_1 + x_2$ for x) to

$$x_1 x_2 + x_2 x_1 - \text{tr}(x_1)x_2 - \text{tr}(x_2)x_1 + \text{tr}(x_1)\,\text{tr}(x_2) - \text{tr}(x_1 x_2).$$

Let us formulate this example in general.

Example 12.7. By the Hamilton-Cayley theorem, any matrix a annihilates its characteristic polynomial $\chi_a(x)$. Newton's formulas (Remark 1.44) imply that in characteristic 0, $M_k(F)$ satisfies the *Hamilton-Cayley* trace identity

$$g_k = x^k + \sum_{j=1}^{k} \left(\sum_{j_1 + \cdots + j_u = j} \alpha_{(j_1, \ldots, j_u)}\,\text{tr}(x^{j_1}) \ldots \text{tr}(x^{j_u}) x^{k-j} \right)$$

where the $\alpha_{(j_1,\ldots,j_u)} \in \mathbb{Q}$ can be computed explicitly. (See also Example 12.21.)

Of course, g_k is a mixed trace polynomial that is homogeneous but not multilinear. Since the Hamilton-Cayley trace identity g_k depends on k, it is nontrivial, and our object here is to prove that every trace identity of $M_k(\mathbb{Q})$ is a consequence of g_k. This is a considerable improvement (and much easier to prove!) than Kemer's theorem, since the minimal identity s_{2k} of $M_k(\mathbb{Q})$ does not generate $\mathrm{id}(M_n(\mathbb{Q}))$ as a T-ideal, cf. Exercise 3.6. However, it still remains a mystery as to how to prove that $\mathcal{M}_n$ is finitely based as a corollary of the Hellman-Procesi-Razmyslov theorem.

Here is the basic set-up, which we shall use throughout:

Remark 12.8. Let $V = F^{(k)}$ and $V^* = \mathrm{Hom}(V, F)$ denote its dual space. Let $e_1, \ldots, e_k \in V$ be a basis of V and $\theta_1, \ldots, \theta_k \in V^*$ its dual basis.

There is a bilinear form $V \times V^* \to F$ denoted by $\langle v, w \rangle$ for $v \in V$ and $w \in V^*$. Thus, we can identify $V \otimes V^*$ with $\mathrm{End}(V) = M_n(F)$, where $v \otimes w$ corresponds to the transformation on V given by $(v \otimes w)(y) = \langle y, w \rangle v$.

For example,

$$(e_i \otimes \theta_j)(e_\ell) = \langle e_\ell, \theta_j \rangle e_i = \delta_{j\ell} e_i,$$

where $\delta_{j\ell}$ is the Kronecker delta, so $e_i \otimes \theta_j$ is identified with the matrix unit e_{ij}.

Remark 12.9. Define $\mathrm{tr}(v \otimes w) = \langle v, w \rangle$. This matches the usual definition of trace on matrices as seen by checking base elements:

$$\mathrm{tr}(e_i \otimes \theta_j) = \langle e_i, \theta_j \rangle = \delta_{ij} = \mathrm{tr}(e_{ij}).$$

Now

$$
\begin{aligned}
(v_1 \otimes w_1)(v_2 \otimes w_2)(v) &= (v_1 \otimes w_1)(\langle v, w_2 \rangle v_2) \\
&= \langle v, w_2 \rangle (v_1 \otimes w_1)(v_2) \\
&= \langle v, w_2 \rangle \langle v_2, w_1 \rangle v_1 \\
&= \langle v_2, w_1 \rangle \langle v, w_2 \rangle v_1 \\
&= \langle v_2, w_1 \rangle (v_1 \otimes w_2)(v).
\end{aligned}
$$

Since this holds for all $v \in V$, we have $(v_1 \otimes w_1)(v_2 \otimes w_2) = \langle v_2, w_1 \rangle (v_1 \otimes w_2)$. By induction on s, it follows that

$$(v_{j_1} \otimes w_{j_1}) \cdots (v_{j_s} \otimes w_{j_s}) = \langle v_{j_2}, w_{j_1} \rangle \cdots \langle v_{j_s}, w_{j_{s-1}} \rangle (v_{j_1} \otimes w_{j_s}). \quad (12.1)$$

12.1.1 The Kostant-Schur trace formula

The theory of trace identities of matrices over a field is based on two main facts. The first is the description of the kernel of the map

$$\varphi_{n,k} : F[S_n] \to \text{End}(T^n(V)),$$

defined in Definition 5.31, where $\dim V = k$.

Theorem 12.10. *If $F[S_n] = \oplus_{\lambda \vdash n} I_\lambda$, the direct sum of its minimal (two-sided) ideals, then*

$$\ker \varphi_{n,k} = \bigoplus_{\lambda \vdash n,\ \lambda_1' \geq k+1} I_\lambda.$$

This classical theorem of Schur and Weyl, a reformulation of Theorem 5.33(i), can be found in more general "hook" form in Theorem 3.20 of [BeRe87].

Let $\bar{\mu} = (1^{k+1})$, then the corresponding ideal $I_{\bar{\mu}} = Fe_{\bar{\mu}}$, where

$$e_{\bar{\mu}} = \sum_{\pi \in S_{k+1}} \text{sgn}(\pi)\pi.$$

Corollary 12.11. *Suppose $\bar{\mu} = (1^{k+1})$ and let $n \geq k + 1$. Then by the Branching Theorem, $\ker \varphi_{n,k}$ is the ideal generated in $F[S_n]$ by $I_{\bar{\mu}}$. Thus, $\ker \varphi_{n,k}$ is spanned over F by the elements*

$$\tau \left(\sum_{\pi \in S_{k+1}} \text{sgn}(\pi)\pi \right) \zeta, \quad \tau, \zeta \in S_n.$$

The second fact is a trace formula due to Kostant [Kos58] which is a generalization of a formula of I. Schur.

Remark 12.12. We have the identification $T^n(V)^* = T^n(V^*)$ via

$$\langle v_1 \otimes \cdots \otimes v_n, w_1 \otimes \cdots \otimes w_n \rangle = \prod_{i=1}^n \langle v_i, w_i \rangle.$$

Theorem 12.13. *Write $\sigma \in S_n$ as a product of disjoint cycles*

$$\sigma = (i_1, \ldots, i_a)(j_1, \ldots, j_b) \cdots (k_1, \ldots, k_c),$$

and let $A_1, \ldots, A_n \in \text{End}(V)$, so that

$$(A_1 \otimes \cdots \otimes A_n) \circ \varphi_{n,k}(\sigma) \in \text{End}(T^n(V)).$$

Then

$$\mathrm{tr}((A_1 \otimes \cdots \otimes A_n) \circ \varphi_{n,k}(\sigma))$$
$$= \mathrm{tr}(A_{i_1} \cdots A_{i_a}) \, \mathrm{tr}(A_{j_1} \cdots A_{j_b}) \cdots \mathrm{tr}(A_{k_1} \cdots A_{k_c}).$$

Here the trace on the left is applied to $\mathrm{End}(T^n(V))$, *while the trace on the right is applied to* $\mathrm{End}(V)$.

Proof. The proof is by employing rather elementary linear algebra to the set-up of Remark 12.8.

Step 1. First, applying *trace* to both sides of (12.1), we obtain

$$\mathrm{tr}((v_{j_1} \otimes w_{j_1}) \cdots (v_{j_s} \otimes w_{j_s})) = \langle v_{j_2}, w_{j_1} \rangle \cdots \langle v_{j_s}, w_{j_{s-1}} \rangle \langle v_{j_1}, w_{j_s} \rangle$$
$$= \prod_{i \in \{j_1, \dots, j_s\}} \langle v_{\sigma(i)}, w_i \rangle, \qquad (12.2)$$

for any cycle $\sigma = (j_1, \dots, j_s)$.

Step 2. If $c \in \mathrm{End}(V)$, then

$$\mathrm{tr}(c) = \sum_{\ell=1}^{k} \langle c e_\ell, \theta_\ell \rangle.$$

This is clear when $c = e_i \otimes \theta_j$, and then we extend it by linearity to all $c \in \mathrm{End}(V)$.

Step 3. Note that $\{e_{i_1} \otimes \cdots \otimes e_{i_n} \mid 1 \leq i_1, \dots, i_n \leq k\}$ is a basis of $T^n(V)$, and $\{\theta_{i_1} \otimes \cdots \otimes \theta_{i_n} \mid 1 \leq i_1, \dots, i_n \leq k\}$ its dual basis.

Thus, if $c \in \mathrm{End}(T^n(V))$, it follows that

$$\mathrm{tr}(c) = \sum_{1 \leq i_1, \dots, i_n \leq k} \langle c(e_{i_1} \otimes \cdots \otimes e_{i_n}), \theta_{i_1} \otimes \cdots \otimes \theta_{i_n} \rangle.$$

Step 4. We need a lemma.

Lemma 12.14. *Let* $\sigma \in S_n$, $A_i = v_i \otimes w_i \in V \otimes V^*$, $i = 1, \dots, n$, *and let* $c = (A_1 \otimes \cdots \otimes A_n) \circ \varphi_{n,k}(\sigma) \in \mathrm{End}(T^n(V))$. *Then*

$$\mathrm{tr}(c) = \prod_{i=1}^{n} \langle v_{\sigma(i)}, w_i \rangle.$$

Proof. By direct calculation,

$$c(e_{j_1} \otimes \cdots \otimes e_{j_n}) = \left(\prod_{i=1}^{n} \langle e_{j_{\sigma(i)}}, w_i \rangle\right)(v_1 \otimes \cdots \otimes v_n),$$

therefore

$$\langle c(e_{j_1} \otimes \cdots \otimes e_{j_n}), \theta_{j_1} \otimes \cdots \otimes \theta_{j_n} \rangle = \prod_{i=1}^{n} (\langle e_{j_{\sigma(i)}}, w_i \rangle \langle v_i, \theta_{j_i} \rangle).$$

Summing over all the k^n tuples $\{(j_1, \ldots, j_n) \mid 1 \leq j_i \leq k\}$ yields $\mathrm{tr}(c)$ on the left. We claim that the right-hand side is

$$\sum_{1 \leq j_i \leq k} \prod_{i=1}^{n} (\langle e_{j_{\sigma(i)}}, w_i \rangle \langle v_i, \theta_{j_i} \rangle) = \prod_{i=1}^{n} \langle v_{\sigma(i)}, w_i \rangle.$$

By linearity, we may assume $v_i = e_{r_i}$ and $w_i = \theta_{s_i}$, $i = 1, \ldots, n$. Since $\langle e_a, \theta_b \rangle = \delta_{a,b}$, it follows that

$$\prod_{i=1}^{n} (\langle e_{j_{\sigma(i)}}, w_i \rangle \langle v_i, \theta_{j_i} \rangle) = \prod_{i=1}^{n} (\langle e_{j_{\sigma(i)}}, \theta_{s_i} \rangle \langle e_{r_i}, \theta_{j_i} \rangle) = \prod_{i=1}^{n} \delta_{j_{\sigma(i)}, s_i} \delta_{r_i, j_i}.$$

The only nonzero summands occur when all $j_i = r_i$, in which case

$$\prod_{i=1}^{n} \delta_{j_{\sigma(i)}, s_i} \delta_{r_i, j_i} = \prod_{i=1}^{n} \delta_{r_{\sigma(i)}, s_i} = \prod_{i=1}^{n} \langle e_{r_{\sigma(i)}}, \theta_{s_i} \rangle = \prod_{i=1}^{n} \langle v_{\sigma(i)}, w_i \rangle.$$

Step 5. Completion of the proof of Theorem 12.13. By linearity, we may assume $A_i = v_i \otimes w_i$, $v_i \in V$, $w_i \in V^*$, $i = 1, \ldots, n$. Then, by Lemma 12.14,

$$\mathrm{tr}(c) = \prod_{i=1}^{n} \langle v_{\sigma(i)}, w_i \rangle.$$

Now $\sigma = (i_1, \ldots, i_a) \cdots (k_1, \ldots, k_c)$, hence

$$\prod_{i=1}^{n} \langle v_{\sigma(i)}, w_i \rangle = \left(\prod_{i \in \{i_1, \ldots, i_a\}} \langle v_{\sigma(i)}, w_i \rangle\right) \cdots \left(\prod_{i \in \{k_1, \ldots, k_c\}} \langle v_{\sigma(i)}, w_i \rangle\right),$$

which by (12.2) equals $\mathrm{tr}(A_{i_1} \cdots A_{i_a}) \cdots \mathrm{tr}(A_{k_1} \cdots A_{k_c})$, as desired. $\qquad \square$

12.1.2 Pure trace polynomials

Let us consider the pure trace polynomials, and then give, in Theorem 12.19, a complete description of the multilinear pure trace identities of the $k \times k$ matrices. The multilinear pure trace polynomials can be described as elements of the group algebras $F[S_n]$ as follows.

Definition 12.15. Write $\sigma \in S_n$ as a product of disjoint cycles

$$\sigma = (i_1, \ldots, i_a)(j_1, \ldots, j_b) \cdots (k_1, \ldots, k_c)$$

and denote

$$M_\sigma(x_1, \ldots, x_n) = \mathrm{Tr}(x_{i_1} \cdots x_{i_a}) \, \mathrm{Tr}(x_{j_1} \cdots x_{j_b}) \cdots \mathrm{Tr}(x_{k_1} \cdots x_{k_c}),$$

the corresponding pure trace monomial. Given $a = \sum_\sigma \alpha_\sigma \sigma \in F[S_n]$, denote the corresponding formal trace polynomial

$$f_a(x_1, \ldots, x_n) = \sum_\sigma \alpha_\sigma M_\sigma(x_1, \ldots, x_n).$$

Remark 12.16. Given a multilinear pure trace monomial $M(x_1, \ldots, x_n)$, there exists $\sigma \in S_n$ such that $M(x_1, \ldots, x_n) = M_\sigma(x_1, \ldots, x_n)$. Given a multilinear pure trace polynomial $f(x_1, \ldots, x_n)$, there exists $a \in F[S_n]$ such that $f(x_1, \ldots, x_n) = f_a(x_1, \ldots, x_n)$.

This vector-space map $a \mapsto f_a$ has the following property, which will be applied later.

Lemma 12.17. Let $a \in F[S_n]$ with corresponding pure trace polynomial f_a, and let $\eta \in S_n$. Then

$$f_{\eta a \eta^{-1}}(x_1, \ldots, x_n) = f_a(x_{\eta(1)}, \ldots, x_{\eta(n)}).$$

Proof. It is enough to check pure trace monomials. Let

$$\sigma = (i_1, \ldots, i_a) \cdots (k_1, \ldots, k_c),$$

so

$$M_\sigma(x_1, \ldots, x_n) = \mathrm{Tr}(x_{i_1} \cdots x_{i_a}) \cdots \mathrm{Tr}(x_{k_1} \cdots x_{k_c}).$$

Now $\eta \sigma \eta^{-1} = (\eta(i_1), \ldots, \eta(i_a)) \cdots (\eta(k_1), \ldots, \eta(k_c))$ and therefore,

$$M_{\eta \sigma \eta^{-1}}(x_1, \ldots, x_n) = \mathrm{Tr}(x_{\eta(i_1)} \cdots x_{\eta(i_a)}) \cdots \mathrm{Tr}(x_{\eta(k_1)} \cdots x_{\eta(k_c)}) =$$

$$= M_\sigma(x_{\eta(1)}, \ldots, x_{\eta(n)}),$$

as desired. $\square$

Corollary 12.18. *Let* $a \in F[S_n]$ *with* $f_a(x)$ *the corresponding trace polynomial, and let* $A_1, \ldots, A_n \in \mathrm{End}(V)$. *Then*

$$f_a(A_1, \ldots, A_n) = \mathrm{tr}((A_1 \otimes \cdots \otimes A_n) \circ \varphi_{n,k}(a)).$$

Proof. Follows at once from Theorem 12.13.   □

We now give a complete description of the multilinear trace identities of $M_k(F)$.

Theorem 12.19. *Let* $f(x) = f_a(x) = f_a(x_1, \ldots, x_n)$ *be a multilinear trace polynomial, where* $a \in F[S_n]$. *Then* $f_a(x)$ *is a trace identity of* $M_k(F)$ *if and only if*

$$a \in \bigoplus_{\lambda \vdash n,\ \lambda_1' \geq k+1} I_\lambda.$$

Proof. Follows easily from the fact that

$$\ker \varphi_{n,k} = \bigoplus_{\lambda \vdash n,\ \lambda_1' \geq k+1} I_\lambda.$$

Indeed, let $\bar{A} = A_1 \otimes \cdots \otimes A_n \in T^n(\mathrm{End}(V))$ $(\dim V = k)$. Now, $f_a(x)$ is a trace identity for $M_k(F) = \mathrm{End}(V)$ if and only if for all such $\bar{A}$, $f_a(\bar{A}) = 0$, and thus, by Corollary 12.18, if and only if $\mathrm{tr}(\bar{A} \circ \varphi_{n,k}(a)) = 0$ for all $\bar{A} \in T^n(\mathrm{End}(V)) \equiv \mathrm{End}(T^n(V))$.

By the non-degeneracy of the trace form, the above happens if and only if $\varphi_{n,k}(a) = 0$, namely $a \in \ker \varphi_{n,k}$. The assertion now follows from the above description of $\ker \varphi_{n,k}$.   □

12.1.3 Mixed trace polynomials

There is an obvious transition between pure and mixed trace identities of an algebra A with a trace, as follows. If $p(x_1, \ldots, x_n)$ is a mixed trace identity of A, then trivially $\mathrm{Tr}(x_0 \cdot p(x_1, \ldots, x_n))$ is a pure trace identity of A. When $A = M_k(F)$, the multilinear converse also holds: Let $f(x_0, x_1, \ldots, x_n)$ be a multilinear pure trace polynomial. Each monomial of f can be written as

$$\mathrm{Tr}(g_0) \mathrm{Tr}(g_1) \cdots \mathrm{Tr}(g_t)$$

for monomials $g_0, \ldots, g_t$, which we can rearrange so that x_0 appears in g_0. Furthermore, performing a cyclic permutation on g_0, we may assume g_0 starts with x_0. Writing $g_0 = x_0 g_0'$, we have

$$\mathrm{Tr}(g_0) \mathrm{Tr}(g_1) \ldots \mathrm{Tr}(g) = \mathrm{Tr}(x_0 g_0' \mathrm{Tr}(g_1) \cdots \mathrm{Tr}(g_t)).$$

Letting p be the sum of these $g'_0 \operatorname{Tr}(g_1) \cdots \operatorname{Tr}(g_t)$, a mixed trace polynomial, we see $f = \operatorname{Tr}(x_0 p)$. For example,

$$\operatorname{Tr}(x_2 x_0 x_3) \operatorname{Tr}(x_1 x_4) = \operatorname{Tr}(x_0 x_3 x_2) \operatorname{Tr}(x_1 x_4) = \operatorname{Tr}(x_0 x_3 x_2 \operatorname{Tr}(x_1 x_4)).$$

The non-degeneracy of the trace on $M_k(F)$ implies the following result.

Proposition 12.20. *Suppose* $f(x_0, x_1, \ldots, x_n) = \operatorname{Tr}(x_0 p(x_1, \ldots, x_n))$. *Then* $f(x_0, \ldots, x_n)$ *is a pure trace identity of* $M_k(F)$ *iff* $p(x_1, \ldots, x_n)$ *is a mixed trace identity of* $M_k(F)$.

We are ready for the main example of a trace identity of matrices.

Example 12.21.

(i) Let $\bar{\mu} = (1^{k+1})$ and

$$e_{\bar{\mu}} = \sum_{\pi \in S_{k+1}} \operatorname{sgn}(\pi) \pi,$$

with the corresponding pure trace polynomial

$$f_{e_{\bar{\mu}}} = f_{e_{\bar{\mu}}}(x_0, x_1, \ldots, x_k).$$

By Theorem 12.19, $f_{e_{\bar{\mu}}}$ is a trace identity of $M_k(F)$. Moreover, it is the multilinear pure trace identity of the lowest degree, and any other such identity of degree $k + 1$ is a scalar multiple of $f_{e_{\bar{\mu}}}$. Thus, $f_{e_{\bar{\mu}}}$ is called the *fundamental pure trace identity*.

(ii) Let $g_k(x)$ denote the trace identity of $M_k(F)$ given in Example 12.7, and take its multilinearization

$$\tilde{g}_k = \tilde{g}_k(x_1, \ldots, x_k).$$

Then $\tilde{g}_k$ is a multilinear mixed trace identity of $M_k(F)$. Hence,

$$\operatorname{Tr}(x_0 \tilde{g}_k(x_1, \ldots, x_k))$$

is a multilinear pure trace identity of $M_k(F)$; by (i), it is a scalar multiple of $f_{e_{\bar{\mu}}}$.

12.2   Finite Generation of Trace T-Ideals

In this section, we prove the fundamental theorem of trace identities, due independently to Razmyslov [Raz74b], Helling[Hel74], and Procesi [Pro76]. Note that the variables $x_0, \ldots, x_k$ are replaced by $x_1, \ldots, x_{k+1}$, to enable us to notate permutations more easily.

Theorem 12.22.

(i) *The T-ideal of the pure trace identities of the $k \times k$ matrices is generated by the single trace polynomial $f_{e_{\bar{\mu}}}$, where $\bar{\mu} = (1^{k+1})$ and*

$$e_{\bar{\mu}} = \sum_{\pi \in S_{k+1}} \operatorname{sgn}(\pi)\pi.$$

(ii) *The T-ideal of the mixed trace identities of the $k \times k$ matrices is generated by the two elements $f_{e_{\bar{\mu}}}$ and $\tilde{g}_k$ (cf. Example 12.21(iii)).*

Proof. We follow [Pro76].

(i). In view of the multilinearization, it suffices to prove that any multilinear pure trace identity of $M_k(F)$ is a consequence of $f_{e_{\bar{\mu}}}$. By Corollary 12.11, it suffices to prove this assertion for the pure trace polynomials that correspond to the elements

$$\tau e_{\bar{\mu}} \zeta = \tau \left(\sum_{\pi \in S_{k+1}} \operatorname{sgn}(\pi)\pi \right) \zeta, \quad \tau, \zeta \in S_n,$$

where $n \geq k + 1$. Now $\tau e_{\bar{\mu}} \zeta = \tau(e_{\bar{\mu}} \zeta \tau)\tau^{-1} = \tau(e_{\bar{\mu}} \eta)\tau^{-1}$ where $\eta = \zeta \tau$. Hence, we have the following reduction.

Reduction. It suffices to prove the assertion for the trace polynomials that correspond to the elements $e_{\bar{\mu}} \eta$, where $\eta \in S_n$ and $n \geq k + 1$.

The essence of the argument is the following lemma.

Lemma 12.23. *Let $n \geq k + 1$ and let $\eta \in S_n$. Then there exist permutations $\sigma \in S_{k+1}$ and $\gamma \in S_n$ such that $\eta = \sigma \cdot \gamma$, and each cycle of γ contains at most one of the elements $1, 2, \ldots, k + 1$.*

Proof. On the contrary, assume for example that both 1 and 2 appear in the same cycle of η:

$$\eta = (1, i_1, \ldots, i_r, 2, j_1, \ldots, j_s) \cdots .$$

Then

$$(1, 2) \cdot \eta = (1, i_1, \ldots, i_r)(2, j_1, \ldots, j_s) \cdots .$$

Continue by applying the same argument to $(1, 2) \cdot \eta$, and the proof of the lemma follows by induction. $\qquad \square$

Let us prove the theorem. For any $\sigma \in S_{k+1}$,

$$e_{\bar{\mu}}\sigma = \sum_{\pi \in S_{k+1}} \text{sgn}(\pi)\pi\sigma = \text{sgn}(\sigma)\sum_{\rho \in S_{k+1}} \text{sgn}(\rho)\rho = \pm e_{\bar{\mu}}.$$

In view of the lemma, it follows that $e_{\bar{\mu}} \cdot \eta = \pm e_{\bar{\mu}} \cdot \gamma$. Thus, it suffices to show that the trace polynomial corresponding to $e_{\bar{\mu}} \cdot \gamma$ is obtained from $f_{e_{\bar{\mu}}}$.

Consider $\pi \in S_{k+1}$ and $\pi\gamma$, with their corresponding trace monomials $M_\pi(x_1, \ldots, x_{k+1})$ and $M_{\pi\gamma}(x_1, \ldots, x_n)$, given that the cycle decomposition of γ is:

$$(1, i_2, \ldots, i_r)(2, j_2, \ldots, j_s) \cdots (k+1, \ell_2, \ldots, \ell_t)(a_1, \ldots, a_u) \cdots (b_1, \ldots, b_v).$$

A straightforward calculation shows that the cycle decomposition of $\pi\gamma$ is obtained from the cycle decomposition of π as follows:

First write the cycle decomposition of π; then replace 1 by the string $1, i_1, \ldots, i_r$, replace 2 by the string $2, j_1, \ldots, j_s$, $\cdots$, replace $k+1$ by the string $k+1, \ell_1, \ldots, \ell_t$, and finally multiply (from the right) by the remaining cycles $(a_1, \ldots, a_u) \cdots (b_1, \ldots, b_v)$.

It follows that

$$M_{\pi\gamma}(x_1, \ldots, x_n) = M_\pi(h_1, \ldots, h_{k+1}) \cdot \text{Tr}(x_{a_1} \cdots x_{a_u}) \cdots \text{Tr}(x_{b_1} \cdots x_{b_v})$$

where $h_1 = x_1 x_{i_2} \cdots x_{i_r}$, $h_2 = x_2 x_{j_2} \cdots x_{j_s}$, $\ldots$, $h_{k+1} = x_{k+1} x_{\ell_2} \cdots x_{\ell_t}$.

As a consequence, we deduce that the trace polynomial that corresponds to $e_{\bar{\mu}} \cdot \gamma$ is

$$F(h_1, \ldots, h_{k+1}) \cdot \text{Tr}(x_{a_1} \cdots x_{a_u}) \cdots \text{Tr}(x_{b_1} \cdots x_{b_v}),$$

and the proof of (i) of the theorem is now complete.

(ii). Let $h = h(x_1, \ldots, x_n)$ be a mixed trace identity of $M_k(F)$. Then $\text{Tr}(hx_{n+1})$ is a pure trace identity of $M_k(F)$. By (i), $\text{Tr}(hx_{n+1})$ is a linear combination of terms of the form $af_{e_{\bar{\mu}}}(h_1, \ldots, h_{k+1})$, where the h_is are ordinary monomials, and the as are pure trace monomials which behave like scalars. Since $\text{Tr}(hx_{n+1})$ is linear in x_{n+1}, we may assume x_{n+1} is of degree 1 in each such summand. Thus, we have the following two cases.

Case 1. x_{n+1} appears in a; then $a = \text{Tr}(hx_{n+1})$, so

$$af_{e_{\bar{\mu}}}(h_1, \ldots, h_{k+1}) = \text{Tr}(hf_{e_{\bar{\mu}}}(h_1, \ldots, h_{k+1})x_{n+1}).$$

Case 2. x_{n+1} appears in, say, h_{k+1} : $h_{k+1} = q_1 x_{n+1} q_2$. By applying cyclic permutations in the monomials of $\tilde{g}_k(h_1, \ldots, h_{k+1})$, it follows that

$$a\tilde{g}_k(h_1, \ldots, h_{k+1}) = \text{Tr}(aq_2\tilde{g}_k(h_1, \ldots, h_k)q_1 x_{n+1}).$$

It follows that $\mathrm{Tr}(hx_{n+1})$ is the following sum:

$$\mathrm{Tr}(h(x_1,\ldots,x_n)x_{n+1}) =$$

$$= Tr\left(\left[\sum a'\tilde{g}_k(h_1,\ldots,h_{k+1}) + \sum aq_2\tilde{g}_k(h_1,\ldots,h_k)q_1\right]x_{n+1}\right).$$

Part (ii) now follows, because of the nondegeneracy of the trace. $\qquad\square$

Noting by Example 12.21(ii) that f_{e_μ} is a consequence of $\tilde{g}_k$, we conclude that every trace identity of $M_k(F)$, and in particular every polynomial identity, is a formal consequence of the Hamilton-Cayley polynomial. See Exercise 1 for a direct proof of the Amitsur-Levitzki theorem from the Hamilton-Cayley polynomial.

12.3 Trace Codimensions

Having seen the benefits of trace identities, it makes sense to incorporate them into the S_n-representation theory. Suppose $\{X\}$ is infinite; let $T_n \subseteq T = T\{x\}$ denote the subspace of the multilinear pure trace polynomials in $x_1,\ldots,x_n$, and similarly let $T'_n \subseteq T' = T'\{x\}$ denote the subspace of the multilinear mixed trace polynomials.

Remark 12.24. The map $p(x_1,\ldots,x_n) \mapsto \mathrm{Tr}(x_0p(x_1,\ldots,x_n))$, yields an isomorphism $T'_n \approx T_{n+1}$, as we saw above.

The pure trace identities of $M_k(F)$ form a submodule $\mathrm{Tid}_{k,n} \subseteq T_n$. By Theorem 12.19, we have the $F[S_n]$-module isomorphism

$$T_n/\mathrm{Tid}_{k,n} \approx \bigoplus_{\lambda \vdash n,\ \lambda'_1 \le k} I_\lambda.$$

Call the S_n-character of $T_n/\mathrm{Tid}_{k,n}$ the *n-th pure trace cocharacter of* $M_k(F)$ and denote it $\chi_n^{Tr}(M_k(F))$. The preceding arguments show that $\chi_n^{Tr}(M_k(F))$ equals the $F[S_n]$-module $\bigoplus_{\lambda \vdash n,\ \lambda'_1 \le k} I_\lambda$ (under conjugation). It is well known that the character of I_λ under conjugation equals the character $\chi^\lambda \otimes \chi^\lambda$, where $\otimes$ indicates here the Kronecker (inner) product of characters. This proves the following result.

Theorem 12.25. *Notation as above,*

$$\chi_n^{Tr}(M_k(F)) = \bigoplus_{\lambda \vdash n,\ \lambda'_1 \le k} \chi^\lambda \otimes \chi^\lambda.$$

Definition 12.26. $t_n(M_k(F)) = \dim(T_n/\operatorname{Tid}_{k,n})$ is the *(pure) trace codimension* of $M_k(F)$.

Corollary 12.27. *Notation as above, we have*

$$t_n(M_k(F)) = \sum_{\lambda \vdash n,\ \lambda'_1 \leq k} (f^\lambda)^2,$$

where $f^\lambda = \deg \chi^\lambda$.

We can now prove a codimension result.

Lemma 12.28.
$$c_{n-1}(M_k(F)) \leq \sum_{\lambda \vdash n,\ \lambda'_1 \leq k} (f^\lambda)^2.$$

Proof. The assertion follows from the nondegeneracy of the trace form. Define $g : V_{n-1} \to T_n$ by

$$g(p(x_1, \ldots, x_{n-1})) = \operatorname{Tr}(p(x_1, \ldots, x_{n-1})x_n).$$

Clearly g is one-to-one. Now $p(x_1, \ldots, x_{n-1})$ is a (non) identity of $M_k(F)$ if and only if $g(p(x_1, \ldots, x_{n-1}))$ is a (non) trace identity of $M_k(F)$, so the assertion follows. $\qquad\square$

Remark 12.29.

(i) Applying Regev's comparison of the cocharacter theories to some deep results of Formanek [For84], it could be shown that when n goes to infinity, $c_{n-1}(M_k(F))$ is asymptotically equal to $t_n(M_k(F))$:

$$c_{n-1}(M_k(F)) \sim t_n(M_k(F)).$$

(ii) It can be shown that as n goes to infinity,

$$t_n(M_k(F)) \sim c\left(\frac{1}{n}\right)^{(k^2-1)/2} k^{2n},$$

where c is the constant

$$c = \left(\frac{1}{\sqrt{2\pi}}\right)^{k-1} \left(\frac{1}{2}\right)^{(k^2-1)/2} 1!2!s(k-1)!k^{(k^2+4)/2}.$$

Combining (i) and (ii) yields the precise asymptotics of the codimensions $c_n(M_k(F))$.

Remark 12.30. This theory has been formulated and proved in exactly the same way for algebras without 1, cf. Exercise 3.

Although the theory has been described here in characteristic 0, the characteristic p version has been developed by Donkin [Do94] and Zubkov [Zu96].

Let us conclude this chapter by noting that Kemer [Kem96] has linked the codimension theory to the n-good words in the sense of Definition 5.25. Namely, let A be the algebra of m generic $n \times n$ matrices, and let c_m denote the codimension of the space of multilinear trace polynomials of length m. Kemer proved c_m is also the number of n-good words among all words of length m. Hence, using Formanek's methods in [For84], one could obtain explicit formulas for the n-good words. This ties in Shirshov's theory with the representation theory.

Chapter 13

Exercises

Chapter 1

1. Prove Hall's identity (Example 1.6(iii).) (Hint: $\text{tr}[a, b] = 0$, for any $a, b \in M_2(F)$, so apply the characteristic polynomial of $[a, b]$.)

2. If $f(x_1, \ldots, x_m)$ is homogeneous of degree d_i in each indeterminate x_i, then the multilinearization of f is the multilinear part of

$$f(x_{1,1} + \cdots + x_{1,d_1}, \ldots, x_{m,1} + \cdots + x_{m,d_m}), \qquad (13.1)$$

where $x_{i,j}$ denote distinct indeterminates. Obtain a suitable generalization for arbitrary polynomials (but take care which degrees appear in which monomials).

3. Although the multilinear part of (13.1) is an identity whenever f is an identity, find an example for which a homogeneous component of (13.1) need not be an identity even when f is an identity. (Hint: Take a finite field.)

4. Any finite ring satisfies an identity $x^n - x^m$ for suitable $n, m \in \mathbb{N}$. (Hint: The powers of any element a eventually repeat.) Multilinearizing, show any finite ring satisfies a symmetric identity.

5. For the purposes of this exercise, "algebraic" means with coefficients in $\mathbb{Z}$. Define the *algebraic radical* of an algebra A to be $\{r \in A : ra$ is algebraic, for all $a \in A.\}$ This contains the nilradical $\text{Nil}(A)$, equality holding when A contains a non-algebraic subfield. (Hint: Passing to $A/\text{Nil}(A)$, one may assume $\text{Nil}(A) = 0$; in particular, A is semiprime. Then, passing to prime homomorphic images, one may assume A is prime; taking central fractions, one may assume A is simple. But any ideal must contain the center, which is not algebraic.)

309

6. Suppose A is a PI-algebra which is affine over a finite field. Definitions as in Exercise 5, show the algebraic radical of A satisfies the identity $(x^m - x^n)^k$ for suitable k and $m > n$ in $\mathbb{N}$. (Hint: Since Nil(A) is nilpotent, one may follow the steps of the previous exercise to assume A is prime; then either A has algebraic radical 0 or A is a matrix algebra over a field; appeal to Exercise 4.)

7. If the algebraic radical of an algebra A is 0, then the homogeneous components of any identity of A are identities of A.

8. Prove Proposition 1.17: If F is an infinite field, then every identity of an algebra A is a consequence of homogeneous identities. (Hint: Write $f = \sum f_u \in \mathrm{id}(A)$ with each f_u homogeneous of degree u in x_i. Then for all $a \in A$, we have

$$
\begin{aligned}
0 = f(a_1, \ldots, \alpha a_i, \ldots, a_m) \\
= \sum f_u(a_1, \ldots, \alpha a_i, \ldots, a_m) = \sum \alpha^u f_u(a_1, \ldots, a_m);
\end{aligned}
\tag{13.2}
$$

view this as a system of linear equations in the $f_u(a_1, \ldots, a_m)$ whose coefficients are α^u; the matrix of the coefficients is the famous Vandermonde matrix, which is nonsingular, so the system of linear equations has the unique solution $f_u(a_1, \ldots, a_m) = 0$ for each u. Thus, we may assume that f is homogeneous in the i indeterminate, and repeat the procedure for each i.)

9. Here is a refinement of the linearization procedure which is useful in characteristic p. Suppose char(F) $= p$. If f is homogeneous in x_i, of degree n_i that is *not* a p-power, then f can be recovered from a homogeneous component of a partial linearization at x_i. (Hint: Take k such that $\binom{n_i}{k}$ is not divisible by p, and take the $(k, n_i - k)$ homogeneous part of the linearization.)

10. Suppose char(F) $= p$. If $n_i = \deg_i f$ is a p-power, then f cannot be recovered from a homogeneous component of a partial linearization at x_i. (Hint: $\binom{n_i}{k}$ is divisible by p for each $0 < k < p_i$, so the procedure only yields 0.)

11. (Amitsur.) $s_{n+1}(y, xy, x^2 y, \ldots, x^n y) \in \mathrm{id}(M_n(F))$. (Hint: Same argument as in Lemma 1.51.) One can cancel the y on the right and get a two-variable identity of degree $\frac{n^2 + 3n}{2}$.

12. In the spirit of Remark 1.41, use the substitution $e_{11}, e_{12}, e_{22}, \ldots$ to show that $M_n(F)$ cannot satisfy any identity of degree $< 2n$.

13. Show that the only multilinear PI of $M_n(\mathbb{Z})$ of degree $2n$ is $\pm s_{2n}$. (Hint: Substitutions of matrix units as in Remark 1.41.)

14. Suppose $n = n_1 + \cdots + n_t$. Given $\pi \in S_n$ with

$$\pi(1) < \pi(2) < \cdots < \pi(n_1), \quad \pi(n_1+1) < \pi(n_1+2) < \cdots < \pi(n_1+n_2),$$

and so on. Write

$$c_{n_1;\pi} = c_{n_1}(x_{\pi(1)}, \ldots, x_{\pi(n_1)}; y_1, \ldots, y_{n_1});$$
$$c_{n_j;\pi} =$$
$$c_{n_j}(x_{\pi(n_1+\cdots+n_{j-1}+1)}, \ldots, x_{\pi(n_1+\cdots+n_j)}; y_{n_1+\cdots+n_{j-1}+1}, \ldots, y_{n_1+\cdots+n_j})$$
$$\tag{13.3}$$

for each $j > 1$. Noting that the alternator of

$$c_{n_1;\pi} c_{n_2;\pi} \cdots c_{n_t;\pi}$$

is $\mathrm{sgn}(\pi)n_1! \ldots n_t! c_n(x_{\pi(1)}, \ldots, x_{\pi(n)}; y_1, \ldots, y_n)$, show

$$\sum_{\pi \in S_n} \mathrm{sgn}(\pi) c_{n_1;\pi} c_{n_2;\pi} \cdots c_{n_t;\pi} = k c_n(x_1, \ldots, x_n; y_1, \ldots, y_n)$$

where $k > 0$. Obtain the analogous formula for S_n, by specializing each y_i to 1. This observation was critical to Razmyslov's proof of the Amitsur-Levitzki theorem (cf. Exercise 12.1) and also is needed in our verification of Capelli identities for affine algebras (although the notation in Chapter 4 is more user-friendly).

15. The free algebra $A = C\{x_1, \ldots, x_\ell\}$ can be $\mathbb{Z}^{(\ell)}$-graded, by defining the *multi-degree* of a monomial $f \in A$ as $(d_1, \ldots, d_\ell)$, where $d_i = \deg_i f$. Compare this with Example 1.31.

16. Show that Example 1.34 (iii) is a special case of (ii), by defining an appropriate function $d : \{1, \ldots, n\} \to \mathbb{Z}$, and define

$$A_m = \{C e_{ij} : d(i) - d(j) = m\}.$$

17. Given a commutative ring C and a C-module V, write $V^{\otimes 0}$ for C and $V^{\otimes 1}$ for V, and, for $i \geq 2$, write $V^{\otimes i}$ for $V \otimes_C \cdots \otimes_C V$, with i tensor factors. Define the *tensor algebra* $T(V) = \oplus_{i \geq 0} V^{\otimes i}$, endowed with the natural multiplication $V^{\otimes i} \times V^{\otimes j} \to V^{\otimes(i+j)}$ given by $(a, b) \mapsto a \otimes b$; show the Grassman algebra $G(V) = T(V)/I$ where $I = \langle a^2 : a \in V \rangle$.

18. If $V_m = \sum_{i=1}^{m} Ce_i$, then any product of $m+1$ elements of V_m is 0 in $G(V)$. (Hint: This can be checked on the generators e_i; in any product $w = e_{j_1} \ldots e_{j_{m+1}}$ some e_i repeats, so $w = 0$, by Remark 1.38(ii).) It follows that every element of V is nilpotent in $G(V)$. The same argument shows that every element of $G(V)$ with constant term 0 is nilpotent, and thus, the set of such elements comprises the unique maximal ideal of $G(V)$.)

19. Verify the assertion in the paragraph after Lemma 1.51, by checking matrix units.

20. Suppose $T \in A = M_n(F)$, and let $\alpha = \text{tr}(T)$ and $\beta = \det(T)$. Then, viewing T in $\text{End}_F A$, show now that the trace is $n\alpha$ and determinant is β^n. How does this affect Equation(1.19)?

21. (Razmyslov-Bergman-Amitsur) Verify that

$$\text{tr}(w_1)\,\text{tr}(w_2) f(a_1, \ldots, a_t, r_1, \ldots, r_m)$$

$$= \sum_{k=1}^{t} f(a_1, \ldots, a_{k-1}, w_1 a_k w_2, a_{k+1}, \ldots, a_t, r_1, \ldots, r_m)$$

(13.4)

holds identically in $M_n(C)$. (Hint: Consider the transformation $a \mapsto tr(w_1)tr(w_2)$.) This observation arises naturally when we consider

$$\text{End}_F A = A \otimes A^{\text{op}},$$

and also leads to the construction of a central polynomial.

22. Any subalgebra of an algebra with ACC(ann) satisfies ACC(ann). (Hint: Consider the intersection of a chain of left annihilators with the smaller algebra.)

23. Any semiprime algebra with ACC(ann) has a finite number of minimal prime ideals, and their intersection is 0. (Hint: These are the maximal annihilator ideals.)

24. Suppose M is generated by n elements as an H-module, for H commutative. Then $\text{End}_H M$ is a homomorphic image of a subalgebra of $M_n(H)$, so $\text{End}_H M$ satisfies all the identities of $M_n(H)$. Compare with Remark 1.70.

25. (Strengthened version of Proposition 1.76.) If C is Noetherian and $A = C\{a_1, \ldots, a_\ell\}$ is affine and representable in the weak sense of Remark 1.70, then $A \subseteq M_n(K)$ for a suitable commutative affine (thus Noetherian) C-algebra K.

26. Any representable algebra over an infinite field is PI-equivalent to a finite dimensional algebra over a (perhaps larger) field. Contrast with the example

$$\begin{pmatrix} \mathbb{Z}_p & \mathbb{Z}_p[x,y] \\ 0 & \mathbb{Z}_p[x] \end{pmatrix}, \tag{13.5}$$

which satisfies the identity $(x^p - x)[y,z]$.

Chapter 2

1. If all subwords of w of length $|u|$ are cyclically conjugate to u, then w is periodic with period u. (Hint: Start with the initial subword of w of length $|u|$; we may assume this is u. Then the subword of length $|u|$ starting with the second position must be $\delta(u)$; continue letter by letter.)

Uniqueness of the period

2. If $w = vu^k r$ with $|v| < |u|$ and u is not periodic, then any appearance of u as a subword of w must be in one of the k obvious positions. In other words, if $w = v'ur'$, then $|v'| = |v| + i|u|$ for some integer i. (Hint: Otherwise, $u = u'u''$ where one of the obvious occurrences of u ends in u' or begins with u''. But then apply Lemma 2.61 to conclude that u is periodic, a contradiction.)

3. If w is quasiperiodic with period u and has a subword v^2 with $|v| > |u|$, then v is cyclically conjugate to a power of u. (Hint: Cyclically conjugating if necessary, write $v = u^j v'$ with j maximal possible. Then $v^2 = u^j v' u^j v'$. Since w is quasiperiodic, some

$$u^i u' = w = w' v^2 w'' = w' u^j v' u^j v' w''.$$

Then $w' = u^k u''$ where u'' is an initial subword of u, and matching parts shows $u''u = uu''$.)

4. If quasiperiodic words w_1, w_2 with respective periods u_1, u_2 of lengths d_1, d_2 have a common subword v of length $d_1 + d_2$, then u_1 is cyclically conjugate to u_2. (Hint: Assume $d_1 \geq d_2$. Replacing u_1 by a cyclic conjugate, write $v = u_1 u_1'$, where $|u_1'| = d_2$. Since v starts with a cyclic conjugate $\tilde{u}_2$ of u_2, one can write $u_1 = \tilde{u}_2^k u_1''$, for $|u_1''| < d_2$, so v starts with $\tilde{u}_2^k u_1'' \tilde{u}_2$; show $u_1'' = \emptyset$.)

5. (Ambiguity in the periodicity.) The periodic word 12111211 can also be viewed as the initial subword of a periodic word of periodicity 7.

6. Write $v \subset w$ to denote that v is a subword of w.

   (i) If two subwords v_1 and v_2 of u^∞ have the same initial subword of length $|u|$, then one of them is a subword of the other. Put another way, if $|c| \geq |u|$ and d_1 and d_2 are lexicographically comparable words, then either cd_1 or cd_2 is not a subword of u^∞.

   (ii) Occurrences of any subword v of length $\geq |u|$ in u^∞ differ by a multiple of $|u|$.

   (iii) If $v^2 \subset u^\infty$ with $|v| \geq |u|$, then v is cyclically conjugate to a power of u.

   (iv) Suppose $h = wvw' \subset u^\infty$, with v cyclically conjugate to u. Then, for all $k \geq 0$, $wv^k w' \subset u^\infty$. In particular, $ww' \subset u^\infty$ and $wv^2 w' \subset u^\infty$.

Better bounds for Shirshov's Lemma

By the d-*type* of a word w, we mean the initial subword of length d, unless d is the preperiodicity of w, in which case the d-type is w itself.

7. The number of d-types of words that do not contain the k-th power of a word with length $\leq d$, is not greater than $\ell^{d+1}d^2 k$. (Hint: The same counting argument as Proposition 2.73(iii).)

8. Show $\beta(\ell, k, d) \leq \ell^d(\ell-1)d^2 k$. (Hint: To establish d-decomposability, it is enough to prove the repetition of some minimal d-type. But there are no triple overlaps of the same d-type. Hence, we do not need to partition w into subwords of length dk to establish the repetition of d-types of subwords. Overlaps of the same type can be only double, and the number of required subwords has thereby been divided by dk.)

9. A d-critical word w is *minimal* if each terminal subword is preperiodic of preperiodicity d. Suppose w is a minimal d-critical word. Show:

   (i) The terminal subword of w of length $|w| - 1$ is quasiperiodic of periodicity exactly d.

   (ii) $d + 1 \leq |w| \leq 2d$.

   (iii) The number of such words $\leq d(\ell - 1)\ell^d$.

   (Hint: (i) The initial and terminal subwords both are preperiodic of periodicity d. But the only way for this to happen is for the terminal subword to be quasiperiodic (since otherwise cutting off the last letter does not change the periodicity).)

10. A d-critical word w will be called a d-*end* if its initial subword of length $|w| - 1$ is quasiperiodic of periodicity $\leq d$. If w is a minimal d-critical word with the same property, then w will be called a *minimal d-end*. Suppose $|w| > 2\ell^{d+1}d^2 k$, and w is not d-decomposable. Then:

   (i) Some subword in fewer than d letters repeats k times in succession. Thus, the estimate in Corollary 2.74 can be lowered to $2\ell^{d+1}d^2 k$.

   (ii) w contains fewer than $2d\ell\ell^{d+1}$ subwords that are d-ends.

   (iii) w can be decomposed into $d\ell\ell^{d+1}$ preperiodic subwords of preperiodicity $\leq d$.

   (iv) In conclusion, w has the form

$$w = v_0\omega_0 v_1\omega_1 \ldots \omega_{r-1}v_r,$$

   where ω_i are quasiperiodic words of periodicity $\leq d$ and of length $\geq 2d$ each, whereas each v_i does not contain such subwords, and the product of ω_i with the first letter of v_i is not preperiodic of preperiodicity d. Hence, $r < d\ell\ell^{d+1}$ and $\sum |v_i| < d\ell\ell^{d+1}$.

11. (Improved bound for Shirshov's height function.) The height function $\mu(\ell, d) < 3d\ell\ell^{d+1}$. (Hint: Write $\omega_i = u_i^{k_i}s_i$, where s_i is an initial subword of ω_i and $|u_{k_i}| \leq d$.) A closer study of d-ends allows one to improve the bound to $\mu(d, \ell) < 2d\ell\ell^{d+1}$.

12. As in Definition 2.76, given an affine algebra A without 1, a right A-module M and a hyperword h, define when $Mh(A) = 0$. Prove: If $MA^k \neq 0$, $\forall k$, for some right module M, then the set of all right hyperwords h, such that $Mh \neq 0$, has maximal and minimal hyperwords.

13. (Counterexample to a naive converse of Shirshov's Height Theorem.) The algebra $A = \mathbb{C}[\lambda, \lambda^{-1}]$ has the property that for any homomorphism sending λ to an algebraic element, the image of A is f.d., but nevertheless $\{\lambda\}$ is not a Shirshov base.

14. The Capelli identity c_d satisfies a stronger property than the sparse property:

   Given any polynomial $f(x_1, \ldots, x_d; y_1, \ldots, y_m)$ and any substitutions $\bar{x}_1, \ldots, \bar{x}_d, \bar{y}_1, \ldots, \bar{y}_m$ in A, one can rewrite

$$\sum_{\sigma \in S_d} \operatorname{sgn}(\sigma) f(\bar{x}_{\sigma(1)}, \ldots, \bar{x}_{\sigma(d)}, \bar{y}_1 \ldots, \bar{y}_m)$$

as

$$\sum_i \bar{w}_{i0} c_d(\bar{x}_{\sigma(1)}, \ldots, \bar{x}_{\sigma(1)}, \bar{w}_{i1} \ldots, \bar{w}_{im}), \tag{13.6}$$

for suitable words $\bar{w}_j$ in the $\bar{x}$ and $\bar{y}$. (Hint: First, assume f is a monomial $h = w_0 x_1 w_1 \ldots x_d w_d$ for suitable words w_j. But then

$$\sum_{\sigma \in S_d} \mathrm{sgn}(\sigma) h(\bar{x}_{\sigma(1)}, \ldots, \bar{x}_{\sigma(d)}, \bar{y}_1 \ldots, \bar{y}_m)$$
$$= \bar{w}_0 c_d(\bar{x}_{\sigma(1)}, \ldots, \bar{x}_{\sigma(d)}, \bar{w}_1 \ldots, \bar{w}_d), \tag{13.7}$$

as desired.

In general, write $f = \sum \alpha_i h_i$, and perform a similar computation.)

Other nil implies nilpotent results

15. (Schelter.) Suppose A is any prime affine PI-algebra over a commutative Noetherian ring, and $I \lhd A$. A/I has a finite set of prime ideals whose intersection is nilpotent, and any nil subalgebra N/I of A/I is nilpotent. (Hint: Otherwise, if A were a counterexample, there would be a prime ideal $P \lhd A$ maximal with respect to A/P also being a counterexample. Replacing A by A/P, assume the assertion holds in A/P for every nonzero prime ideal P of A. Pass to the characteristic closure $\hat{A}$. Take I_A as in Proposition 2.31. $I_A I$ is also an ideal of $\hat{A}$, which is Noetherian, so by Theorem 1.97, has a finite set of prime ideals $\hat{P}_1, \ldots, \hat{P}_m$ minimal over $I_A I$, such that $\hat{P}_1 \cdots \hat{P}_m \subseteq I_A I$. Let $P_i = A \cap \hat{P}_i$, a prime ideal of A. Apply the theorem to A/P_i and its ideal $(I + P_i)/P_i$, and thus, there are a finite set of prime ideals $Q_{i1}/(I + P_i), \ldots, Q_{it}/(I + P_i)$ in $A/(I + P_i)$ whose intersection is nilpotent, i.e.,

$$(Q_{i1} \cap \cdots \cap Q_{it})^k \subseteq I + P_i.$$

But then, $\prod_{i=1}^{m} (Q_{i1} \cap \cdots \cap Q_{it})^k \subseteq I + P_1 \ldots P_m \subseteq I$.)

16. If A satisfies the identities of $n \times n$ matrices and is affine over a Noetherian commutative ring, then any nil subalgebra N is nilpotent. (Hint: The relatively free algebra of $\mathrm{id}(M_n(A))$ is the algebra of generic matrices, which is a prime algebra. Apply Exercise 15.)

17. If A is affine and integral over a Noetherian subring of its center, then any nil ideal of A is nilpotent. (Hint: Shirshov's Height Theorem 2.3(ii).)

18. If A is affine PI over a commutative Noetherian ring C and satisfies a Capelli identity, then every nil subalgebra of A is nilpotent. (Hint: Apply Exercise 17.)

19. If $A = F\{a_1, \ldots, a_\ell\}$ is an algebra without 1, and $\mathrm{id}(A) \supset \mathrm{id}(M_n(F))$, and each word of deg $\leq n$ is nilpotent, then A is nilpotent. (Hint: The lower nilradical is nilpotent, so one can factor it out and assume A is semiprime.)

Pumping

20. (Pumping words.) Suppose the word

$$w = v_0 c_1 v_1 c_2 v_2 \ldots c_m v_m$$

where the c_i are powers of x_ℓ, and x_ℓ does not appear in the v_j. If d of the v_j have length $\geq d$, then w is d-decomposable with respect to the lexicographic order assigning value 0 to each letter of v_j. (Hint: By induction on the order of w. Discarding the initial segment v_0, assume w starts with c_1. Assuming on the contrary that $v_{j_1}, \ldots, v_{j_d}$ have length $\geq d$, write $w = c'_1 w_1 c'_2 w_2 \ldots w_d c'_{d+1} \ldots$ where each w_u ends with v_{j_u}, and c'_i start with c_i. Write $w_j = w'_j w''_j$, where w''_j has length j. Letting $a_j = w''_j c'_{i_j} w'_{j+1}$, note that $a_j > a_{j+1}$ for $1 \leq j \leq d - 1$. Thus, w is d-decomposable.)

21. By pumping, use an identity of degree d to rewrite a word as a linear combination of words in which all but at most $d - 1$ of the words separating the occurrences of x_ℓ have length $< d$.

Words on hemirings

The following set of exercises, following I. Bogdanov [Bog01] and Belov, enables us to put Shirshov's theory in a general context, applicable to automata. A *generalized hemiring* is an algebra $(S, +, \cdot)$ with the following conditions: $(S, +)$ and $(S, \cdot)$ are semigroups; a semigroup $(S, +)$ has a neutral element 0; multiplication is distributive on both sides over addition; and $0S = S0 = 0$. A generalized hemiring is called a *hemiring* if addition is commutative. (A *semiring* usually is assumed to have a multiplicative unit.) A *(generalized) semialgebra* over a commutative associative semiring Ψ is defined as usual, but not assumed to have a multiplicative unit.

22. The axiom $0S = S0 = 0$ does not follow from the other axioms, even if addition is commutative; however, it does follow for any nilpotent element. (Hint: For any semigroup $(S, +)$, adjoin a nonzero

idempotent a satisfying $xy = a$ for any $x, y \in S$. If $x^n = 0$, then $0 = x \cdot x^{n-1} = x(x^{n-1} + 0) = x^n + x0 = x0$.)

23. There exists a generalized hemiring S with non-commutative addition. Moreover, $(S^n, +)$ is not necessarily commutative. (Hint: Consider any semigroup $(S, \cdot)$; define addition by $x + y = y$, and adjoin 0.)

24. If $(S, +)$ is a group, then $(S^2, +)$ is commutative.

25. An ideal I of a hemiring S is the kernel of some congruence relation iff the condition $a, a + b \in I \Rightarrow b \in I$ is satisfied. Such ideals are called *subtractive*. (Hint: The desired congruence relation is given by $x\theta_I y \iff \exists a, b \in I : x + a = y + b$.) θ_I is called the *Bourne relation*.

26. Suppose C is a commutative and associative ring, and $\Psi \subseteq C$ is a semiring satisfying $\Psi - \Psi = C$. Then the (ℓ-generated) free Ψ-semialgebra $\mathcal{F}_\Psi$ is embedded into an (ℓ-generated) free C-algebra $\mathcal{F}_C$ in a straightforward way and $\mathcal{F}_C = \mathcal{F}_\Psi - \mathcal{F}_\Psi = \mathcal{F}_\Psi - n\mathcal{F}_\Psi$ for any integer $n \geq 1$. Moreover, any subtractive ideal J of $\mathcal{F}_\Psi$ has the form $J = I \cap \mathcal{F}_\Psi$ for some ideal I of $\mathcal{F}_C$.

The following exercises show that long words in a nil hemiring have the same behavior as in a nil ring. By *negative*, we mean "additive inverse."

27. If S is a hemiring satisfying the identity $x^n = 0$, then S^n is a ring (without 1). (Hint: Expanding the brackets in $(s_1 + \cdots + s_n)^n = 0$, we obtain that $s_1 \cdots s_n$ has a negative.)

28. If $(S, +)$ is a (not necessary commutative) monoid in which

$$s_1 + s_2 + \cdots + s_t = s_t + s_{t-1} + \cdots + s_1 = 0,$$

then all the s_i have negatives. (Hint: $s_1 + \cdots + s_{t-1} = s_{t-1} + \cdots + s_1$, since right and left negatives coincide if both exist. Conclude by induction.)

29. If S is a generalized hemiring satisfying the identity $x^n = 0$, then any monomial of the form $s_1 \cdots s_n$ has a negative. (Hint: Expand all brackets in $(s_1 + \cdots + s_n)^n = (s_n + \cdots + s_1)^n = 0$, and apply Exercise 28.)

30. Under the same assumptions, S^n is a ring with commutative addition. (Hint: If $a, b \in S$ and $a', b' \in S^{n-1}$, then

$$ab' + aa' + bb' + ba' = (a + b)(b' + a') = ab' + bb' + aa' + ba'.$$

Add the negatives of ab' from the left and of ba' from the right.)

Suppose $\mathcal{F}_C$ and $\mathcal{F}_\Psi$ are the free C-algebra and free Ψ-semialgebra introduced above.

31. Denote by I_n and I_n^+ the ideals in $\mathcal{F}_C$ generated by $\{r^n : r \in \mathcal{F}_C\}$ and $\{r^n : r \in \mathcal{F}_\Psi\}$, respectively. Then $I = I^+$. (Hint: Denote by h the determinant of the Vandermonde matrix generated by powers of $0, 1, \ldots, n$. If $(x + y)^n = f_0(x, y) + \cdots + f_n(x, y)$ is the expansion of $(x+y)^n$ into a sum of homogeneous components, then $h f_i(x, y) \in I_n^+$, implying $(x - hy)^n \in I_n^+$.)

32. Suppose d_n is a natural number such that each (ℓ-generated) C-algebra satisfying the identity $x^n = 0$ is nilpotent of degree d_n. Then each (ℓ-generated) generalized Ψ-semialgebra satisfying the identity $x^n = 0$ is also nilpotent of degree d_n. (Hint: Suppose S is such a generalized semialgebra. There exists an obvious morphism of semialgebras $\psi : \mathcal{F}_\Psi^n \to S^n$ induced by $x_i \to a_i$, and $I_n \cap \mathcal{F}_\Psi = I_n^+ \cap \mathcal{F}_\Psi \subseteq \psi^{-1}(0)$, since $I_n^+ \cap \mathcal{F}_\Psi$ is the subtractive ideal generated by $\{r^n : r \in \mathcal{F}_\Psi\}$.)

Chapter 3

1. Amitsur's original proof of Theorem 3.38: All values of s_n are nilpotent. Index copies of A according to the n-tuples of its elements, and take the direct product, which is PI-equivalent to A, and use a diagonalization argument.

2. Define the *ring of generic matrices* $\mathbb{Z}\{Y\}_n$ in analogy to Definition 3.22. Show that if $f(y_1, \ldots, y_m) = 0$, then $f(x_1, \ldots, x_m) \in \text{id}(M_n(C))$ for every commutative algebra C. (Hint: As in Remark 3.23.)

3. $\text{id}(M_n(\mathbb{Z})) \subseteq \text{id}(M_n(C))$ for any commutative algebra C. (Hint:

$$M_n(\mathbb{Z}[\Lambda]) \cong M_n(\mathbb{Z}) \otimes_\mathbb{Z} \mathbb{Z}[\Lambda],$$

so $\text{id}(M_n(\mathbb{Z})) = \text{id}(M_n(\mathbb{Z}[\Lambda]))$. Conclude with Exercise 2.)

4. Prove Theorem 3.40. (Hint: Partial linearization gives the identity $\sum_{i=0}^{n-1} x^i y x^{n-i-1}$, and thus,

$$\sum_{i,j} x^i (z y^j) x^{n-i-1} y^{n-j-1},$$

which is $\sum_i x^i z \sum_j y^j x^{n-i-1} y^{n-j-1}$, and thus, $\sum x^{n-1} z y^{n-1}$. If N is the ideal of A generated by all a^{n-1}, then A/N is nilpotent by induction, but $NAN = 0$, so conclude A is nilpotent.) Also see Exercise 12.4.

5. In any characteristic $p \neq 2$, the T-ideal of the Grassman algebra G over an infinite dimensional vector space is generated by the Grassman identity $[[x_1, x_2], x_3]$. (Hint: Applying Proposition 3.36 to Theorem 3.45, reduce any element of $\mathrm{id}(G)$ to the form

$$ f = \sum \alpha_u [x_{i_1}, x_{i_2}] \ldots [x_{i_{2k-1}}, x_{i_{2k}}] x_1^{u_1} \ldots x_m^{u_m} $$

where $i_1 < i_2 < \ldots i_{2k}$. Let f_1 be the sum of terms in which $i_1 = 1$. Then $f = f_1 + f_2$ where x_1 does not appear in any of the commutators used in f_2. Taking u to be the largest value for u_1, and specializing $x_1 \mapsto e_1 e_2 + \cdots + e_{2u-1} e_{2u}$, a central element of G, show any specialization of $x_2, \ldots x_m$ sends $f_1 \mapsto 0$, implying $f_2 \in \mathrm{id}(G)$, so $f_1 = f - f_2 \in \mathrm{id}(G)$. By induction on the number of monomials in f, assume $f = f_1$ or $f = f_2$, and continuing in this way, assume

$$ f = [x_{i_1}, x_{i_2}] \ldots [x_{i_{2k-1}}, x_{i_{2k}}] \sum \alpha_u x_1^{u_1} \ldots x_m^{u_m}, $$

i.e., the same $i_1, i_2, \ldots, i_{2k}$ appears in each term.

If $f \neq 0$, one can obtain the contradiction $f \notin \mathrm{id}(G)$. Indeed, specializing x_j to 1 for all $j \neq i_1, \ldots, i_{2k}$ and renumbering, assume

$$ f = [x_1, x_2] \ldots [x_{2k-1}, x_{2k}] \sum \alpha_u x_1^{u_1} \ldots x_m^{u_m}. $$

Now specialize x_i to a sum of one odd and $u_i - 1$ even base elements. Then, specialize to the product of the commutators of the odd parts; f has reverted to the even part, and the usual Vandermonde argument shows that this product is nonzero, a contradiction. Thus, $f = 0$, as desired.) The result also holds in characteristic 2, as shown in Chapter 7.

6. The standard identity s_{2n} does not generate $\mathrm{id}(M_n(F))$, for any $n \geq 2$. (Hint: Confront Exercise 1.11 with the smallest degree obtained by substituting monomials in x, y into s_4.)

7. ([Raz74c], [Dr81c]) (Very difficult) $\mathrm{id}(M_2(\mathbb{Q}))$ is based by s_4 and Hall's identity (cf. Example 1.7). This is very computational, and is done in several stages, one of which requires passing to the theory of Lie identities. First Razmyslov obtained a slightly longer list for a

base of the Lie identities of the Lie algebra of upper triangular 2×2 matrices, from which he obtained a base for $\mathrm{id}(M_2(\mathbb{Q}))$, and then Drensky showed that all the other identities are consequences of s_4 and Hall's identity. This takes up about eight pages in Razmyslov's book [Raz89]. The proof is streamlined somewhat in [Ba87].

Chapter 4

1. If A has the form (1.20) with $I_1 I_2 = 0$, then $\mathrm{id}(A) \supseteq \mathrm{id}(A/I_1)\,\mathrm{id}(A/I_2)$. Conclude, using Lewin's theorem, that if $\mathrm{id}(W/J(W)) = \mathcal{M}_n$ and $J(W)^k = 0$, then $\mathcal{M}_n^k \subseteq \mathrm{id}(W)$.

2. For any κ, there exists a T-ideal with Kemer index κ and $\delta(W)$ arbitrarily large. (Hint: Take the direct product of an algebra with Kemer index κ and upper triangular matrices of large size.)

3. Fill in the details to Kemer's argument (given below) that any T-ideal in characteristic 0 is finitely based, given Theorem 4.1: Suppose on the contrary there is an infinitely based T-ideal generated by multilinear polynomials $f_1, f_2, f_3, \ldots$, such that $\deg f_{j-1} \le \deg f_j$, and f_j is not a consequence of $\{f_1, \ldots, f_{j-1}\}$ for all j. Assume each f_j is Spechtian, cf. Proposition 3.36. Let $n_j = \deg f_j$, and let T_j denote the T-ideal generated by all

$$f_j(x_1, \ldots, x_{i-1}, [x_i, x_{n_j+1}], x_{i+1}, \ldots, x_{n_j}), \quad 1 \le i \le n_j.$$

Then $f_j \notin T_j$, and by hypothesis $T_j = \mathrm{id}(A)$ for some f.d. algebra A. Thus, there are semisimple or radical substitutions $a_1, \ldots, a_{n_j}$ such that $f_j(a_1, \ldots, a_{n_j}) \ne 0$. Taking n_j larger than the index of nilpotence of $J(A)$, some substitution a_i must be semisimple, say in some simple component A_k of $\bar{A} = A/J(A)$. But $A_k = F \oplus [A_k, A_k]$ (since $[A_k, A_k]$ are the matrices of trace 0), and

$$f_j(a_1, \ldots, a_{i-1}, [A_k, A_k], a_{i+1}, \ldots, a_{n_j})$$

is a specialization of $T_j(A)$ and thus, is 0. This implies

$$f_j(a_1, \ldots, a_{n_j}) = 0,$$

contradiction.

4. Suppose A is a PI-basic algebra without 1, and in the terminology of Remark 4.7, $e_0 A = A$. Then $A = J(A)$; in this case, $\beta(A) = 0$,

and $\gamma(A)$ equals the index of nilpotence of $J(A)$, so Kemer's Second Lemma clearly holds. Use this observation to prove Kemer's Theorem for algebras without 1.

5. Suppose $h = f_{A(1,...,t)}$ is a t-alternating polynomial, and $I_1, \ldots, I_\mu$ are a partition of $\{1, \ldots, t\}$, i.e., $\{1, \ldots, t\}$ is their disjoint union. Then there are permutations $\sigma_1 = 1, \sigma_2, \ldots, \sigma_m \in S_t$ for some m, such that

$$h = \sum_{j=1}^{m} \mathrm{sgn}(\sigma_j) \sigma_j f_{A(I_1)...A(I_s)}. \tag{13.8}$$

(Hint: Writing each I_k as $\{i_1, \ldots, i_{t_k}\}$, let S_{I_k} denote the group of permutations in $i_1, \ldots, i_{t_k}$, viewed naturally as a subgroup of S_t. Since the $I_1, \ldots, I_\mu$ are disjoint, S_{I_k} commutes with $S_{I_{k'}}$ for each k, k', so $H = S_{I_1} \ldots S_{I_\mu}$ is a subgroup of S_t. Take a transversal $\sigma_1 = 1, \sigma_2, \ldots, \sigma_m$ of H in S_t, i.e.,

$$S_t = \cup \sigma_i H.$$

Then match the terms of (13.8) permutation by permutation.)

6. Define algebras A, A' to be Γ-*equivalent*, written $A \sim_\Gamma A'$, if

$$\mathrm{id}(A') \cap \Gamma = \mathrm{id}(A) \cap \Gamma.$$

Thus, PI-equivalent means Γ_∞-equivalent. Also, an algebra A is Γ-*full* if A is full with respect to a polynomial in Γ. A f.d. algebra A is Γ-*minimal* (among all f.d. algebras Γ-equivalent to A) if the ordered pair (t_A, s_A) is minimal (lexicographically) among all such algebras. In other words, if $A' \sim_\Gamma A$, and A' is f.d. with ordered pair (t', s'), then either $t' > t_A$, or $t' = t_A$ with $s' \geq s_A$. A Γ-*basic* algebra is a Γ-minimal, Γ-full subdirectly irreducible f.d. algebra.

Define $\beta_\Gamma(A) = 0$ whenever $\Gamma \subseteq \mathrm{id}(A)$; if $\Gamma \not\subseteq \mathrm{id}(A)$, then $\beta_\Gamma(A)$ is the largest t such that for any s there is an s-fold t-alternating polynomial $f(X_1, \ldots, X_s; Y) \in \Gamma \setminus \mathrm{id}(A)$. Likewise, $\gamma_\Gamma(A) = 0$ whenever $\Gamma \subseteq \mathrm{id}(A)$; if $\Gamma \not\subseteq \mathrm{id}(A)$, then $\gamma_\Gamma(A)$ is the largest s such that some $(s-1)$-fold $(\beta_\Gamma(A)+1)$-alternating polynomial $f(X_1, \ldots, X_{s-1}; Y) \in \Gamma \setminus \mathrm{id}(A)$. The pair $(\beta_\Gamma(A), \gamma_\Gamma(A))$ is called the *relative Kemer index*.

Show $\beta_\Gamma(A) \leq \beta(A)$; $\gamma_\Gamma(A) \leq \gamma(A)$. Prove the relative versions of Kemer's first and second lemmas in this setting.

Internal characteristic coefficients

7. If A is affine and satisfies c_{n+1}, then the obstruction to integrality of order n has nilpotence index not greater than $1 + 2^2 + \cdots + n^2 = \frac{n(n+1)(2n+1)}{12}$. (Hint: Use Theorem 4.82. For characteristic p, one also needs the Donkin-Zubkov results.)

8. Using Exercise 7, conclude that if $\mathcal{M}_n(A)$ is the set of evaluations of identities of $n \times n$ matrices on A, then $\mathcal{M}_n(A)^k = 0$, for some k. Using Lewin's embedding Theorem 1.72, conclude that A satisfies all identities of $m \times m$ matrices for some m. Since m depends only on k, this proves that any algebra satisfying a Capelli identity also satisfies the identities of $n \times n$ matrices for some n.

Chapter 5

1. Fill in the $u \times v$ rectangular tableau λ with the numbers

$$
\begin{array}{cccc}
1 & 2 & \cdots & v \\
v+1 & v+2 & \cdots & 2v \\
\vdots & \vdots & \ddots & \vdots \\
(u-1)v+1 & (u-1)v+2 & \cdots & uv
\end{array}
$$

and compute its semi-idempotents e_{T_λ} and e'_{T_λ}. In case $u = v = 2$, one of them is the multilinearization of S_2^2, but the other corresponds to the multilinearization of

$$
x_1^2 x_3^2 - x_3 x_1^2 x_3 - x_1 x_3^2 x_1 + x_3^2 x_1^2,
$$

which can be rewritten as

$$
x_1[x_1, x_3^2] - x_3[x_1^2, x_3].
$$

2. Let $\lambda \vdash n$ and let $E_\lambda \in I_\lambda$ be the corresponding central idempotent. Then $I_\lambda \subseteq \mathrm{id}(A)$ if and only if $E_\lambda \in \mathrm{id}(A)$. (Hint: $I_\lambda = F[S_n]E_\lambda$.)

3. In Theorem 3.38, if $\mathrm{char}(F) = 0$, k is bounded by a function of d. (Hint: One considers only a finite number of semi-idempotents of $\mathbb{Q}[S_d]$, and takes the maximal k corresponding to them all.)

4. If $n = uv$ is as in Example 5.47, then $c_n(x_1, x_1^2, \ldots, x_1^n, y_1, \ldots, y_n) \in \mathrm{id}(A)$. (Hint: Take $p = e'_{T_\lambda}$, where λ is the rectangular tableau v^u of

Example 5.47 which can be written explicitly as

$$p(x_1, \ldots, x_n) = \sum_{\sigma \in C} \sum_{\tau \in R} \mathrm{sgn}(\sigma) x_{\sigma\tau(1)} \cdots x_{\sigma\tau(n)},$$

where $C = C_{T_\lambda}$ and $R = R_{T_\lambda}$. In view of Corollary 5.52 and Remark 1.27, one wants to take the alternator of p^*. However, this could be 0. So look for a clever substitution of p^* that will preserve the column permutations (which create alternating terms) while neutralizing the row transformations.

Towards this end, for each $1 \le k \le n$, write $k = i + (j-1)u$, i.e., k is in the (i,j) position in the tableau T_λ, and substitute $x_k \mapsto x_1^i$ and $y_k \mapsto x_1^{(j-1)u} y_k$. Since x_1^i now repeats throughout the i row, this neutralizes the effect of row permutations, and

$$p^*(x_1, \ldots, x_1, x_1^2, \ldots, x_1^2, \ldots; y_1, \ldots, y_u, x_1 y_{u+1}, \ldots, x_1^{(v-1)u} y_n)$$

$$= v! \sum_{\sigma \in C} \mathrm{sgn}(\sigma) x_1^{\sigma(1)} y_1 x_2^{\sigma(2)} y_2 \ldots x_k^{\sigma(k)} y_k \ldots x_n^{\sigma(n)} y_n.$$

Since these terms respect the sign of the permutation, one can apply the alternator of Remark 1.27 (but treating these power of x_1 as separate entities) to get the identity

$$u! v! \sum_{\sigma \in \pi} \mathrm{sgn}(\pi) x_1^{\pi(1)} y_1 x_2^{\pi(2)} y_2 \ldots x_k^{\pi(k)} y_k \ldots x_n^{\pi(n)} y_n,$$

which is $u! v! c_{2n}(x_1, x_1^2, \ldots, x_1^n, y_1, \ldots, y_n)$.)

5. If $n = uv$ are as in Theorem 5.48 and $f(x_1, \ldots, x_n; y)$ is any n-alternating polynomial, then

$$f(w_1 x_1 w_2, w_1 x_1^2 w_2, \ldots, w_1 x_1^n w_2, y_1, \ldots, y_n) \in \mathrm{id}(A),$$

for any words $w_1, w_2 \in C\{X\}$. (Hint: Apply Corollary 1.26 to Exercise 4.)

6. Using Exercise 4, prove that $c_n(x_1^{k_1}, x_1^{k_2}, \ldots, x_1^{k_n}; y) \in \mathrm{id}(A)$. (Hint: An induction argument.)

Kemer then uses an induction argument on the tableaux to prove that A satisfies c_n for n as in Proposition 5.48. This is a much better bound than the one given in our text.

7. If A_i satisfies the identities of $n_i \times n_i$ matrices for $i = 1, 2$, then $A_1 \otimes A_2$ satisfies the identities of $n_1 n_2 \times n_1 n_2$ matrices. (Hint: Reduce to generic matrices via Exercise 3.3, and embed these into matrix algebras, for which the assertion is easy.)

Chapter 6

1. Define the Grassman involution for T-ideals over an arbitrary infinite field F. In other words, extend the $(*)$-action naturally to homogeneous polynomials. Extend the $(*)$-action linearly to any T-ideal by noting that any polynomial can be written canonically as a sum of homogeneous polynomials.

2. Show $G \otimes G \sim_{PI} M_{1,1}$; $M_{k,\ell} \otimes G \sim_{PI} M_{k+\ell,k+\ell}$ for $k, \ell \neq 0$, and
$$M_{k_1,\ell_1} \otimes M_{k_2,\ell_2} \sim_{PI} M_{k_1 k_2 + \ell_1 \ell_2, k_1 \ell_2 + k_2 \ell_1}.$$
Conclude that the PI-degree of the tensor product of two algebras can be more than the product of their PI-degrees.

3. (For those who know ultraproducts.) Any relatively free T-prime algebra U_A can be embedded into an ultraproduct of copies of A. (Hint: Take the filter consisting of those components on which the image of some element is 0.)

4. An algebra A is called *Grassman representable* if A can be embedded into $M_n(G)$ for suitable n and for a suitable Grassman algebra over a suitable commutative ring. Show the Grassman envelope of a f.d. superalgebra is Grassman representable. Conclude that any relatively free PI-algebra U in characteristic 0 is Grassman representable. (Hint: $U = U_A$ where A is the Grassman envelope of a f.d. superalgebra, and construct U by means of generic linear combinations of a base of A.)

5. Call a superalgebra *left superNoetherian* if it satisfies the ACC on graded left ideals. (This is a weaker condition than left Noetherian.) Every affine left superNoetherian PI-superalgebra is superrepresentable, in the sense that there is a graded injection into a f.d. superalgebra. (Hint: Follow the proof of Anan'in's Theorem 1.91, using homogeneous elements at each stage.)

6. Suppose C is a commutative affine superalgebra over a field F, and $R \supset C$ is a PI-superalgebra that is f.g. as a C-module. Then R is superrepresentable. (Hint: R is spanned by homogeneous elements, so is left superNoetherian; apply Exercise 5.)

7. For any relatively free algebra $\mathcal{A}$ in an infinite number of indeterminates, $J(\mathcal{A})$ is contained in every maximal T-ideal, and thus is the intersection of the maximal T-ideals of $\mathcal{A}$. (Hint: Any maximal T-ideal M is contained in a maximal ideal, which clearly contains $J(\mathcal{A})$, so $M + J(\mathcal{A})$ is a T-ideal containing M and, thus, equals M.)

Chapter 7

The extended Grassman algebra (for char$(F) = 2$)

1. $[a, b] = \varepsilon_a \varepsilon_b ab$ for all words a, b in G^+. (Hint: Write $a = e_i a_1$ and proceed by induction, since

$$[a, b] = [e_i a_1, b] = e_i[a_1, b] + [e_i, b]a_1.) \tag{13.9}$$

2. Conclude from Exercise 1 that if v is a word in G^+ and v' is a permutation of the letters of v, then $v^2 = (v')^2$.

3. Every identity of G^+ is a consequence of the Grassman identity; one such identity is $(x + y)^4 - x^4 - y^4$. Conclude the identity $(x + y)^{2^k} - x^{2^k} - y^{2^k}$ for all $k \geq 2$. (Compare with Exercises 7, 8.)

Identities of Grassman algebras

4. In any algebra satisfying the Grassman identity, the terms $[v_1, v_2]wv_1$ and $v_1[v_1, v_2]w$ are equal. (Hint: $[v_1, v_2][w, v_1] = 0$.)

5. If a, b are odd elements of G, then $(a+b)^m = 0$ for any $m > 2$. (Hint: In each word of $(a + b)^m$, either a or b repeats.)

6. In characteristic p, G satisfies the superidentity $(y + z)^p = y^p$ for y even and z odd. (Hint: Lemma 7.29.)

7. Any Grassman algebra in characteristic $p > 2$ satisfies the *Frobenius identity* $(x+y)^p - x^p - y^p$. (Hint: Writing $a = a_0 + a_1$ and $b = b_0 + b_1$, for $a, b \in G$, use Exercise 6 to show

$$(a + b)^p = (a_0 + b_0)^p = a_0^p + b_0^p = a^p + b^p.)$$

8. Derive Exercise 7 directly from the Grassman identity. (This is possible, in view of Exercise 3.5.)

9. The Frobenius identity implies $(x + y)^{p^k} - x^{p^k} - y^{p^k}$ for all $k \geq 1$.

Non-finitely based T-ideals

10. In characteristic 2, use the extended Grassman algebra G^+, together with the relations $x_i^3 = 0$. Then x^2 is central, and define

$$\tilde{Q}_n = [[e, t], t] \prod_{i=1}^{n} x_i^2([t, [t, f]][[e, t], t])^3[t, [t, f]].$$

Carry out the analog to the proof that was given in characteristic $p > 2$, to prove $\{\tilde{Q}_n : n \in \mathbb{N}\}$ generates a non-finitely based T-ideal in characteristic 2.

11. The construction given in the text can be generalized to non-finitely generated T-ideals for varieties of Lie and special Jordan algebras (associative algebras are a special case of alternative algebras), with virtually the same proofs. Here are the modifications needed for the Jordan case in characteristic $p > 2$.

Let $Q(x, y) = x^{p-1}y^{p-1}[x, y]$, let $\mathfrak{r}[x]$ denote right multiplication by x, and let $\{x, y, z\} = (xy)z - x(yz)$ be the associator. The polynomials are considered as polynomials in operators. The relation $t^{4q+1} = 0$ is replaced by $t^{8q+1} = 0$.

$$
Q_n = \{e, t^2, t^2\} \prod_{i=1}^{n} Q(\mathfrak{r}[x_i], \mathfrak{r}[y_i])\mathfrak{r}[x_i]^2
$$

$$
\cdot \left(\mathfrak{r}(\{e, t^2, t^2\})\mathfrak{r}[\{t^2, t^2, f\}] \right)^{q-1} \mathfrak{r}[\{t^2, t^2, f\}]
$$

(13.10)

A non-finitely based T-space in the affine case

The following exercises present Shchigolev's example of a non-finitely based T-space in the affine case in characteristic $p > 2$, despite Belov's proof that T-ideals are finitely based in this case. Although it requires considerable preparation, the example (given in Exercise 16) has a surprisingly simple formulation.

12. The Grassman identity implies the identities

$$
[x_1^{m_1}x_2^{m_2}, x_1^{n_1}x_2^{n_2}] = c[x_1, x_2]x_1^{m_1+n_1-1}x_2^{m_2+n_2-1}
$$

where $c = m_1 n_2 - n_1 m_2$. (Hint: In view of Corollary 3.46, one may check this identity for G. Write $x_i = y_i + z_i$ where y_i are even and z_i are odd. By Lemma 7.29(i),

$$
(y_1 + z_1)^{m_1}(y_2 + z_2)^{m_2} = (y_1^{m_1} + m_1 y_1^{m_1-1}z_1)(y_2^{m_2} + m_2 y_2^{m_2-1}z_2)
$$

$$
= y_1^{m_1}y_2^{m_2} + m_1 m_2 y_1^{m_1-1}y_2^{m_2-1}z_1 z_2
$$

$$
+ m_1 y_1^{m_1-1}y_2^{m_2}z_1 + m_2 y_1^{m_1}y_2^{m_2-1}z_2.
$$

(13.11)

Since the first two terms on the RHS are central, and recalling any term with z_i repeated is 0, note that

$$[(y_1 + z_1)^{m_1}(y_2 + z_2)^{m_2}, (y_1 + z_1)^{n_1}(y_2 + z_2)^{n_2}]$$
$$= [m_1 y_1^{m_1-1} y_2^{m_2} z_1 + m_2 y_1^{m_1} y_2^{m_2-1} z_2, n_1 y_1^{n_1-1} y_2^{n_2} z_1 + n_2 y_1^{n_1} y_2^{n_2-1} z_2]$$
$$= m_1 n_2 [y_1^{m_1-1} y_2^{m_2} z_1, y_1^{n_1} y_2^{n_2-1} z_2] + m_2 n_1 [y_1^{m_1} y_2^{m_2-1} z_2, y_1^{n_1-1} y_2^{n_2} z_1]$$
$$= 2c z_1 z_2 y_1^{m_1+n_1-1} y_2^{m_2+n_2-1}$$
$$= c[x_1, x_2] x_1^{m_1+n_1-1} x_2^{m_2+n_2-1}$$

(13.12)

by Lemma 7.29(iv).)

13. The Grassman identity implies the identities

$$[f(x_1, x_2), g(x_1, x_2)] = c[x_1, x_2] x_1^{m_1+n_1-1} x_2^{m_2+n_2-1}$$

for arbitrary homogeneous polynomials f, g in x_1, x_2 where $m_i = \deg_i f$, $n_i = \deg_i g$, and c is a multiple of

$$m_1 n_2 - n_1 m_2 = \left\| \begin{matrix} m_1 & n_1 \\ m_2 & n_2 \end{matrix} \right\|.$$

(Hint: Writing f and g as sums of monomials, it suffices to assume both f and g are monomials. One may assume all occurrences of x_1 in f (and likewise in g) appear before all occurrences of x_2. Otherwise, if say $f = h_1 x_2 x_1 h_2$, then

$$f = h_1 x_1 x_2 h_2 + h_1[x_2, x_1]h_2,$$

so $[f, g] = [h_1 x_1 x_2 h_2, g] + [h_1[x_2, x_1]h_2, g]$, and the last term is 0 by Lemma 7.19(iv). Continuing this process of moving the occurrences of x_1 to the left of x_2, conclude by the previous exercise.)

In the subsequent exercises, define

$$q_n(x_1, x_2) = x_1^{p^n-1}[x_1, x_2] x_2^{p^n-1};$$

thus, $P = q_1$.

14. Writing $x_i = y_i + z_i$ for y_i even and z_i odd, the polynomial q_n is equivalent to $2z_1 z_2 y_1^{p^n-1} y_2^{p^n-1}$ on G. (Hint: Lemma 7.35.)

15. Assume char$(F) = p > 2$. The T-space generated by homogeneous substitutions (in x_1 and x_2) of total degree p^n in $\{q_j : j < n\}$ is a consequence of the Grassman identity. (Hint: Appeal to Remark 7.2. Specialize $x_1 \mapsto f$ and $x_2 \mapsto g$. Writing $m_i = \deg_i f$ and $n_i = \deg_i g$, we see $p^j(m_i + n_i) = \deg_i q_j(f, g) = p^n$, implying $m_i + n_i = p^{n-j} \equiv 0 \pmod{p}$. But Exercise 13 says $[f, g]$ has coefficient a multiple of

$$\left\| \begin{matrix} m_1 & n_1 \\ m_2 & n_2 \end{matrix} \right\|,$$

which is 0 since its rows are dependent modulo p.)

16. (Shchigolev [Shch00].) If $\text{char}(F) = p$, then the T-space generated by $\{q_n : n \in \mathbb{N}\}$ is not finitely based as T-space, modulo the Grassman identity.

    (Hint: Evaluated in the Grassman algebra G, q_n is equivalent to $2z_1 z_2 y_1^{p^n-1} y_2^{p^n-1}$, which obviously can take nonzero values. Thus, it suffices to show for all $j < n$, and any specializations of q_j, that the part of degree p^n in each of x_1 and x_2 is identically 0 in G; this would yield the assertion, since the $q_j, j < n$, could not be used to cancel out q_n.

    As in Remark 7.2, it suffices to check this for word substitutions (which was done in Exercise 15) and for partial multilinearizations. Since only two indeterminates are involved, one can compute the coefficients, which turn out to be 0 (mod p) by means of a more intricate version of the computations of Exercise 7.13.)

Another Theorem of Kemer

The following exercises give the highlights of Kemer's suprising theorem [Kem93] that every PI-algebra A of characteristic p satisfies a suitable standard identity s_n.

17. If $A \otimes G$ satisfies the symmetric identity $\tilde{s}_n$, then A satisfies s_n, so it suffices to prove every PI-algebra satisfies $\tilde{s}_n$ for suitable n.

18. Let $F[\varepsilon]_0 = F[\varepsilon_i : i \in \mathbb{N}]_0$ denote the commutative polynomial algebra without 1 modulo the relations $\varepsilon_i^2 = 0$, cf. Definition 7.7. Then A satisfies $\tilde{s}_n$, iff $A \otimes F[\varepsilon]_0$ satisfies x_1^n. Conclude that A satisfies $\tilde{s}_n$ iff A satisfies the same multilinear identities of a suitable nil algebra (without 1) of bounded index. (Hint: Recall $\tilde{s}_n$ is the multilinearization of x_1^n.)

19. For any t, A satisfies $\tilde{s}_n$ iff $M_t(A)$ satisfies $\tilde{s}_m$ for some m. (Hint: Using Exercise 18, translate to nil of bounded index, which is locally nilpotent; do this argument generically.)

20. Define the k-*Engel polynomial* f_k inductively, as $[f_{k-1}, x_2]$, where $f_0 = x_1$. (Thus, $f_2 = [[x_1, x_2], x_2]$.) If $\text{char}(F) = p$ and $k = p^j$, then f_k is the homogeneous component of the partial linearization of x_1^k. Conclude in this case that any (associative) algebra satifying an Engel identity satisfies some $\tilde{s}_n$.

21. (Difficult) Prove by induction on the Kemer index that every associative algebra in characteristic p satisfies a suitable Engel identity.

This is a tricky computational argument that takes about five pages in [Kem93].

Chapter 8

1. Any prime PI-ring R with Noetherian center C is a f.g. C-module. (Hint: Take any evaluation $c \neq 0$ of an n^2-alternating central polynomial and note that $R \approx Rc$, which is contained in a f.g. C-module, by Theorem 1.6.)

2. An uncountable class of non-finitely presented monomial algebras satisfying ACC on ideals: Consider all monomial algebras generated by three elements a, b, c, supported by ba^m, $a^n c$, $ba^q c$, where $m, n \in \mathbb{N}$ are arbitrary, but q is in a specified infinite subset of $\mathbb{N}$ whose complement is also infinite. There are uncountably many such sets, and an uncountable number of such algebras, each of which satisfies ACC on ideals. But there are only countably many finitely presented algebras, so we are left with uncountably many non-finitely presented ones. (Details are as in Example 1.81.)

3. If H is a subgroup of G of index n, then H contains a normal subgroup of index $\leq n!$. (Hint: Consider the homomorphism $G \to S_n$ by identifying S_n with the permutations of the cosets $g_i H$ of H, sending $a \in G$ to the permutation given by $g_i H \mapsto ag_i H$.)

4. Prove the analog of Lemma 8.10 for prime enveloping algebras of restricted Lie algebras.

5. Prove Bachturin's theorem, that the (unrestricted) enveloping algebra of a Lie algebra L in characteristic p satisfies a PI iff L has an Abelian ideal of finite codimension and the adjoint representation of L is algebraic. (Hint: Define $L^p = \{a^p : a \in L\}$, and take $H = \sum_{i \in \mathbb{N}} L^{p^i} \subseteq U(L)$, a restricted Lie algebra, whose restricted enveloping algebra $u(H) = U(L)$ is prime. By Exercise 4, H has an Abelian ideal of finite codimension, which translates to the required conditions on L.)

Chapter 9

1. Define the following relation on the set of all growth functions:

   $f \succ g$, if there exist c and k such that $cf(kn) \geq g(n)$ for all n. This gives rise to an equivalence relation: $f \equiv g$, if $f \succ g$ and $g \succ f$. This equivalence relation does not depend on the choice of generators; this is a general fact about growth in algebraic structures.

Algebras having linear growth

The object of the next few exercises is to prove highlights of the Small-Stafford-Warfield theorem, that any algebra of GK dimension 1 is PI. We follow Pappacena-Small-Wald [PaSmWa].

2. (Pappacena-Small-Wald.) If $A = F\{a_1,\ldots,a_\ell\}$ with GKdim$(A) = 1$, then any simple A-module M is f.d. over F. (Hint, using some structure theory: Factoring A by the annihilator of M, one may assume M is also a faithful A-module, i.e., A is a primitive algebra. Write $M = Aw$, and let $D = \mathrm{Hom}_A(M, M)$. If M has m linearly independent elements $b_1,\ldots,b_m$ over D, take $s_i \in A$, $1 \le i \le m$, such that $s_i b_j = 0$ for $j \le i$ and $s_i b_i = b_1$. Take n such that each $b_i \in V^n$, where $V = \sum_{i=1}^\ell Fa_i$, and let L_k be the left annihilator of $b_1,\ldots,b_k$ for each $k \le n$. Note that $L_k w = M$ and the L_k are independent over F. If M is infinite-dimensional over F, then comparing growth rates, show $m \le n$. But this means M has dimension at most n over D, i.e., $A = M_n(D)$ is simple Artinian. By the Artin-Tate lemma, the center Z of A is affine over F and thus, f.d. over F; hence, one may assume $Z = F$. It remains to show that A has PI-class n. But this is immediate by splitting A in the same way as in the standard proofs of Kaplansky's theorem.)

3. Suppose P is a maximal ideal of $A = F\{a_1,\ldots,a_\ell\}$, and A/P has linear growth rate N. Let n be the PI-class of A/P. Then any word of length $(n+1)N$ in the a_i is a linear combination of smaller words.

4. Notation as in Exercise 3, show $n \le \frac{1}{2}(N^2 + \sqrt{N^4 + 4N^2 - 4N})$. (Hint: The words of length $(n+1)N - 1$ span R/P, which has dimension n^2, so $(n+1)N^2 - N \ge n^2$.)

5. If A is semiprimitive affine of GK dimension 1, then A is PI, and its PI-class does not exceed the function of Exercise 4. (Hint: Apply Exercises 2 and 4 to each primitive homomorphic image.)

6. If GKdim$(A) = 1$ and A is affine with no nonzero nil ideals, then A is PI, of PI-class not exceeding the function of Exercise 4. (Hint: Expand the base field to an uncountable field; then one may assume A is semiprimitive and apply Exercise 5.)

7. (Small-Warfield [SmWa84].) Define a *principal left annihilator* to be the left annihilator of an element. If A is prime with $\mathrm{GKdim}(A) = 1$, then every chain of principal left annihilators of A has length bounded by the linear growth rate of A. (Hint: Given principal left annihilators $L_1 \supset L_2$, show $L_1 a = 0$ but $L_2 a \neq 0$ for some $a \in A$; $L_2 a$ is a faithful A-module, and thus has linear growth rate at least 1; continuing in this way, show the linear growth rate of A is at least the length of any chain of principal left annihilators.)

8. If A is semiprime affine with $\mathrm{GKdim}(A) = 1$, then A is PI. (Hint: One may assume A is prime. By Exercise 7, A satisfies the ACC on principal left annihilators, so, by [Row88, Proposition 2.6.24], A satisfies the conditions of Exercise 6.)

9. (Small-Warfield.) If R is prime of GK dimension 1, then R is a f.g. module over a central subring isomorphic to the polynomial ring $F[\lambda]$, and in particular, R is Noetherian. (Hint: Let $C = \mathrm{Cent}(R)$. R is PI, by Exercise 8. If C is a field, then R is simple, so assume C is not a field. Then C has GK dimension 1, implying C is integral over $F[\lambda]$, and thus, f.g. over $F[\lambda]$, by Exercise 8.1.)

Examples of PI-algebras with unusual growth

These examples are stronger than those appearing in prior expositions.

10. Define the algebra $A^{\infty} = F\{x, y\}/\langle y \rangle^2$, i.e., y occurs at most once in each nonzero word. Then $d_n(A^{\infty}) = n(n + 3)/2$. (Hint: $d_n = n + 1$.)

11. Define the algebra $A^n = F\{x, y\}/I_y^n$, where I_y^n is the ideal generated by monomials that do not contain subwords of the form $yx^k y$, $k < n$. Then $2^{\lfloor k/n \rfloor} \leq d_k(A^n) = d_k(A^{\infty}) = k + 1$. (Hint: There are $2^{\lfloor k/n \rfloor}$ words having y in the positions not divisible by n.)

12. Suppose ψ, φ are functions satisfying

$$\lim_{n \to \infty} \frac{\ln \psi(n)}{n} = 0, \qquad \lim_{n \to \infty} \left(\varphi(n) - \frac{n(n+3)}{2} \right) = \infty.$$

Then there exist an algebra A and two infinite subsets $\mathcal{K}$ and $\mathcal{L}$ of $\mathbb{N}$, such that:

(i) $d_n(A) > \psi(n), \forall n \in \mathcal{K}$.

(ii) $d_n(A) < \varphi(n), \forall n \in \mathcal{L}$.

If $\psi(n) < Cn^k$, then there exists a PI-algebra A satisfying these conditions.

(Hint: Let $\alpha : \ell_1 < k_1 < \ell_2 < k_2 < \ldots < \ell_i < k_i < \ell_{i+1} < \ldots$ be a sequence of positive integers, and let $I(\alpha)$ denote the ideal generated by the sets $\mathcal{W}_1$ and $\mathcal{W}_2$ defined as follows:

$$\mathcal{W}_1 = \{w : \ell_i \le |w| < k_i \quad \text{and} \quad \exists k < \ell_i : yx^k y \subset w\},$$
$$\mathcal{W}_2 = \{w : k_i \le |w| < \ell_{i+1} \quad \text{and} \quad \exists k < \ell_{i+1} : yx^k y \subset w\}.$$

Define $A(\alpha) = F\{x,y\}/I(\alpha)$. If $\ell_i \le n < k_i$, then $d_n(A(\alpha)) = d_n(A^{\ell_i})$, and, if $k_i \le n < \ell_{i+1}$, $d_n(A(\alpha)) = d_n(A^{\infty}) = n + 1$. Exercise 11 allows one to choose ℓ_{i+1} so that

$$d_{\ell_{i+1}-1}(A(\alpha)) < \varphi(\ell_{i+1} - 1).$$

A similar argument allows one to choose k_i so that

$$d_{k_i-1}(A(\alpha)) = d_{k_i-1}(A^{\ell_i}) > \psi(k_i - 1).$$

Hence, $A(\alpha)$ satisfies (1) and (2).

The proof of the last assertion is analogous. Note that $F\{x,y\}/\langle y \rangle^k$ is a PI-algebra; the number of words with y occuring k times, such that the distance between each two occurences of y is greater than some given large enough number, has growth n^k.)

13. In Exercise 12, let $\psi(n) = e^{\sqrt{n}}$, and $\varphi(n) = n(n+3)/2 + \ln(n)$. Then the growth function of A has a slow start and lags behind φ, but then begins to grow very fast and outstrips ψ.

Essential Shirshov bases

14. Suppose u is a non-periodic word, of length n. Then $h = u^{\infty}$ is a hyperword of periodicity n. Let A_u be the monomial F-algebra whose generators are letters of u, with relations $v = 0$ iff v is not a subword of h. Prove that $A_u \sim_{\text{PI}} M_n(F[\lambda])$. (Hint: Construct the homomorphism $A_u \to M_n(F[\lambda])$ obtained by taking the cyclic graph of length n. The edges of the cycle are labeled with the letters in u. Identify the edge between i and $i+1$ with λ times the matrix unit $e_{i,i+1}$. For example, if the letter x_1 appears on the edges connecting 1 and 2, 3 and 4, and 7 and 8, then x_1 is sent to $\lambda(e_{12} + e_{34} + e_{78})$. This gives an injection of A_u into $M_n(F[\lambda])$. Hence, $\text{id}(A) \supseteq \text{id}(F[\lambda])$. If the ground field is infinite, conclude the reverse inclusion using the surjection $A_u \to M_n(F)$ sending $\lambda \mapsto 1$.)

15. Let A_u be as in Exercise 14. The set of non-nilpotent words in A_u is the set of words that are cyclically conjugate to powers of u.

16. Every prime monomial PI-algebra A is isomorphic to some A_u as defined in Exercise 14. (Hint: Use Shirshov's Height Theorem, and show A has bounded height over some word.)

17. Suppose A is an algebra with generators $a_s \succ \cdots \succ a_1$, and M is a f.g. right A-module spanned by $w_k \succ \cdots \succ w_1$. Put $w_1 \succ a_s$, and $a_1 \succ x$. Make $\tilde{A} = A \oplus M$ into an algebra via the "trivial" multiplication

$$0 = w_i w_j = a_i w_j \quad \forall i, j.$$

Let A'' denote $\tilde{A} * F\{x\}/I$, where the ideal I is generated by the elements xw_i. If $MA^k \neq 0$ for all k, then the minimal hyperword in the algebra A'' is preperiodic of preperiodicity $\leq n + 1$. (Hint: By Theorem 2.79.)

18. The set of lexicographically minimal words in a PI-algebra A of PI-class n has bounded height over the set of words of length $\leq n$. (Hint: Let $d = 2n$, the PI-degree of A. As A has bounded height over the set of words of degree not greater than d, it is enough to prove that if $|u|$ is a nonperiodic word of length not greater than n, then the word u^k is a linear combination of lexicographically smaller words, for k sufficiently big. This is done in several steps.

   (i) Consider the right A-module N defined by the generator v and relations $vw = 0$ whenever $w \prec u^\infty$ (i.e., when w is smaller than some power of u). The correspondence $t : vs \to vus$ is a well-defined endomorphism of N; hence, N is an $A[t]$-module. The goal is to prove that $Nt^k = 0$ for some k.

   (ii) If $Mt^k \in MJ(\text{Ann } M)$, where $J(\text{Ann } M)$ is the Jacobson radical of the annihilator, then $Mt^{\ell k} \in MJ(\text{Ann } M)^\ell$ and $Nt^{\ell k} = 0$, for ℓ sufficiently big. Hence, viewing N as a module over $A[t]/J(\text{Ann } M)$, one may assume that $J(\text{Ann } M) = 0$.

   (iii) Reduce the proof to the case when M is a faithful module over a semiprime ring B.

   (iv) The elements from $Z(B) \setminus \{0\}$ all have annihilator 0. Localize with respect to them, and, by considering an algebraic extension of $Z(B)$, reduce to the case that N is a matrix algebra over a field.

(v) By applying Example 17, get a minimal nonzero right hyper-word vu^∞ and hence, a contradiction to Corollary 2.80 of the Independence Theorem.)

19. A set of words S in a relatively free algebra $\mathcal{A}$, is an essential Shirshov base iff, for each word u of length not greater than the PI-degree of $\mathcal{A}$, S contains a word that is cyclically conjugate to some power of u. (Hint: Since $(uv)^n = u(vu)^{n-1}v$, if S contains two cyclically conjugate words, then one of them can be deleted from the base. Hence, sufficiency is a consequence of Exercise 18. To prove necessity, suppose u is non-cyclic of length $\leq n$. By Exercise 14, $A_u \sim_{PI} A$, implying A_u is a quotient algebra of $\mathcal{A}$. The image of S must contain a non-nilpotent element. Conclude by Exercise 15.)

20. (Converse to Shirshov's Height Theorem.) Suppose A is a C-algebra. We say we are making $a \in A$ *generically integral* of degree m if we replace A by

$$A \otimes_C C[\lambda_1, \ldots, \lambda_{m-1}] / \langle a^m - \sum_{i=0}^{m-1} \lambda_i a^i \rangle,$$

where $\lambda_1, \ldots, \lambda_{m-1}$ are new commuting indeterminates over C. Repeating this process for each a in a given subset W of A, we say W has been made *generically integral* of degree m. Also we say W is a *Kurosch set* for A if A becomes f.g. over C when we make W generically integral. Show W is an essential Shirshov base iff W is a Kurosch set. (Hint: ($\Rightarrow$) is easy. Conversely, first note that if an ideal I of A has bounded essential height over W and A/I has bounded essential height over W, then A also has bounded essential height over W. Take I to be the evaluations of μ-Kemer polynomials for large enough μ; apply Zubrilin traces (Section 4.7.1). Corollary 4.93 guarantees that the induction terminates.)

21. Reformulate Exercise 20 as follows: W is an essential Shirshov base of A iff, for any algebra homorphism $\varphi : A \to B$ such that $\varphi(W)$ is integral (over the center of B), we must have $\varphi(A)$ integral.

22. Suppose A is an affine PI-algebra, $I \triangleleft A$, and $W \subset A$ such that $(W + I)/I$ is an essential Shirshov base in A/I. Then there exists a finite set of elements $\{v_1, \ldots, v_p\} \subseteq I$, such that $W \cup \{v_1, \ldots, v_p\}$ is an essential Shirshov base in A. (Hint: Let A_k be the module spanned by words from W of degree $\leq k$. A_k is finite-dimensional for all k. By the definition of the essential height, there exists $\mu \in \mathbb{N}$ and a finite set S_k such that A_k is spanned by elements of essential height μ with respect to $S_k \cup W$, modulo a finite linear combination (which

depends on k) of elements from I. It remains to choose k sufficiently large.)

23. If $I \lhd A$ and A/I both have bounded essential height over (the image of) W, then A also has bounded essential height over W.

24. (Essential height in the graded case.) Let A be a finitely generated graded associative algebra without 1 of PI-degree d, and $B \subset A$ be a finite set of homogeneous elements that generate A as an algebra. If $\bar{A} = A/\langle B^m \rangle$ is nilpotent, then A has bounded essential height over B. (Hint: Use Exercise 2.20 to show A is spanned by elements of the form

$$v_0 b_0^{j_0} v_1 b_1^{j_1} \ldots b_{s-1}^{j_{s-1}} v_s$$

where, for all i, $b_i \in M$, $|v_i| < m$, and $j_i \geq d$, and not more than $d-1$ words v_i have length $\geq d$, and not more than $d-1$ of the v_i are equal.)

25. (Berele's proof of Proposition 9.31.) By Theorem 6.53, there is a nilpotent ideal J such that A/J is the subdirect product of relatively free T-prime algebras $A_1, \ldots, A_q$ of T-prime ideals. Then $J^s = 0$ by Lemma 4.54. It remains to compute GKdim(A) when A is relatively free T-prime. But these algebras were determined in Corollary 6.62, so it remains to compute the GK dimension for the three kinds of prime T-ideals, cf. Definition 6.34. Note that $n \leq [d/2]$ by the Amitsur-Levitzki Theorem.

Case 1. If $I = \mathcal{M}$, the GK dimension is $(\ell - 1)n^2 + 1$, by Procesi's result.

Case 2. If $I = \mathcal{M}'$, the GK dimension is $(\ell - 1)n^2 + 1$. To see this, one displays the relatively free algebra in terms of generic matrices over the Grassman algebra.

Case 3. If $I = \mathcal{M}_{j,k}$ where $j + k = n$, the relatively free algebra is contained in that of Case 2, so the GK dimension is bounded by that number. (Note: Berele [Ber93] computed the GK dimension for $I = \mathcal{M}_{j,k}$ to be precisely $(\ell - 1)(j^2 + k^2) + 2$.)

The multivariate Poincare-Hilbert series

26. (Suggested by Berele.) Show the multivariate Poincaré–Hilbert series satisfies the properties described in Lemma 9.40 and the ensuing theory, and conclude that Belov's theorem about the rationality of the Hilbert series of a relatively free PI-algebra holds in this more

general context. (Hint: Over an infinite field, every T-ideal is homogeneous with respect to each variable, so the relatively free algebra is multi-graded. Every Kemer polynomial generates a homogeneous module with respect to formal traces, in view of Theorem J.)

Reduction modulo p

The following exercises show how to reduce a representable affine PI-algebra A modulo p, for arbitrarily large primes p (depending on A). There are examples of Schelter and Drensky to show that this does not work for all primes p, even when A is the algebra of generic matrices.

27. Let M be a graded A-module. The Hilbert series of $M \otimes \mathbb{Q}$ and $M \otimes \mathbb{Z}/p$ coincide iff the additive group of M is p-torsion free. (Hint: The dimensions of the corresponding homogeneous components are equal.)

28. A Noetherian module M is p-torsion free for all sufficiently large primes $p \in \mathbb{N}$. (Hint: Let $\{p_i\}$ be an infinite ascending sequence of primes such that M is not p_i-torsion free, and let M_i be the corresponding nonzero torsion module. Then $M_i \cap M_j = 0$ for $i \neq j$, implying

$$\bigoplus_i M_i$$

is an infinite direct sum of submodules, contrary to M Noetherian.)

29. If p is sufficiently large, then $H_{M \otimes \mathbb{Q}} = H_{M \otimes \mathbb{Z}/p}$.

30. If A is Noetherian, then the minimal dimensions of representations for $A \otimes \mathbb{Q}$ and of $A \otimes \mathbb{Z}/p$ are the same, for sufficiently large p.

31. Let A be a relatively free finitely generated $\mathbb{Z}$-algebra. Then for all sufficiently large p, the algebra $A \otimes \mathbb{Z}/p$ can be represented in matrices of bounded degree, and A is p-torsion free. (Hint: First show that a finitely generated relatively free $\mathbb{Z}$-algebra A is p-torsion free for all sufficiently large p: It suffices to construct a nonzero T-ideal I such that $H_{I \otimes \mathbb{Q}} = H_{I \otimes \mathbb{Z}/p}$ and $pI = I \cap pA$, for all sufficiently large p; then $H_A = H_I + H_{A/I}$. If the algebra has T-ideals with intersection zero, use Noetherian induction. Otherwise, use Kemer polynomials, cf. the proofs of Lemma 8.4 and Theorem 8.2, to construct an ideal I of the trace ring which is also a Noetherian module over A. Apply the previous exercises to I, and induction on the Kemer index to A/I.

A closer argument is needed to obtain representability of bounded degree. The algebra $\mathbb{Q} \otimes A$ is representable, and consequently this is true for its quotient algebra $A' = A/N$ relative to the torsion module N (which is a T-ideal). Let $\tilde{A}$ denote the trace ring of A. Then $\tilde{A}$ contains a common ideal with A.

$pI = I \cap p\tilde{A} = I \cap pA$ for all sufficiently large p; hence, $I \otimes \mathbb{Z}/p$ can be embedded into $\tilde{A}/p\tilde{A}$ and into A/pA. Thus, we have a map $\pi_1 : A \to \tilde{A} \otimes \mathbb{Z}/p$ with $I \cap \ker \pi_1 = pI$. Clearly, $\ker \pi_1 \supset pA$.

Using Noetherian induction for T-ideals, assume that the basic assertion holds for all homomorphic images with respect to T- ideals. Hence, there is a map π_2' of A/I to a Noetherian algebra B and a corresponding map $\pi_2 : A \to B$ with $\ker \pi_2 = I + pA$.

Now consider the map $\pi_1 \oplus \pi_2 : A \to (\tilde{A} \otimes \mathbb{Z}/p) \oplus B$. Then, using the modularity law,

$$\ker(\pi_1 \oplus \pi_2) = (\ker \pi_1) \cap (I + pA) = (I \cap \ker \pi_1) + pA,$$

which is $pI + pA = pA$. Thus, $A \otimes \mathbb{Z}/p$ is representable. The degree is bounded, since the dimension of B is bounded (by induction) and the dimension of $\tilde{A} \otimes \mathbb{Z}/p$ also is obviously bounded.)

32. Hypotheses as in Exercise 31, $H_{A \otimes \mathbb{Z}/p} = H_{A \otimes \mathbb{Q}}$ for all sufficiently large p, and the algebras $A \otimes \mathbb{Z}/p$ and $A \otimes \mathbb{Q}$ have the same set of bases consisting of words in generators.

    (Hint: For p large enough, A is p-torsion free, so the Hilbert series for $A \otimes \mathbb{Q}$ and $A \otimes \mathbb{Z}/p$ coincide. It suffices to show that the same sets of words are linearly dependent in these two algebras. In one direction, suppose $\sum \alpha_i v_i = 0$, $\alpha_i \in \mathbb{Z}$ for words v_i. This relation remains nontrivial in $A \otimes \mathbb{Z}/p$. The other direction now follows from a dimension count.)

33. Let C be a commutative Noetherian ring, S a Noetherian subring, M a Noetherian C-module, and $N \subseteq M$ an S-submodule. Then for all sufficiently large p, the image of S in $C \otimes \mathbb{Z}/p$ is isomorphic to $S \otimes \mathbb{Z}/p$, and the image of N in $M \otimes \mathbb{Z}/p$ is isomorphic to $N \otimes \mathbb{Z}/p$. (Hint: The degrees of defining relations for N and S are determined by the number of generators, by the degrees of the corresponding relations in M and C, and by lengths of words in generators expressing the generators of N and S; the bounds do not depend on the base field.)

Chapter 10

1. (Regev.) If A satisfies the Capelli identity c_{m+1} and B satisfies c_{n+1}, then $A \otimes B$ satisfies c_{mn+1}. (Hint: By Proposition 10.4.)

2. In the group algebra $F[S_n] = \oplus I_\lambda$, the element

$$E_\lambda = \sum_{\sigma \in S_n} \chi^\lambda(\sigma)\sigma \in I_\lambda$$

is the (central) unit in I_λ.

3. Recall the polynomials p^* from Definition 5.8. Taking E_λ as in Exercise 2, define the *Amitsur-Capelli polynomial*

$$E_\lambda^* = \sum_{\sigma \in S_n} \chi^\lambda(\sigma)x_{\sigma(1)}y_1 x_{\sigma(2)}y_2 \cdots x_{\sigma(n)}y_n,$$

a natural generalization of the Capelli polynomial. If $E_\lambda^* \in \mathrm{id}(A)$, then $f^* \in \mathrm{id}(A)$ for all $f \in I_\lambda$. (Hint: Since $f = aE_\lambda$, $a \in F[S_n]$, it suffices to show that for $\sigma \in S_n$, $(\sigma E_\lambda)^* \in \mathrm{id}(A)$. Let $\pi \in S_{2n}$ be the permutation of Lemma 5.9. Then

$$\begin{aligned} (\sigma E_\lambda)^*((\sigma E_\lambda)(x_1,\ldots,x_n,x_{n+1},\ldots,x_{2n}))\pi \\ = \sigma(E_\lambda(x_1,\ldots,x_{2n})\pi) = \sigma(E_\lambda^*) \in \mathrm{id}(A).) \end{aligned} \tag{13.13}$$

4. Let $\lambda \vdash n$ be a partition and let E_λ^* be the corresponding Amitsur-Capelli polynomial. Then $E_\lambda^* \in \mathrm{id}(A)$, if and only if the two-sided ideal $I_\mu \subseteq \mathrm{id}(A)$ for every partition μ containing λ. (Hint: Exercise 3, Corollary 5.22, and Theorem 5.50(ii).)

Chapter 11

1. Show the three basic Propositions 1.18, 1.17, and 1.19 also hold in the wider context of Chapter 11, when we define the *degree* of a x_i in a monomial to be the number of times the symbol x_i appears.

2. Suppose $\mathcal{V}$ is a variety of associative algebras (perhaps with additional structure) that has the property that if $A \in \mathcal{V}$, then $A/\mathrm{Nil}(A)$ is in $\mathcal{V}$ and satisfies the identity s_n. Then every algebra in $\mathcal{V}$ satisfies s_n^k for suitable k. (Hint: Consider the direct product $\tilde{A}$ of copies of A, indexed over all n-tuples of elements of A. $\tilde{A}/\mathrm{Nil}(\tilde{A})$ satisfies s_n, and $\mathrm{Nil}(\tilde{A})$ is nil of bounded index, so A satisfies s_n^k for some k. Conclude as in Exercise 3.1.)

3. $(x+y)^{-1} - y^{-1}(x^{-1} + y^{-1})^{-1}x^{-1}$ is a trivial rational identity of any ring.

4. (Hua's identity.) Verify the "trivial" rational identity

$$(x^{-1} + (y^{-1} - x)^{-1})^{-1} - (x - xyx).$$

(Hint: Substitute x^{-1} for x and $(y^{-1} - x)^{-1}$ for y in Exercise 3.)

5. (Burnside.) Any finitely generated periodic subgroup of $M_n(F)$ containing a base of $M_n(F)$ is finite. (Hint: If $g = \sum \alpha_i g_i$, then $\text{tr}(gg_i) = \sum \alpha_i(\text{tr}(gg_i))$, which has only a finite number of possible solutions since gg_i is a root of 1.)

6. (Procesi.) Any finitely generated torsion PI-group G is finite. (Hint: One may replace A by the F-subalgebra generated by G, which is f.d., and thus Noetherian. Take $I \lhd A$ maximal with respect to the image of G in A/I not being finite. Show I is a prime ideal; hence A/I is a matrix algebra, so use Exercise 5.)

Chapter 12

1. Prove the Amitsur-Levitzki Theorem for $M_n(F)$ as a formal consequence of the Hamilton-Cayley trace identity. (Hint: Linearize and then substitute commutators. Then take the $2n$-alternator, cf. Definition 1.27, and dispose of the superfluous terms using Remark 1.48.)

2. Reformulate the Razmyslov-Zubrilin theory of Theorem 2.55 in terms of trace identities.

3. The trace identity theory goes through for algebras without 1.

4. (Improved in Exercise 8.) If A is an algebra without 1 satisfying the identity x^n, then A is nilpotent of index $\leq 2n^2 - 3n + 2$. (Hint: A satisfies the Hamilton-Cayley trace identity, where formally we take $\text{tr}(a) = 0$ for all $a \in A$. Hence, A satisfies all the identities of $n \times n$ matrices, including s_{2n}, whose Young diagram is a $1 \times 2n$ column. But A also satisfies the symmetric identity $\tilde{s}_n$, whose Young diagram is an $n \times 1$ row. By the Branching Theorem, any Young diagram containing a row of length n or a column of length $2n$ must be an identity of A. This means the largest possible non-identity has $(n - 1)(2n - 1) = 2n^2 - 3n + 1$ boxes.)

5. Define a theory of Young diagrams analogous to that of Chapter 5, by means of Definition 12.15 and Theorem 12.19. Use the trick of passing from mixed trace identities to pure trace identities by adding an indeterminate x_0.

6. Define an operation φ on $F[S_n]$ by $\varphi(\pi) = \mathrm{sgn}(\pi)\pi$. Under the correspondence of Definition 12.15, show

$$\varphi(\mathrm{Tr}(x_{i_1} \cdots x_{i_{m_1}}) \mathrm{Tr}(x_{j_1} \cdots x_{j_{m_2}}) \cdots \mathrm{Tr}(x_{k_1} \cdots x_{k_{m_t}}))$$
$$= \pm \mathrm{Tr}(x_{i_1} \cdots x_{i_{m_1}}) \mathrm{Tr}(x_{j_1} \cdots x_{j_{m_2}}) \cdots \mathrm{Tr}(x_{k_1} \cdots x_{k_{m_t}}). \qquad (13.14)$$

7. Using the correspondence of Exercise 5, show φ of Exercise 6 sends a Young tableau T_λ to its transpose.

8. (Razmyslov's improved version of Theorem 3.40.) If A is an algebra without 1 satisfying the identity x^n, then A is nilpotent of index $\leq n^2$. (Hint: Use the Young theory with traces of Exercise 5, adding one extra variable x_0. Any non-identity must not contain a column of length $n + 1$, nor, by Exercise 7, a row of length $n + 1$, so has length at most n^2.)

9. (Kuzmin's lower bound.) The T-ideal (without 1) $\mathcal{I}$ of $\mathbb{Q}\{x, y\}$ generated by X^n does not imply nilpotence of degree $\frac{n^2+n-2}{2}$. (Hint: It is enough to show that

$$yx^1 yx^2 y \ldots yx^{n-1} \notin \mathcal{I}.$$

Note that $\mathcal{I}$ is generated by values of the symmetric polynomials $\tilde{s}_n$, cf. Equation (1.6). Given a function $\pi : \{0, \ldots, k\} \to \mathbb{N}$, identify π with the monomial

$$w_\pi = x^{\pi(0)} yx^{\pi(1)} yx^{\pi(2)} y \ldots yx^{\pi(n-1)};$$

it suffices to show $w_{(1)} \notin \mathcal{I}$. Let V be the T-space generated by $\{w_\pi : \pi \in S_n\}$; let $\mathcal{J}$ be the T-space generated by all terms of the following form:

(i) x^j for all $j \geq n$.

(ii) w_π unless π is $1 : 1$.

(iii) $w_\pi - w_{\pi\tau}$ when π is $1 : 1$, for all transpositions τ.

The intuition for $n = k$: viewing these as reduction procedures, one may eliminate any monomial w_π unless π is a permutation of $\{0, \ldots, n-1\}$, and then use (iii) to transpose the powers of x. However, one permits $k \leq n$ to facilitate the induction procedure.

Since $w_{(1)} \notin \mathcal{J}$, it suffices to show that $\mathcal{I} \cap V \subseteq \mathcal{J}$.

Take $a = \tilde{s}_n(u_1, \ldots, u_n) \in \mathcal{I} \cap V$. Rearranging the u_i in decreasing order according to $\deg_y u_i$, assume $\deg_y u_i = 0$ for all $i \geq k$. It suffices to prove $a \in \mathcal{J}$; $k \leq n - 1$ since $\deg_y a = n - 1$.

Note that $\mathcal{J}$ is invariant under the derivations $x^i \frac{\partial}{\partial x}$, so one can build a from terms of the form $\tilde{s}_n(v_1, \ldots, v_{k-1}, x, x, \ldots, x)$; thus, assume a has this form. Prove the stronger assertion:

$$\sum_{j_1 + \cdots + j_k = n+1-k} x^{j_1} v_1 x^{j_2} v_2 \ldots v_{k-1} x^{j_k} \in J.$$

This is seen by induction on k, using (iii) above to rearrange the terms in ascending order, and then breaking off the initial term and applying induction on k.)

Chapter 14

Lists of Theorems and Examples

14.1 Major Theorems

This is a list of the major results which are treated in this book; it does not include the major structure theorems (such as Kaplansky's Theorem and Posner's Theorem), which are quoted separately in Section 1.6. Background results are listed by name only.

Chapter 1.

1. Theorem 1.49. (Amitsur-Levitzki.) The standard identity s_{2n} is an identity of $M_n(C)$. (Rosset's proof; alternate proof given in Exercise 12.1.)

2. Theorem 1.55. (Razmyslov.) There is a 1:1 correspondence between multilinear central polynomials of $M_n(C)$ and multilinear 1-weak identities that are not identities.

3. Proposition 1.58. The identity of algebraicity.

4. Theorem J. A Hamilton-Cayley type equation involving evaluations of polynomials.

5. Theorem 1.72. (Lewin.) If the algebras $F\{X\}/I_i$ are representable, then $F\{X\}/I_1 I_2$ is representable.

6. Theorem 1.79. Any nil subalgebra N of a representable affine algebra is nilpotent, of bounded nilpotence index. ("Representable" is superfluous, by Theorem 2.57.)

7. Theorem 1.80. The Jacobson radical of a representable affine algebra is nilpotent. (Improved in Theorem 2.57 and in Exercises 2.15 ff.)

343

8. Theorem 1.91. (Anan'in.) Every affine, left Noetherian PI-algebra is representable.

9. Theorem 1.98. Suppose $\mathcal{I}$ satisfies the ACC. Then for any $I \in \mathcal{I}$, there are only a finite number of primes $P_1, \ldots, P_t$ in $\mathcal{I}$ minimal over I, and some finite product of the P_i is contained in I.

Chapter 2.

1. Theorem 2.3. (Shirshov's Theorem.) Let A be an affine PI-algebra of PI-degree d. The set of words of length $\leq d$ in the generators is a Shirshov base for A; if these words are integral, then A is f.g. as a module.

2. Lemma 2.10. (Shirshov's Lemma.) For any ℓ, k, d, there is $\beta = \beta(\ell, k, d)$ such that any d-indecomposable word w of length $\geq \beta$ in ℓ letters must contain a nonempty subword of the form u^k, with $|u| \leq d$.

3. Theorem 2.17. Any hyperword h is either d-decomposable or is preperiodic of periodicity $< d$.

4. Theorem 2.32. (Amitsur.) The Jacobson radical of an affine PI-algebra is locally nilpotent. (Superceded by the Razmyslov-Kemer-Braun Theorem.)

5. Corollary 2.45. (Zubrilin.) Suppose a polynomial $h(x_1, \ldots, x_n; y)$ A-alternates in the variables $x_i, 1 \leq i \leq n$. Then any algebra A satisfying the Capelli identity c_{n+1} also satisfies:

$$\sum_{j=0}^{n} (-1)^j z^{n-j} \delta_j(h(x_1, \ldots, x_n; y)) \tag{14.1}$$

(Most effective when used in conjunction with Theorem 4.82; also see Exercise 12.2.)

6. Propositions 2.38 and 2.39. Properties of the trace ring of a representable PI-algebra, respectively in characteristic 0 and positive characteristic.

7. Theorem 2.56. (Kemer.) Every affine PI-algebra satisfies a Capelli identity. (Theorem 2.60 gives an alternate proof of Theorem 2.56, via sparse identities.)

8. Theorem 2.57. (Razmyslov-Kemer-Braun.) The Jacobson radical of an affine PI-algebra is nilpotent.

9. Theorem 2.56. Every affine PI-algebra satisfies a Capelli identity. (Theorem 2.60 gives an alternate proof of Theorem 2.56, via sparse identities.)

10. Theorem 2.74. Improved bound for Shirshov's function $\beta(\ell, k, d)$. (Improved further in Exercises 2.8 and 2.11.)

11. Theorem 2.79. If a minimal nonzero hyperword on A is not preperiodic, then the values of its initial hyperwords are linearly independent. This yields a quick proof of a theorem of Ufnarovski'i (Corollary 2.81).

Chapter 3.

1. Theorem 3.25. For every field F and every n, the *generic division algebra* $UD(n, F)$ is indeed a division algebra.

2. Theorem 3.38. (Amitsur.) Every algebra A of PI-degree d satisfies a power s_{2n}^k of the standard identity, for $2n \leq d$. (Alternate proof in Exercise 1. A general version is given in Exercise 11.2.)

3. Theorem 3.40. (Dubnov-Ivanov-Nagata-Higman.) Any nil $\mathbb{Q}$-algebra A (without 1) of bounded index n is nilpotent. (Razmyslov's nilpotence bound of n^2 is given in Exercise 12.8.)

4. Corollary 3.46. Each multilinear identity of the Grassman algebra G (over an infinite dimensional vector space over a field) is a consequence of the Grassman identity $[[x_1, x_2], x_3]$. (Proved in characteristic 0; Exercise 3.5 handles arbitrary characteristic.)

Chapter 4.

1. Theorem 4.1; Theorem 4.66. (Kemer.) Suppose $\mathrm{char}(F) = 0$. Any affine PI-algebra is PI-representable; in other words, its T-ideal is the set of identities of a suitable f.d. algebra over a field.

2. Corollary 4.67. (Kemer.) Any relatively free PI-algebra (not necessarily affine) in characteristic 0 is Grassman representable, i.e., into matrices over a Grassman algebra.

3. Theorem 4.69. Any T-ideal of $F\{x_1, \ldots, x_\ell\}$ is finitely based (assuming $\mathrm{char}(F) = 0$). (A quicker but less intuitive proof is given in Exercise 4.3; the non-affine version is Theorem 6.52.)

4. Theorem 4.82. (Razmyslov-Zubrilin.) Suppose $f(x_1, \ldots, x_{2n+1}; y)$ is an RZ-polynomial for A. Then

$$f(x_1, \ldots, x_n, x_{n+1}, \ldots, x_{2n}, x_{2n+1}; y)$$
$$-f(x_{n+1}, \ldots, x_{2n}, x_1, \ldots, x_n, x_{2n+1}; y)$$
$$(14.2)$$

is an identity of A.

5. Exercise 4.8. (Improved in Proposition 6.84.) Any algebra satisfying a Capelli identity also satisfies the identities of $n \times n$ matrices, for some n.

6. Theorem 4.89. If c_d vanishes with respect to all word substitutions of length $\leq d$ in the d alternating variables $x_1, \ldots, x_d$, then $c_{2d-1} \in \mathrm{id}(A)$.

7. Corollary 4.93. Suppose A is an affine algebra, of Kemer index κ. Taking d such that A satisfies the Capelli identity c_d, let $\mu \geq d$ be arbitrary and let I be the ideal generated by all evaluations of μ-Kemer polynomials, where each designated indeterminate is specialized to a word in A of length $\leq d + 1$. Then $\mathrm{index}(A/I) < \kappa$.

Chapter 5.

1. Theorem 5.18. (Young.) Decomposition of the group algebra $F[S_n]$ in terms of the semi-idempotents of Young diagrams.

2. Theorem 5.21. Branching theorem for characters.

3. Theorem 5.23. Let $\sigma \in S_n$ and

$$\sigma \mapsto (P_\lambda, Q_\lambda)$$

in the Robinson-Schensted-Knuth correspondence, so $\lambda \vdash n$, and let $d = \ell(\lambda)$. Then d is the length of a maximal chain $1 \leq i_1 < \cdots < i_d \leq n$ such that $\sigma(i_1) > \cdots > \sigma(i_d)$. In other words, $\ell(\lambda)$ is the length of a maximal descending subsequence in σ.

Consequently, if f^λ is the number of standard tableaux of shape λ, then

$$n! = \sum_{\lambda \vdash n} (f^\lambda)^2.$$

The following three theorems provide extra important combinatoric information concerning the f^λ.

4. Theorem 5.27. The Young-Frobenius formula.

5. Theorem 5.29. The Hook formula.

6. Theorem 5.34. The Schur-Weyl theorem.

7. Theorem 5.38. (The Regev-Latyshev exponential bound theorem.) If A satisfies a PI of degree d, then the codimension

$$c_n(A) \leq (d-1)^{2n}.$$

8. Theorem 5.42. (Regev.) The tensor product of two two PI-algebras is also PI.

9. Theorem 5.50. (Amitsur.) Let A be a PI-algebra and let $I \subseteq F[S_n]$ be a two-sided ideal in $F[S_n] = V_n$.

   (i) Let $f = f(x) \in I$ and assume for $n \leq m \leq 2n - 1$ that the elements of the two-sided ideal generated by f in $F[S_m]$ are identities of A. Then $f^* \in \text{id}(A)$.

   (ii) Assume $I \subseteq \text{id}(A)$ and that for each $f \in I$, $f^* \in \text{id}(A)$.

   Then, for all $m \geq n$, all the elements of the two-sided ideal generated by I in $F[S_m]$ are identities of A: $F[S_m]IF[S_m] \subseteq \text{id}(A)$.

10. Theorem 5.51. (Kemer-Regev-Amitsur.) Suppose A has PI-degree d. Then there are u, v, depending on d, satisfying the following:

    (i) $F[S_m]e_{T_{\lambda_0}}F[S_m] \subseteq \Gamma_m$, for any tableau T_{λ_0} of $\lambda_0 = (v^u)$.

    (ii) Γ_m contains every two-sided ideal of S_m corresponding to a shape containing the rectangle (v^u).

11. Theorem 5.58. (Sparse identity.) Let A be a PI-algebra (not necessarily affine) of characteristic 0. There exists an integer n and nonzero $g \in F[S_n]$ that generates a two-sided ideal in $F[S_m]$ for all $m \geq n$. The identity g can be calculated explicitly.

Chapter 6.

1. Theorem 6.10. Let G denote the infinite dimensional Grassman algebra. Given a multilinear polynomial

$$p = p(x_1, \ldots, x_n) = \sum_{\sigma \in S_n} \alpha_\sigma x_{\sigma(1)} \cdots x_{\sigma(n)}$$

and $I \subseteq \{1, \ldots, n\}$, define

$$p*_I = \sum_{\sigma \in S_n} \alpha_\sigma \varepsilon(\sigma, I) x_{\sigma(1)} \cdots x_{\sigma(n)}.$$

The following assertions are equivalent:

(i) $p \in \mathrm{id}(A_0)$.

(ii) $\tilde{p}*_I$ is a superidentity of $A_0 \otimes G$ for some subset $I \subseteq \{1, \ldots, n\}$.

(iii) $\tilde{p}*_I$ is a superidentity of $A_0 \otimes G$ for every subset $I \subseteq \{1, \ldots, n\}$.

2. Theorem 6.13, Theorem 6.14. (Kemer.) The correspondence between superidentities and identities of the Grassman envelope.

3. Theorem 6.20. (Kemer's affine trick.) For any algebra A of characteristic 0, $\mathrm{id}_2(A \otimes G) = \mathrm{id}_2(\tilde{A})$, for a suitable affine superalgebra $\tilde{A}$.

4. Exercise 6.4. Any relatively free PI-algebra (not necessarily affine) in characteristic 0 is Grassman representable, i.e., into matrices over a Grassman algebra.

5. Exercise 6.5. "Super" version of Anan'in's theorem.

6. Theorem 6.31. (Wedderburn Principal supertheorem.) Any f.d. superalgebra R over an algebraically closed field has a semisimple supersubalgebra $\bar{R}$ such that $R = \bar{R} \oplus J(R)$.

7. Theorem 6.32. "Super" version of Theorem J.

8. Theorem 6.33. (Kemer.) Every affine superalgebra is PI$_2$-equivalent to a finite dimensional superalgebra.

9. Theorem 6.37. "Super" version of Lewin's theorem.

10. Theorem 6.51. (Kemer) The ACC holds for T_2-ideals of affine superalgebras.

11. Theorem 6.52. Kemer's solution to Specht's problem: The ACC holds for T-ideals, over any field of characteristic 0.

12. Theorem 6.53. (Kemer.) Any T-ideal J of $\mathcal{A}$ has a radical $\sqrt{J}$ that is a finite intersection of prime T-ideals, and $\sqrt{J}^k \subseteq J$ for some k.

13. Theorem 6.64. (Kemer.) A complete list of the prime T-ideals of $F\{X\}$, for any field F of characteristic 0:

    (i) $\mathrm{id}(M_n(F))$;

    (ii) $\mathrm{id}(M_{k\ell})$;

    (iii) $\mathrm{id}(M_n(G))$.

14. Theorem 6.70. (Kemer.) Every verbal ideal I has only a finite number of verbal primes $P_1, \ldots, P_t$ minimal over I, and some finite product of the P_i is contained in I.

15. Theorem 6.75. (Kemer) Any algebra (not necessarily affine) of characteristic 0 satisfying s_n satisfies a Capelli identity. (A better estimate for n is indicated in Exercises 5.4 ff.)

Chapter 7.

1. Theorem 7.5. (Kemer-Belov.) Any PI-algebra in characteristic $p > 0$ satisfies the identities of $M_m(\mathbb{Z}/p)$, for suitable m. A key step is Exercise 7.21 (Kemer), that every PI-algebra A of characteristic p satisfies a standard identity.

2. Theorem 7.27; Theorem 7.38. (Shchigolev.) In nonzero characteristic, the T-space generated by $\{P_n : n \in \mathbb{N}\}$ is not finitely based.

3. Theorem 7.45. (Belov.) The T-ideal generated by the polynomials $\{Q_n : n \in \mathbb{N}\}$ is not finitely based in characteristic $p > 2$. (The case for $p = 2$ is given in Exercise 7.10.)

4. Exercise 7.16. (Shchigolev.) An affine PI-algebra with a non-finitely based as T-space, modulo the Grassman identity. (Nevertheless, every T-ideal must be finitely based.)

Chapter 8.

1. Theorem 8.2. Every affine Noetherian PI-algebra A (over any commutative Noetherian ring) is finitely presented.

2. Exercise 8.1. Any prime PI-ring R with Noetherian center C is a f.g. C-module.

3. Theorem 8.9. (Isaacs-Passman.) If $\mathrm{char}(F) = 0$ and $F[G]$ satisfies a PI, then G has a normal Abelian subgroup of finite index.

4. Exercise 8.5. (Bachturin.) The (unrestricted) enveloping algebra of a Lie algebra L in characteristic p satisfies a PI iff L has an Abelian ideal of finite codimension and the adjoint representation of L is algebraic.

Chapter 9.

1. Theorem 9.11. (Bergman's gap.) If $GKdim(A) > 1$, then $GKdim(A) \geq 2$.

2. Exercise 9.8. (Small-Stafford-Warfield.) If A is affine of GK dimension 1, then A is PI. (Treated in the semiprime case.)

3. Exercise 9.9. (Small-Warfield.) If A is prime affine of GK dimension 1, then A is f.g. over a central polynomial ring, and thus Noetherian.

4. Theorem 9.19. (Amitsur-Drensky.) $GKdim(A)$ is bounded by the Shirshov height, for any affine algebra A.

5. Proposition 9.31. (Berele.) $GKdim(A)$ is bounded by a function of the Kemer index.

6. Theorem 9.26. (Belov-Borisenko-Latyshev.) $GKdim(A)$ equals the essential height, for any representable affine algebra.

7. Exercise 9.18. The set of lexicographically minimal words in a PI-algebra A of PI-class n has bounded height over the set of words of length $\leq n$.

8. Exercise 9.19. A set of words S in a relatively free algebra $\mathcal{A}$ is an essential Shirshov base iff, for each word u of length not greater than the PI-degree of $\mathcal{A}$, S contains a word that is cyclically conjugate to some power of u.

9. Exercise 9.20. A converse to Shirshov's theorem.

10. Theorem 9.36. (Bell.) Any generating set of a prime affine PI-algebra can be expanded to a rational generating set.

11. Theorem 9.44. (Belov.) In characteristic 0, every relatively free, affine PI-algebra has a rational Hilbert series. (A multivariate version is given in Exercise 9.26.)

12. Theorem 9.46. If A is affine and its graded algebra is Noetherian PI, then A has a rational generating set.

13. Exercise 9.31. If A is relatively free and $\mathbb{Z}$-affine, then for all sufficiently large p, the algebra $A \otimes \mathbb{Z}/p$ can be represented by matrices of bounded degree, and A is p-torsion free.

Chapter 10.

1. Theorem 10.5. (Regev.) The cocharacters of a PI-algebra over a field F of characteristic zero are supported on a hook of Young diagrams

2. Theorem 10.9. Schur's double centralizer theorem.

3. Theorem 10.14. Description of the Weyl module.

4. Theorem 10.17. (Weyl.) The $GL(V) \times S_n$ decomposition of $T^n(V)$.

5. Theorem 10.19. The dimension of the Weyl module.

6. Theorem 10.22. (Drensky, Berele.) The multiplicities m_λ in the homogeneous cocharacters are the same as in the multilinear cocharacters.

Chapter 11.

1. Exercise 11.6. (Procesi.) Any finitely generated torsion PI-group is finite.

Chapter 12.

1. Theorem 12.13. Suppose $\sigma = (i_1, \ldots, i_a)(j_1, \ldots, j_b) \cdots (k_1, \ldots, k_c) \in S_n$, and let $A_1, \ldots, A_n \in \operatorname{End}(V)$. Then

$$\operatorname{tr}((A_1 \otimes \cdots \otimes A_n) \circ \varphi_{n,k}(\sigma))$$
$$= \operatorname{tr}(A_{i_1} \cdots A_{i_a}) \operatorname{tr}(A_{j_1} \cdots A_{j_b}) \cdots \operatorname{tr}(A_{k_1} \cdots A_{k_c}).$$

2. Theorem 12.19. A multilinear trace polynomial $f_a(x_1, \ldots, x_n)$, for $a \in F[S_n]$, is a trace identity of $M_k(F)$ if and only if

$$a \in \bigoplus_{\lambda \vdash n,\ \lambda_1' \geq k+1} I_\lambda.$$

3. Theorem 12.22. (Razmyslov, Helling, and Procesi.) The fundamental theorem of trace identities:

   The T-ideal of the pure trace identities of the $k \times k$ matrices is generated by the single trace polynomial $f_{e_{\bar{\mu}}}$, where

$$e_{\bar{\mu}} = \sum_{\pi \in S_{k+1}} \operatorname{sgn}(\pi)\pi.$$

   The T-ideal of the mixed trace identities of the $k \times k$ matrices is generated by the two elements $f_{e_{\bar{\mu}}}$ and $\tilde{g}_k$ (cf. Example 12.21(iii)).

4. Theorem 12.25. $\chi_n^{Tr}(M_k(F)) = \bigoplus_{\lambda \vdash n,\ \lambda_1' \leq k} \chi^\lambda \otimes \chi^\lambda$.

14.2   Major Examples and Counterexamples

The following is a list of interesting examples appearing in the text. The major counterexamples of Chapter 7 are listed as theorems, as is the universal division algebra described in Chapter 3.

1. Example 1.7. Easy examples of PI-algebras.

2. Exercise 1.4. Any finite ring satisfies an identity $x^n - x^m$, for suitable $n, m \in \mathbb{N}$.

3. Example 1.13. The symmetric polynomial $\tilde{s}_n$ is the multilinearization of x^n.

4. Exercise 1.11. (Amitsur.) $s_{n+1}(y, xy, x^2y, \ldots, x^ny) \in \mathrm{id}(M_n(F))$.

5. Exercise 1.12. (Amitsur-Levitzki.) $M_n(F)$ cannot satisfy any identity of degree $< 2n$.

6. Exercise 1.14. The Capelli polynomial as an alternator of Capelli polynomials of smaller size.

7. Example 1.34. The matrix superalgebras.

8. Definition 1.35. The Grassman algebra.

9. Example 1.83. (Lewin.) Existence of nonrepresentable affine PI-algebras.

10. Exercise 1.26. Existence of affine PI-algebra which is not PI-equivalent to a f.d. algebra.

11. Exercise 2.13. Counterexample to a naive converse of Shirshov's Theorem.

12. Exercise 2.23. A generalized hemiring with non-commutative addition.

13. Exercise 3.7. (Razmyslov-Drensky.) $\mathrm{id}(M_2(\mathbb{Q}))$ is based by s_4 and Hall's identity.

14. Example 4.28. The algebra of upper triangular $n \times n$ matrices has low Kemer index, namely $(1, n)$, but does not satisfy the standard identity s_{2n-1}.

15. Example 4.36. A non-full, subdirectly irreducible algebra.

16. Example 4.50. The "generic" algebra of a PI-minimal f.d. algebra.

17. Example 4.73. Examples of K-admissible algebras.

18. Example 5.5. The Grassman T-ideal is not preserved under the right action of permuatations.

19. Example 5.6. The Capelli polynomial c_n is

$$(s_n(x_1, \ldots, x_n)x_{n+1} \cdots x_{2n})\pi,$$

for suitable $\pi \in S_{2n}$.

20. Example 5.20. The standard identity $s_n = \sum_{\sigma \in S_n} \mathrm{sgn}(\sigma)\sigma$ is the (central) unit in the two-sided ideal I_λ, where $\lambda = (1^n)$.

21. Example 5.47. A polynomial corresponding to a rectangular frame. (But Exercise 5.1 shows one can get nonequivalent identities by filling the frame differently.)

22. Example 5.57. Although the standard identity is not sparse, the Capelli identity is sparse.

23. Example 6.27. A supersimple superalgebra that is not simple.

24. Exercise 6.2. The tensor product of various T-prime T-ideals.

25. Example 6.76. The relatively free superalgebra $\tilde{G}$ with respect (only) to the graded identities of (6.1) of Remark 6.5.

26. Exercise 7.3. All identities of the extended Grassman algebra G^+ are consequences of the Grassman identity.

27. Exercise 7.7. The Frobenius identity $(x + y)^p - x^p - y^p$ holds in Grassman algebras of characteristic $p > 2$.

28. Exercise 8.2. An uncountable class of non-finitely presented monomial algebras satisfying ACC on ideals.

29. Example 8.6. If G is any group that contains an Abelian group A of finite index, then $F[G]$ is a f.g. module over the the commutative algebra $F[A]$.

30. Example 9.4. The Hilbert series of the (commutative) polynomial algebra.

31. Example 9.10. The GK dimension of the polynomial algebra.

32. Example 9.13. The GK dimension of block triangular matrix rings.

Chapter 15

Some Open Questions

Central Polynomials

1. What are the central polynomials of minimal degree for $M_n(\mathbb{Q})$?

   Although Razmyslov's method in principle gives all the central polynomials for matrices in terms of 1-weak identities, this question stubbornly remains.

2. For p prime, does $M_p(\mathbb{Q})$ have a noncentral polynomial f such that f^p is central? This easily-formulated question is equivalent to division algebras of degree p being cyclic, a long-standing open question. Thus, it is known to be true for $p \leq 3$, but is open for all other p. It is easy to see for $p \geq 2$ that f cannot be multilinear.

3. Describe the central polynomials (if any) for $M_n(G)$ for $n > 1$.

   The standard methods for $M_n(F)$ fail because the grading throws us off balance; for example, we lose the Capelli identity in characteristic 0. On the other hand, $[x_1, x_2]$ is central for G.

T-Ideals

1. What are the cocharacters of $\mathrm{id}(M_n(\mathbb{Q}))$? for $\mathrm{id}(M_n(G))$? $\mathcal{M}_{k,\ell}$?

   This is perhaps the most studied question in the theory, and considerable advances have been made using multivariate Hilbert series, as described at the end of Chapter 10. Nevertheless, the solution still seems far away.

2. What is the best asymptotic formula for the codimensions of a T-ideal? (Some major advances have been made recently by Giambruno, Regev, and Zaicev, cf. the bibliography.)

3. Describe the identities of minimal degree of $M_n(G)$.

4. Find a set of identities that generate the T-ideal $\mathrm{id}(M_n(G))$.

5. Find a set of trace identities that generate $\mathrm{id}(M_n(G))$.

   In positive characteristic we still are far from the full picture. For example, the following question surprising remains open:

6. Is the T-ideal (in an infinite number of indeterminates) of $M_n(F)$ finitely based, when F is an infinite field of characteristic p? (This is known to have a negative answer for f.d. Lie algebras.)

7. Does Specht's conjecture hold for T-ideals generated by multilinear polynomials? (This is true in characteristic p.)

8. Classify the prime T-ideals over an infinite field of characteristic > 0. (This is extremely difficult, with the best results provided by Kemer [Kem96]).

9. (Kemer.) Is Remark 6.66 true in positive characteristic?

10. (L'vov.) Is every PI-algebra a homomorphic image of a torsion-free PI-algebra? (This would enable one to transfer several theorems directly from characteristic 0 to characteristic p.)

11. (Procesi's problem.) Let $C\{Y\}_n$ denote the algebra of generic matrices over C, cf. Definition 3.22. There is a natural homomorphism $\Phi_p \mathbb{Z}\{Y\}_n \to (\mathbb{Z}/p\mathbb{Z})\{Y\}_n$, obtained by taking mod p, and Schelter and Kemer found examples where $\ker \Phi_p$ properly includes $p\mathbb{Z}\{Y\}_n$. Fixing n, is $\ker \Phi_p = p\mathbb{Z}\{Y\}_n$ for large enough p? (Belov has showed this is true in the affine case, i.e., when there are a fixed finite number of Y.)

12. What are the prime ideals of the group algebra of the infinite symmetric group, in positive characteristic?

13. What are the T-prime varieties in the class of algebras having a formal trace function for which $\mathrm{tr}(1) = 0$?

GK Dimension and Essential Height

1. Is there an example of a prime Goldie ring of GK dimension 2 that is neither primitive nor PI?

2. Find the best possible formulas for GK dimension of affine PI-algebras.

3. Find the best possible formulas for the essential (Shirshov) height of affine PI-algebras. (Recall that this is the same as for GK dimension for representable PI-algebras.)

4. Is the difference between the essential height and the GK dimension bounded? By 1?

5. Is there any natural criterion weaker than representability that will guarantee that a PI-algebra has finite GK dimension?

6. What can one say about the (minimal) kernel of a PI-algebra into a representable algebra?

Nonassociative Algebras

1. Does every affine alternative algebra satisfy all identities of a f.d. alternative algebra? What about Jordan algebras? Find some workable (perhaps asymptotic) analog of Lewin's theorem.

2. Does the local Specht problem have an affirmative answer for Lie algebras? (Iltyakov [Ilt91] obtained an affirmative answer in characteristic 0 for Lie algebras satisfying a Capelli identity.)

Bibliography

[ADF85]   Almkvist, G., Dicks, W., and Formanek, E., *Hilbert series of fixed free algebras and noncommutative classical invariant theory*, J. Algebra **93** (1985), no. 1, 189–214.

[Am57]   Amitsur, S. A., *A generalization of Hilbert's Nullstellensatz*, Proc. Amer. Math. Soc. **8** (1957), 649–656.

[Am66]   Amitsur, S. A., *Rational identities and applications to algebra and geometry*, J. Algebra **3** (1966), 304–359.

[Am70]   Amitsur, S. A., *A noncommutative Hilbert basis theorem and subrings of matrices*, Trans. Amer. Math. Soc. **149** (1970), 133–142.

[Am71]   Amitsur, S. A., *A note on PI-rings*, Israel J. Math. **10** (1971), 210–211.

[Am72]   Amitsur, S. A., *On central division algebras*, Israel J. Math. **12** (1972), 408–420.

[Am74]   Amitsur, S. A., *Polynomial identities*, Israel J. Math. **19** (1974), 183–199.

[AmL50]   Amitsur, S. A. and Levitzki, J., *Minimal identities for algebras*, Proc. Amer. Math. Soc. **1** (1950), 449–463.

[AmPr66]   Amitsur, S. A. and Procesi, C., *Jacobson-rings and Hilbert algebras with polynomial identities*, Ann. Mat. Pura Appl. **71** (1966), 61–72.

[AmRe82]   Amitsur, S. A. and Regev, A., *PI-algebras and their cocharacters*, J. Algebra **78** (1982), no. 1, 248–254.

[An77]   Anan'in, A. Z., *Locally residually finite and locally representable varieties of algebras*, Algebra i Logika **16** (1977), no. 1, 3–23.

[An92]   Anan'in, A. Z., *The representability of finitely generated algebras with chain condition*, Arch. Math. **59** (1992), 1–5.

[Ani82a]   Anick, D., *The smallest singularity of a Hilbert series*, Math. Scand. **51** (1982), 35–44.

[Ani82b]   Anick, D., *Non-commutative graded algebras and their Hilbert series*, J. Algebra **78** (1982), no. 1, 120–140.

[Ani88]   Anick, D., *Recent progress in Hilbert and Poincare series*, Lect. Notes Math., vol. 1318 (1988), 1–25.

[ArSc81]   Artin, M. and Schelter, W., *Integral ring homomorphisms*, Advances in Math. **39** (1981), 289–329.

[Ba74]   Bakhturin, Yu. A., *Identities in the universal envelopes of Lie algebras*, Australian Math. Soc. **18** (1974), 10–21.

[Ba87]   Bakhturin, Yu. A., *Identical relations in Lie algebras*. Translated from the Russian by Bakhturin. VNU Science Press, b.v., Utrecht, (1987).

[Ba89]   Bakhturin, Yu. A. and Ol'shanskiĭ, A. Yu., *Identities*. (Russian) Current problems in mathematics. Fundamental directions **18**, Itogi Nauki i Tekhniki, Akad. Nauk SSSR, Vsesoyuz. Inst. Nauchn. i Tekhn. Inform., Moscow, (1988), 117–240, in Russian.

[Ba91]   Bakhturin, Yu. A., *Identities*, Encyclopedia of Math. Sci. **18** Algebra II, Springer-Verlag, Berlin-New York (1991), 107–234.

[BGGKPP]   Bandman, T., Grunewald, F., Gruel, G. M., Kunyavski, B., Pfister, G., and Plotkin, E., *Two-variable identities for finite solvable groups*, C.R. Acad. Sci. Paris, Ser. I **337** (2003), 581–586.

[Bei86]   Beĭdar, K. I., *On A. I. Mal'tsev's theorems on matrix representations of algebras*. (Russian) Uspekhi Mat. Nauk **41** (1986), no. 5, (251), 161–162.

[BeMM96]   Beĭdar, K. I., Martindale, W. S., and Mikhalev, A. V., *Rings with Generalized Identities*, Pure and Applied Mathematics **196**, Dekker, New York (1996).

[Bell03]   Bell, J. P., *Examples in finite Gelfand-Kirillov dimension*, J. Algebra **263** (2003), no. 1, 159–175.

[Bell04]   Bell, J. P., *Hilbert series of prime affine PI-algebras*, Israel J. Math. **139** (2004), 1–10.

[Bel88a]   Belov, A., *On Shirshov bases in relatively free algebras of complexity n*, Mat. Sb. **135** (1988), no. 3, 373–384.

[Bel88b]   Belov, A., *The height theorem for Jordan and Lie PI-algebras*,
                in: Tez. Dokl. Sib. Shkoly po Mnogoobr. Algebraicheskih Sistem,
                Barnaul (1988), pp. 12–13.

[Bel89]   Belov, A., *Estimations of the height and Gelfand-Kirillov dimen-
                sion of associative PI-algebras*, in: Tez. Dokl. po Teorii Ko-
                letz, Algebr i Modulei. Mezhdunar. Konf. po Algebre Pamyati
                A.I.Mal'tzeva, Novosibirsk (1989), p. 21.

[Bel92]   Belov, A., *Some estimations for nilpotency of nil-algebras over a
                field of an arbitrary characteristic and height theorem*, Commun.
                Algebra **20** (1992), no. 10, 2919–2922.

[Bel97]   Belov, A., *Rationality of Hilbert series with respect to free alge-
                bras*, Russian Math. Surveys **52** (1997), no. 10, 394–395.

[Bel99]   Belov, A., *About non-Spechtian varieties*, Fundam. Prikl. Mat. **5**
                (1999), no. 6, 47–66.

[Bel2000]   Belov, A., *Counterexamples to the Specht problem*, Sb. Math. **191**
                (2000), no. 3-4, 329–340.

[Bel2002]   Belov, A., *Algebras with polynomial idenitities: Representa-
                tions and combinatorial methods*, Doctor of Science Dissertation,
                Moscow (2002).

[BBL97]   Belov, A., Borisenko, V. V., and Latyshev, V. N., *Monomial
                algebras. Algebra 4*, J. Math. Sci. (New York) **87** (1997), no. 3,
                3463–3575.

[BelIv03]   Belov, A. Ya. and Ivanov, I. A., *Construction of semigroups with
                some exotic properties*, Comm. Algebra **31** (2003), no. 2, 673–696.

[Ber82]   Berele, A., *Homogeneous polynomial identities*, Israel J. Math.
                42 (1983), no. 3, 258–272.

[Ber85a]   Berele, A., *On a theorem of Kemer*, Israel J. Math. **51** (1985),
                no. 1-2, 121–124.

[Ber85b]   Berele, A., *Magnum PI*, Israel J. Math. **51** (1985), no. 1-2, 13–19.

[Ber88]   Berele, A., *Trace identities and $\mathbb{Z}/2\mathbb{Z}$ graded invariants*, Trans.
                Amer. Math. Soc. **309** (1988), 581–589.

[Ber93]   Berele, A., *Generic verbally prime PI-algebras and their GK-
                dimensions*, Comm. Algebra **21** (1993), no. 5, 1487–1504.

[Ber94]   Berele, A., *Rates of growth of p.i. algebras*, Proc. Amer. Math.
                Soc. **120** (1994), no. 4, 1047–1048.

[Ber04]   Berele, A., *Colength sequences for matrices*, J. Algebra (to appear).

[BerBer]   Berele, A. and Bergen, K., *PI algebras with Hopf algebra actions*, J. Algebra **214** (1999), 636–651.

[BeRe87]   Berele, A. and Regev, A., *Hook Young diagrams with applications to combinatorics and to representations of Lie superalgebras*, Advances in Math. **64** (1987), no. 2, 118–175.

[BeP92]   Bergen, J. and Passman, D. S., *Delta methods in enveloping algebras of Lie superalgebras*, Trans. Amer. Math. Soc. **334** (1992), no. 1, 295–280.

[Bir35]   Birkhoff, G., *On the structure of abstract algebras*, Proc. Cambridge Phil. Soc. **31** (1935), 433–431.

[Bog01]   Bogdanov, I., *Nagata-Higman's theorem for hemirings*, Fundam. Prikl. Mat. **7** (2001), no. 3, 651–658 (in Russian).

[BLH88]   Bokut', L. A., L'vov, I. V., and Harchenko, V. K., *Noncommutative rings*, in: Sovrem. Probl. Mat. Fundam. Napravl. Vol. 18, Itogi Nauki i Tekhn., All-Union Institute for Scientificand Technical Information (VINITI), Akad. Nauk SSSR, Moscow (1988), 5–116.

[BoKr76]   Borho, W. and Kraft, H., *Uber die Gelfand-Kirillov dimension*, Math. Ann. **220** (1976), no. 1, 1–24.

[Br81]   Braun, A., *A note on Noetherian PI-rings*, Proc. Amer. Math. Soc. **83** (1981), 670–672.

[Br82]   Braun, A., *The radical in a finitely generated PI-algebra*, Bull. Amer. Math. Soc. **7** (1982), no. 2, 385–386.

[Br84]   Braun, A., *The nilpotency of the radical in a finitely generated PI-ring*, J. Algebra **89** (1984), 375–396.

[BrVo92]   Braun, A. and Vonessen, N., *Integrality for PI-rings*, J. Algebra **151** (1992), 39–79.

[Che94a]   Chekanu, G. P., *On independence in algebras*, Proc. Xth National Conference in Algebra (Timişoara, 1992), Timişoara (1994), 25–35.

[Che94b]   Chekanu, G. P., *Independency and quasiregularity in algebras*, Dokl. Akad. Nauk **337** (1994), no. 3, 316–319; translation: Russian Acad. Sci. Dokl. Math. **50** (1995), no. 1, 84–89.

[ChKo93]   Chekanu, G. P. and Kozhuhar', E. P., *Independence and nilpo-
             tency in algebras*, Izv. Akad. Nauk Respub. Moldova Mat. (1993),
             no. 2, 51–62, 92–93, 95.

[ChUf85]   Chekanu, G. P. and Ufnarovski'i, V. A., *Nilpotent matrices*, Mat.
             Issled. no. 85, Algebry, Kotsa i Topologi (1985), 130–141, 155.

[CuRei62]   Curtis, C. W. and Reiner, I., *Representation Theory of Finite
             Groups and of Associative Algebras*, Interscience Publishers, New
             York, London, Wiley (1962).

[De22]   Dehn, M., *Uber die Grundlagen der projektiven Geometrie und
             allgemeine Zahlsysteme*, Math. Ann. **85** (1922), 184–193.

[Do92]   Donkin, S., *Invariants of several matrices*, Invent. Math. **110**
             (1992), no. 2, 389–401.

[Do93]   Donkin, S., *Invariant functions on matrices*, Math. Proc. Cam-
             bridge Philos. Soc. **113** (1993), no. 1, 23–43.

[Do94]   Donkin, S., *Polynomial invariants of representations of quivers*,
             Comment. Math. Helv. **69** (1994), no. 1, 137–141.

[Dr74]   Drensky, V., *About identities in Lie algebras*, Algebra and Logic
             13 (1974), no. 3, 265–290, 363–364.

[Dr81a]   Drensky, V., *Representations of the symmetric group and vari-
             eties of linear algebras*, Mat. Sb. **115** (1981), no. 1, 98–115.

[Dr81b]   Drensky, V., *Codimensions of T-ideals and Hilbert series of rel-
             atively free algebras*, C.R. Acad. Bulgare Sci. **34** (1981), no. 9,
             1201–1204.

[Dr81c]   Drensky, V., *A minimal basis for identities of a second order ma-
             trix over a field of characteristic 0*, Algebra and Logic **20** (1981),
             282–290.

[Dr84]   Drensky, V., *Codimensions of T-ideals and Hilbert series of rela-
             tively free algebras*, J. Algebra **91** (1984), no. 1, 1–17.

[Dr95]   Drensky, V., *New central polynomials for the matrix algebra*, Is-
             rael J. Math. **92** (1995), no. 1-3, 235–248.

[Dr00]   Drensky, V., *Free Algebras and PI-algebras: Graduate Course in
             Algebra*, Springer-Verlag, Singapore (2000).

[DrFor04]   Drensky, V. and Formanek, E., *Polynomials Identity Rings*, CRM
             Advanced Courses in Mathematics, Birkhäuser, Basel (2004).

[DrKa84]   Drensky, V. and Kasparian, A., *A new central polynomial for* 3×3 *matrices*, Comm. Algebra **13** (1984), no. 3, 745–752.

[DrRa93]   Drensky, V. and Rashkova, T. G., *Weak polynomial identities for the matrix algebras*, Comm. Algebra **21** (1993), no. 10, 3779–3795.

[Fl95]   Flavell, P., *Finite groups in which every two elements generate a soluble group*, Invent. Math. **121** (1995), 279–285.

[For72]   Formanek, E., *Central polynomials for matrix rings*, J. Alg. **23** (1972), 129–132.

[For82]   Formanek, E., *The polynomials identities of matrices*, Algebraists' Homage: Papers in Ring Theory and Related Topics (New Haven, Conn., 1981), Contemp. Math. **13** (1982), 41–79.

[For84]   Formanek, E., *Invariants and the ring of generic matrices*, J. Algebra **89** (1984), no. 1, 178–223.

[For87]   Formanek, E., *The invariants of* $n \times n$ *matrices*, Invariant Theory, Lecture Notes in Math. **1278**, Springer, Berlin-New York (1987) 18–43.

[For91]   Formanek, E., *The Polynomial Identities and Invariants of* $n \times n$ *matrices*, CBMS Regional Conference Series in Mathematics, 78, AMS (1991).

[For02]   Formanek, E., *The ring of generic matrices*, J. Algebra **258** (2002), no. 1, 310–320.

[ForL76]   Formanek, E. and Lawrence, J., *The group algebra of the infinite symmetric group*, Israel J. Math. **23** (1976), no. 3-4, 325–331.

[GeKi66a]   Gelfand, I. M. and Kirillov, A. A., *Sur les corps lies aux algebras envelopantes des algebras de Lie*, Publs. Math. Inst. Hautes Etud. Sci. **31** (1966), 509–523.

[GeKi66b]   Gelfand, I. M. and Kirillov, A. A., *On division algebras related to enveloping algebras of Lie algebras*, Dokl. Akad. Nauk SSSR **167** (1966), no. 3, 503–505.

[GRZ00]   Giambruno, A., Regev, A., and Zaicev, M., *Simple and semisimple Lie algebras and codimension growth*, Trans. Amer. Math. Soc. **352** (2000), no. 4, 1935–1946.

[GiZa03a]   Giambruno, A. and Zaicev, M., *Codimension growth and minimal superalgebras*, Trans. Amer. Math. Soc. **355** (2003), no. 12, 5091–5117.

[GiZa03b]   Giambruno, A. and Zaicev, M., *Asymptotics for the standard and the Capelli identities*, Israel J. Math. Soc. **135** (2003), 125–145.

[GiZa03c]   Giambruno, A. and Zaicev, M., *Minimal varieties of algebras of exponential growth*, Advances Math. **174** (2003), no. 2, 310–323.

[GoWa89]   Goodearl, K. R. and Warfield, R. B., *An Introduction to Noncommutative Noetherian Rings*, London Math. Soc. Student Texts, vol. 16 (1989).

[Gol64]   Golod, E. S., *On nil-algebras and residually finite p-groups*, Izv. Akad. Nauk SSSR **28** (1964), no. 2, 273–276.

[Gov72]   Govorov, V. E., *On graded algebras*, Mat. Zametki **12** (1972), no. 2, 197–204.

[Gov73]   Govorov, V. E., *On dimension of graded algebras*, Mat. Zametki, **14** (1973) no. 2, 209–216.

[Gre80]   Green, J. A., *Polynomial representations of* GL_n, Lecture Notes in Math., vol. 830, Springer-Verlag, Berlin-New York (1980).

[Gri87]   Grishin, A. V., *The growth index of varieties of algebras and its applications*, Algebra i Logika **26** (1987), no. 5, 536–557.

[Gri99]   Grishin, A. V., *Examples of T-spaces and T-ideals in Characteristic 2 without the Finite Basis Property*, Fundam. Prikl. Mat. **5** (1) (1999), no. 6, 101–118 (in Russian).

[GuKr02]   Gupta, C. K. and Krasilnikov, A. N., *A simple example of a nonfinitely based system of polynomial identities*, Camm. Algebra **36** (2002), 4851–4866.

[Ha43]   Hall, M., *Projective planes*, Trans. Amer. Math. Soc. **54** (1943), 229–277.

[Hel74]   Helling, H., *Eine Kennzeichnung von Charakteren auf Gruppen und Assoziativen Algebren*, Comm. in Alg. **1** (1974), 491–501.

[Her68]   Herstein, I. N., *Noncommutative Rings*, Carus Math. Monographs **15**, Math. Assoc. Amer. (1968).

[Her71]   Herstein, I. N., *Notes from a Ring Theory Conference*, CBMS Regional Conference Series in Math. **9**, Amer. Math. Soc. (1971).

[Hig56]   Higman, G., *On a conjecture of Nagata*, Proc. Cam. Phil. Soc. **52** (1956), 1–4.

[Ilt91]   Iltyakov, A. V., *Finiteness of basis identities of a finitely generated alternative PI-algebra*, Sibir. Mat. Zh. **31** (1991), no. 6, 87–99; English translation: Sib. Math. J. **31** (1991), 948–961.

[Ilt92]   Iltyakov, A. V., *On finite basis identities of identities of Lie algebra representations*, Nova J. Algebra Geom. **1** (1992), no. 3, 207–259.

[Ilt03]   Iltyakov, A. V., *Polynomial identities of Finite Dimensional Lie Algebras*, monograph (2003).

[IsPa64]   Isaacs, I. M. and Passman, D. S., *Groups with representations of bounded degree*, Canadian J. Math. **16** (1964), 299–309.

[Jac75]   Jacobson, N., *PI-Algebras, an introduction*, Springer Lecture Notes in Math. **441**, Springer, Berlin-New York (1975).

[Jac80]   Jacobson, N., *Basic Algebra II*, Freeman, New York (1980).

[Kap48]   Kaplansky, I., *Rings with a polynomial identity*, Bull. Amer. Math. Soc. **54** (1948), 575–580.

[Kap49]   Kaplansky, I., *Groups with representations of bounded degree*, Canadian J. Math. **1** (1949), 105–112.

[Kap50]   Kaplansky, I., *Topoloigcal representation of algebras. II*, Trans. Amer. Math. Soc. **66** (1949), 464–491.

[Kap70]   Kaplansky, I., *"Problems in the theory of rings" revisited*, Amer. Math. Monthly **77** (1970), 445–454.

[Kem78]   Kemer, A. R., *Remark on the standard identity*, Math. Notes **23** (1978), no. 5, 414–416.

[Kem80]   Kemer, A. R., *Capelli identities and the nilpotency of the radical of a finitely generated PI-algebra*, Soviet Math. Dokl. **22** (1980), no. 3, 750–753.

[Kem81]   Kemer, A. R., *Nonmatrix varieties with polynomial growth and finitely generated PI-algebras*, Ph.D. Dissertation, Novosibirsk (1981).

[Kem87]   Kemer, A. R., *Finite basability of identities of associative algebras (Russian)*, Algebra i Logika **26** (1987), 597–641; English translation: Algebra and Logic **26** (1987), 362–397.

[Kem88]   Kemer, A. R., *The representability of reduced-free algebras*, Algebra i Logika **27** (1988), no. 3, 274–294.

[Kem90]   Kemer, A. R., *Identities of Associative Algebras*, Transl. Math. Monogr., vol. 87, Amer. Math. Soc. (1991).

[Kem91]   Kemer, A. R., *Ideals of Identities of Associative Algebras*, Amer. Math. Soc. Translations of monographs **87** (1991).

[Kem91b]   Kemer, A. R., *Identities of finitely generated algebras over an infinite field*, Math-USSR Izv. **37** (1991), 69–96.

[Kem93]   Kemer, A. R., *The standard identity in characteristic p: a conjecture of I.B. Volichenko*, Israel J. Math. **81** (1993), 343–355.

[Kem95]   Kemer, A. R., *Multilinear identities of the algebras over a field of characteristic p*, Internat. J. Algebra Comput. **5** (1995), no. 2, 189–197.

[Kem96]   Kemer, A. R., *Remarks on the prime varieties*, Israel J. Math. **96** (1996), part B, 341–356.

[Kn73]   Knuth, D. E., *The Art of Computer Programming. Volume 3. Sorting and Searching*, Addison-Wesley Pub. (1991).

[Kol81]   Kolotov, A. T., *Aperiodic sequences and growth functions in algebras*, Algebra i Logika **20** (1981), no. 2, 138–154.

[Kos58]   Kostant, B., *A theorem of Frobenius, a theorem of Amitsur-Levitzki and cohomology theory*, J. Mathematics and Mechanics **7** (1958), no. 2, 237–264.

[KrLe00]   Krause, G. R. and Lenagan, T. H., *Growth of Algebras and Gelfand-Kirillov Dimension*, Amer. Math. Soc. Graduate Studies in Mathematics **22** (2000).

[Kuz75]   Kuzmin, E. N., *About Nagata-Higman Theorem*, Proceedings dedicated to the 60th birthday of Academician Iliev, Sofia (1975), 101–107 (in Russian).

[Lat63]   Latyshev, V. N., *Two remarks on PI-algebras*, Sibirsk. Mat. Z. **4** (1963), 1120–1121.

[Lat73]   Latyshev, V. N., *On some varieties of associative algebras*, Izv. Akad. Nauk SSSR **37** (1973), no. 5, 1010–1037.

[Lat77]   Latyshev, V. N., *Nonmatrix varieties of associative rings*, Doctor of Science Dissertation, Moscow (1977).

[Lat88]   Latyshev, V. N., *Combinatorial Ring Theory. Standard Bases*, Moscow University Press, Moscow (1988), (in Russian).

[LeBPr90]   Le Bruyn, L. and Procesi, C., *Semisimple representations of quivers*, Trans. Amer. Math. Soc. **317** (1990), no. 2, 585–598.

[Lev46]   Levitzki, J., *On a problem of Kurosch*, Bull. Amer. Math. Soc. **52** (1946), 1033–1035.

[Lev50]   Levitzki, J., *A theorem on polynomial identities*, Proc. Amer.
             Math. Soc. **1** (1950), 449–463.

[Lew74]   Lewin, J., *A matrix representation for associative algebras. I and
             II*, Trans. Amer. Math. Soc. **188** (1974), no. 2, 293–317.

[Lic89]   Lichtman, A. I., *PI-subrings and algebraic elements in enveloping
             algebras and their field of fractions*, J. Algebra. **121** (1989), no. 1,
             139–154.

[Lor88]   Lorenz, M., *Gelfand-Kirillov Dimension and Poincaré Series*,
             University of Granada, unpublished lecture notes (1988).

[MacD95]   MacDonald, I. G., *Symmetric Functions and Hall Polynomials*,
             2nd ed., with contributions by A. Zelevinsky, Oxford Mathemat-
             ical Monographs, The Clarendon Press, Oxford University Press
             (1995).

[Mal36]   Mal'tsev, A. I., *Unterschungen aus dem Gebiete des mathemati-
             cischen Logik*, Mat. Sb. 1 **43** (1936), 323–336.

[Mar79]   Markov, V. T., *On generators of T-ideals in finitely generated
             free algebras*, Algebra i Logika **18** (1979), no. 5, 587–598.

[Mar88]   Markov, V. T., *Gelfand-Kirillov dimension: nilpotency, repre-
             sentability, nonmatrix varieties*, in: Tez. Dokl. Sib. Shkola po
             Mnogoobr. Algebraicheskih Sistem, Barnaul (1988), 43–45.

[Neu67]   Neumann, H., *Varieties of Groups*, Springer-Verlag, Berlin-New
             York (1967).

[Oa64]   Oates, S. and Powell, M. B., *Identical relations in finite gorups*,
             J. Algebra **1** (1964), no. 1, 11–39.

[PaSmWa]   Pappacena, Ch. J., Small, L. W., and Wald, J., *Affine semiprime
             algebras of GK dimension one are (still) PI*, Glasg. Math. J. **45**
             (2003), no. 2, 243–247.

[Pas71]   Passman, D. S., *Group rings satisfying a polynomial identity II*,
             Pacific J. Math. **39** (1971), 425–438.

[Pas72]   Passman, D. S., *Group rings satisfying a polynomial identity*, J.
             Algebra **20** (1972), 103–117.

[Pas77]   Passman, D. S., *The Algebraic Structure of Group Rings*, Wiley,
             New York (1977).

[Pas89]   Passman, D. S., *Crossed products and enveloping algebras satis-
             fying a polynomial identity*, IMCP **1** (1989), 61–73.

[Pchl84]   Pchelintzev, S. V., *The height theorem for alternate algebras*, Mat. Sb. **124** (1984), no. 4, 557–567.

[Plo04]   Plotkin, B., *Notes on Engel groups and Engel-like elements in groups. Some generalizations*, (2004) preprint (15 pp.).

[Pro73]   Procesi, C., *Rings with Polynomial Identities*, M. Dekker, New York (1973).

[Pro76]   Procesi, C., *The invariant theory of $n \times n$ matrices*, Advances in Math. **19** (1976), 306–381.

[Raz72]   Razmyslov, Yu. P., *On a problem of Kaplansky*, Math. USSR. Izv. **7** (1972), 479–496.

[Raz74a]   Razmyslov, Yu. P., *The Jacobson radical in PI-algebras*, Algebra and Logic **13** (1974), no. 3, 192–204.

[Raz74b]   Razmyslov, Yu. P., *Trace identities of full matrix algebras over a field of characteristic zero*, Math. USSR Izv. **8** (1974), 724–760.

[Raz74c]   Razmyslov, Yu. P., *Existence of a finite basis for certain varieties of algebras*, Algera i Logika **13** (1974), 685–693; English translation: Algebra and Logic **13** (1974), 394–399.

[Raz78]   Razmyslov, Yu. P., *On the Hall-Higman problem*, Izv. Akad. Nauk SSSR Ser. Mat. **42** (1978), no. 4, 833–847.

[Raz89]   Razmyslov, Yu. P., *Identities of Algbras and their Representations*, Nauka, Moscow (1989).

[RaZub94]   Razmyslov, Yu. P. and Zubrilin, K. A., *Capelli identities and representations of finite type*, Comm. Algebra **22** (1994), no. 14, 5733–5744.

[Reg78]   Regev, A., *The representations of S_n and explicit identities for P.I. algebras*, J. Algebra **51** (1978), no. 1, 25–40.

[Reg80]   Regev, A., *The Kronecker product of S_n-characters and an $A \otimes B$ theorem for Capelli identities*, J. Algebra **66** (1980), 505–510.

[Reg81]   Regev, A., *Asymptotic values for degrees associated with strips of Young diagrams*, Advances in Math. **41** (1981), 115–136.

[Reg84]   Regev, A., *Codimensions and trace codimensions of matrices are asymptotically equal*, Israel J. Math. **47** (1984), 246–250.

[ReS93]   Resco, R. and Small, L., *Affine Noetherian algebras and extensions of the base field*, Bulletin London Math. Soc. **25** (1993), 549–552.

[Row80]   Rowen, L. H., *Polynomial Identities in Ring Theory*, Acad. Press Pure and Applied Math., vol. 84, New York (1980).

[Row88]   Rowen, L. H., *Ring Theory II*, Acad. Press Pure and Applied Math., vol. 128, New York (1988).

[Sag01]   Sagan, B. E., *The symmetric group: representations, combinatorial algorithms, and symmetric functions*, 2nd ed. Graduate text in mathematics **203**, Springer, Berlin-New York (2001).

[Sch76]   Schelter, W., *Integral extensions of rings satisfying a polynomial identity*, J. Algebra **40** (1976), 245–257; errata op. cit. **44** (1977), 576.

[Sch78]   Schelter, W., *Noncommutative affine PI-algebras are catenary*, J. Algebra **51** (1978), 12–18.

[Shch00]   Shchigolev, V. V., *Examples of infinitely basable T-spaces*, Mat. Sb. **191** (2000), no. 3, 143–160; translation: Sb. Math. **191** (2000), no. 3-4, 459–476.

[Shch01]   Shchigolev, V. V., *Finite basis property of T-spaces over fields of characteristic zero*, Izv. Ross. Akad. Nauk Ser. Mat. **65** (2001), no. 5, 191–224; translation: Izv. Math. **65** (2001), no. 5, 1041–1071.

[Shes83]   Shestakov, I. P., *Finitely generated special Jordan and alternate PI-algebras*, Mat. Sb. **122** (1983), no. 1, 31–40.

[Shir57a]   Shirshov, A. I., *On some nonassociative nil-rings and algebraic algebras*, Mat. Sb. **41** (1957), no. 3, 381–394.

[Shir57b]   Shirshov, A. I., *On rings with identity relations*, Mat. Sb. **43** (1957), no. 2, 277–283.

[SmStWa84] Small, L. W., Stafford, J. T., and Warfield, R., *Affine algebras of Gelfand Kirillov dimension one are PI*, Math. Proc. Cambridge Phil. Soc. **97** (1984), 407–414.

[SmWa84]   Small, L. W. and Warfield, R., *Prime affine algebras of Gelfand-Kirillov dimension one*, J. Algebra **91** (1984), 386–389.

[SmoVi03]   Smoktunowicz, A. and Vishne, U., *Primitive algebras of GK-dimension 2*, preprint (2003).

[Sp50]   Specht, W., *Gesetze in Ringen I*, Math. Z. **52** (1950), 557–589.

[St99]   Stanley, R. P., *Enumerative Combinatorics. Vol. 2*, Cambridge Studies in Advanced Math. **62**, Cambridge Univ. Press (1999).

[Ufn78]   Ufnarovski'i, V. A., *On growth of algebras*, Vestn. Moskovskogo
               Universiteta Ser. 1, no. 4 (1978), 59–65.

[Ufn80]   Ufnarovski'i, V. A., *On Poincare series of graded algebras*, Mat.
               Zametki **27** (1980), no. 1, 21–32.

[Ufn85]   Ufnarovski'i, V. A., *The independency theorem and its conse-
               quences*, Mat. Sb., **128** (1985), no. 1, 124–13.

[Ufn89]   Ufnarovski'i, V. A., *On regular words in Shirshov sense*, in: Tez.
               Dokl. po Teorii Koletz, Algebr i Modulei. Mezhd. Konf. po Alge-
               bre Pamyati A.I. Mal'tzeva, Novosibirsk (1989), 140.

[Ufn90]   Ufnarovski'i, V. A., *On using graphs for computing bases, growth
               functions and Hilbert series of associative algebras*, Mat. Sb. **180**
               (1990), no. 11, 1548–1550.

[VaZel89]   Vais, A. Ja. and Zelmanov, E. I., *Kemer's theorem for finitely gen-
               erated Jordan algebras*, Izv. Vyssh. Uchebn. Zved. Mat. (1989),
               no. 6, 42–51; translation: Soviet Math. (Iz. VUZ) **33** (1989),
               no. 6, 38–47.

[Vas99]   Vasilovsky, S. Yu., $\mathbb{Z}_n$-*graded polynomial identities of the full ma-
               trix algebra of order n*, Proc. Amer. Math. Soc. **127** (1999), 3517–
               3524.

[Vau70]   Vaughan-Lee, M. R., *Varieties of Lie algebras*, Quart. J. of Math.
               21 (1970), 297–308.

[Vis99]   Vishne, U., *Primitive algebras with arbitrary GK-dimension*, J.
               Algebra **211** (1999), 150–158.

[Wag37]   Wagner, W., *Uber die Grundlagen der projektiven Geometrie und
               allgemeine Zahlsysteme*, Math. Z. **113** (1937), 528–567.

[Wil91]   Wilson, J. S., *Two-generator conditions for residually finite
               groups*, Bull. London Math. Soc. **23** (1991), 239–248.

[Za90]   Zalesskiĭ, A. E., *Group rings of inductive limits of alternating
               groups*, Algebra i Analiz **2** (1990), no. 6, 132–149; translation:
               Leningrad Math. J. **2** (1991), no. 6, 1287–1303.

[Za96]   Zalesskiĭ, A. E., *Modular group rings of the finitary symmetric
               group*, Israel J. Math. **96** (1996), 609–621.

[ZaMi73]   Zalesskiĭ, A. E. and Mihalev, A. V., *Group rings*, Current Prob-
               lems in Mathematics, Vol. 2, Akad. Nauk SSSR Vseojuz. Inst.
               Naučn. i Tehn. Informacii, Moscow (1973), 5–118.

[Zel88]   Zelmanov, E. I., *On nilpotency of nilalgebras*, Lect. Notes Math. **1352** (1988), 227–240.

[ZelKos90]   Zelmanov, E. I. and Kostrikin, A. I., *A theorem about sandwich algebras*, Trudy Mat. Inst. Akad. Nauk SSSR **183** (1990), 106–111.

[ZSlShSh78] Zhevlakov, K. A., Slin'ko, A. M., Shestakov, I. P., and Shirshov, A. I., *Rings, that are Nearly Associative*, Nauka, Moscow (1978), in Russian; English translation by Harry F. Smith, Acad. Press Pure and Applied Mathematics, vol. 104, New York (1982).

[Zu96]   Zubkov, A. N., *On a generalization of the Razmyslov-Procesi theorem*, Algebra i Logika **35**, (1996), no. 4, 433–457, 498; transl. Algebra and Logic **35** (1996), no. 4, 241–254.

[Zub95a]   Zubrilin, K. A., *Algebras satisfying Capelli identities*, Sb. Math. **186** (1995), no. 3, 53–64.

[Zub95b]   Zubrilin, K. A., *On the class of nilpotency of obstruction for the representability of algebras satisfying Capelli identities*, Fundam. Prikl. Mat. **1** (1995), 409–430.

[Zub97]   Zubrilin, K. A., *On the largest nilpotent ideal in algebras satisfying Capelli identities*, Sb. Math. **188** (1997), 1203–1211.

Index